CHEMICAL PROCESS ECONOMICS

CHEMICAL PROCESSING AND ENGINEERING

An International Series of Monographs and Textbooks

EDITORS

Lyle F. Albright
Purdue University
Lafayette, Indiana

R. N. Maddox
Oklahoma State University
Stillwater, Oklahoma

John J. McKetta
University of Texas
at Austin
Austin, Texas

Volume 1: Chemical Process Economics, Second Edition, Revised and Expanded
by John Happel and Donald G. Jordan

IN PREPARATION

Petroleum Economics and Engineering: An Introduction, by H. K. Abdel-Aal and Robert Schmelzlee

Thermodynamics of Fluids: An Introduction to Equilibrium Theory, by K.C. Chao and R.A. Greenkorn

Chemical Reactions as a Separation Mechanism—Sulfur Removal, edited by Billy L. Crynes

Continuum Mechanics of Viscoelastic Fluids, by Ronald Darby

Petroleum Refining: Technology and Economics, by James H. Gary and Glenn E. Handwerk

Control of Air Pollution Sources, edited by J. M. Marchello

Gas-Solids Handling in the Process Industry, edited by J. M. Marchello and Albert Gomezplata

Gas Cleaning for Air Quality Control, edited by J. M. Marchello and J. J. Kelly

Computers in Process Control, by Ulrich Rembold, Mahesh Seth, and Jeremy S. Weinstein

Solvent Extraction in Hydrometallurgical Processing, by G. M. Ritcey and A. W. Ashbrook

ADDITIONAL VOLUMES IN PREPARATION

Chemical Process Economics

SECOND EDITION

revised and expanded

JOHN HAPPEL
Department of Chemical Engineering
Columbia University in the City of New York

DONALD G. JORDAN
Consulting Chemical Engineer
Reno, Nevada

MARCEL DEKKER, INC.   New York

Copyright © 1975 by Marcel Dekker, Inc. All rights reserved.

Neither this book nor any part may be reproduced or transmitted in any form or by any means, electronic or mechanical, including photocopying, microfilming, and recording, or by any information storage and retrieval system, without permission in writing from the publisher.

MARCEL DEKKER, INC.

270 Madison Avenue, New York, New York 10016

LIBRARY OF CONGRESS CATALOG CARD NUMBER: 74-76718

ISBN: 0-8247-6155-3

Current printing (last digit):
10 9 8 7 6 5 4 3 2 1

PRINTED IN THE UNITED STATES OF AMERICA

FOREWORD

    This Chemical Engineering and Processing Series will be a welcome addition to the literature in that it provides new material on a series of topics important to the chemical engineering aspects of the process industries.

    Today we are in great need of carrying new scientific and engineering developments through to the application state. Unfortunately chemical engineering had a lull in attention to new information via university research having to do with the real world of the process industries. It is believed this situation has turned around both in research and teaching and this series of books should assist in the educational effort in this direction. The bridge between science and engineering application is a continuing challenge to chemical engineers and has been taken up by the editors of this series.

    One important characteristic of the book or series is the background and ability of the authors and editors, their discernment of what is important and what should be left out, their dedication to the task of communicating the essence of ideas to the reader and their preparation of material from the viewpoint of the user overrides most other considerations. I wish to commend Marcel Dekker, Inc., for the leadership they have chosen and the editors in turn for prospective lists of authors.

    I wish the editors and authors well in their efforts to increase the use of technical knowledge for the benefit of process industries and hence the people of our nation.

DONALD L. KATZ

PREFACE

This is the second edition of Happel's 1958 text Chemical Process Economics. In the years since, the chemical industry has undergone many changes, and is now greatly different from what it was in 1958. It was felt that a new edition of the text should be written so as to reflect the impact of these changes, and to improve and modernize the original material.

The purpose of this book is to provide a working tool to assist the student or chemical engineer in applying technical information to the economic design and operation of chemical process plants.

Bridging the gap between theory and practice is often a matter of experience acquired over a number of years. Even then, methods developed from experience all too often must be re-evaluated in the light of changing economic conditions if optimum plant designs are to result. The approach here represents an attempt to provide a consistent and reasonably concise method for the solution of those problems involving economic alternatives.

The material presented formed a major part of the subject matter of the course in plant design formerly given to fourth-year students in chemical engineering at New York University. The study of economic balances furnishes a natural extension to the subjects of heat and material balances which are the necessary prerequisite to development of a plant design project. The optional design variables which cannot be established by technical considerations are resolved by means of the criterion of design for maximum profitability. The student is thus immediately able to develop a feeling for the usefulness of the technical knowledge which he has gained as well as for the limitation of computations without regard for the dollar sign.

The quantitative viewpoint is emphasized, although it is realized that the broad subject of chemical engineering economics cannot be fitted into any rigid set of formulas. There are many excellent books covering such subjects as administration, market research, and plant location, which are not treated here.

The material in this text is divided roughly into five parts.

1. Chapter 1 is an addition to the original work. The authors' experience, extending now over 30 years, teaches that the structure and procedures of the chemical business are not well known. Students and young engineers in particular are unlikely to be knowledgable about the business, profit making, aspect of the chemical industry. This chapter was added to provide some needed background material.

2. Chapters 2, 3, and 4 are devoted to a development of the problem of making economic evaluations of any situation, either parts of a process plant or the complete project. The proper fixed charge to apply to capital investment is developed as a function of such variables as tax rate, minimum acceptable rate of return, and anticipated plant life. The presentation starts with the simplest situations and proceeds to more complicated formulations and techniques which may be employed if there are sufficient data available. Chapter 4 presents a discussion of more advanced mathematical techniques that may be used to solve more complex problems, particularly those involving more than one variable.

3. Chapters 5 and 6 provide specific information required to apply the techniques previously developed. Methods are suggested for establishing the various items entering into the determination of profitability. Chapter 5 gives a general survey of methods and techniques used in cost estimation for equipment and other items. Only one chapter is devoted to cost estimation because this subject is treated in other books, and also because the authors' long experience in this field teaches them that detailed cost estimation should be left to professional cost estimators.

Chapter 6 is devoted to a discussion of the basis for determination of acceptable rates of return. This subject is one in which much difference of opinion exists, and many engineers will prefer to use different approaches to establish the information needed. Management and company policy will often be important in this area.

4. Chapters 7 and 8 are devoted to applications of the treatment to process problems. First, situations are considered involving overall plant evaluation and economic balances which include an entire process. This is followed by a more detailed examination of specific problems in the economic design of individual component parts of process plants. One aspect of this is the extension of techniques showing how various rules of thumb, or approximation design

PREFACE

procedures, may be justified and new ones developed, depending on circumstance. Appendix C provides a number of such approximation design procedures.

5. Appendixes A, B, and C contain short discussions on subjects that are pertinent to, but not a principal part of, the text.

Appendix A is a short explanation of the mathematical basis and derivation of the compound interest formulas used in the text.

Appendix B is a short essay on inflation. This subject is now much more important and much more widely discussed than at the time of the first edition of this book. Because so much of the work of process economics involves periods of time of as much as 10 years, and in that period the forces of inflation can seriously disrupt economic studies, it was thought that a short discussion on inflation would be helpful.

Appendix C is a collection of rapid approximate design procedures which are suitable for preliminary equipment sizing and process evaluation. The authors have found by experience that such approximately, but largely correct, methods are enormously helpful in rapid economic and technical evaluation.

Numerous problems are provided at the conclusions of Chapters 2, 3, 5, 6, 7, and 8. These problems are not difficult in principle, but for many of them the time consumed in calculation is significant. It is quite difficult to make up suitable student problems in this field because of the excessive burden of computation and the difficulty of obtaining the necessary data. For the more complex problems, such as those treated in Chapter 4, the amount of computation is so heavy that the task of solving such problems is unusually time consuming and difficult. For this reason, no problems are provided at the end of Chapter 4. There are several illustrative examples that can be studied.

As this work is completed, January 1975, the industrial world is in turmoil reflecting the various forces of serious monetary inflation, rapidly changing prices, and a deepening world wide economic recession. The data used in the statement and solution of process-economic problems are changing rapidly and reliable realistic solutions are difficult to obtain, even by those practicing professionally in the field.

In this text there are inconsistencies in the economic data presented, particularly in the examples and in the problems. The student is advised to study the principles presented, to use the data herein only as a guide, and to solve his problems with his own data within the framework of the methods discussed in the text.

New York

JOHN HAPPEL
DONALD G. JORDAN

CONTENTS

FOREWORD iii

PREFACE v

Chapter 1
THE CHEMICAL INDUSTRY:
ITS PRESENT AND FUTURE PROSPECTS 1

1.00 Introduction 1
1.01 Some Characteristics of the Chemical Industry 3
1.02 Raw Materials, Energy, and Manpower 9
1.03 The Prices of Chemical Products 16
1.04 Research and Development in the
       Chemical Industry 21
1.05 The Financial Aspects of the Chemical Business 26
1.06 Future Prospects 31
       References 34

Chapter 2
PRINCIPLES OF ECONOMIC EVALUATION 35

2.00 Basic Concepts 35
2.01 Gross Income 36
2.02 Capital Invested in Facilities and Working Capital ... 36
2.03 Income Taxes Levied by Governmental Agencies 37
2.04 Allowable Depreciation for Accounting
       and Income Tax 38
2.05 Minimum Acceptable Return Rate 43
2.06 Methods for Evaluating Investment Opportunities ... 44
2.07 The Capital Investment 46
2.08 The Manufactured Cost of the Product 46
2.09 The Net Profit 47
2.10 Cash Position; the Cash Position Diagram 47

2.11 Rate of Return on Investment	48
2.12 Payout Time	49
2.13 Incremental Rate of Return	50
2.14 The Venture Profit Concept	54
2.15 Components of a Process Unit	62
2.16 Applications to Plant Operation	72
Summary	81
References	81
Problems	82

Chapter 3
EXPANDED ECONOMIC EVALUATION EQUATIONS 89

3.00 The Time Value of Money	89
3.01 The Venture Worth Method	91
3.02 Extended Venture Profit Equation	92
3.03 The Venture Worth Equation	94
3.04 Rate of Return	111
3.05 The Discounted Cash Flow Method	112
3.06 Income-Tax Credit Summations	116
3.07 Declining Gross Return	129
3.08 Termination of Operations	133
3.09 Maximum in Venture Worth as a Criterion of Project Attractiveness	136
3.10 Components of Process Units	138
3.11 Limited Availability of Capital	143
Summary	143
References	143
Problems	146

Chapter 4
SPECIAL MATHEMATICAL TECHNIQUES 149

4.00 Introduction	149
4.01 General Formulation of Optimum	151
4.02 Complex Relationships: The Use of Special Methods	162
4.03 Region Elimination	163
4.04 One Variable at a Time— With Graphical Differentiation	167
4.05 The Method of Steepest Ascents or Descents	179
4.06 Optimization Problems Involving Restrictions	184
4.07 Linear Programming	195
4.08 Dynamic Programming	203
References	210

CONTENTS   xi

    Chapter 5
NOTES ON COST ESTIMATION   213

5.00 Introduction   213
5.01 Three Steps in Cost Estimation   215
5.02 Equipment Cost Estimation   216
5.03 Precision of Preliminary Cost Estimate   222
5.04 A Compilation of Equipment Cost   222
5.05 Estimating the Final Plant Cost   232
5.06 Sources of Cost Data   242
5.07 The Manufacturing Cost of the Product   244
     Summary   254
     References   255
     Problems   256

    Chapter 6
RISK, RETURN RATE, AND INVESTMENT RECOVERY   259

6.00 Introduction   259
6.01 Risks   259
6.02 Gross Sales Return, $R_k = Q(s-c)$ 261
6.03 Total New Capital Investment, I   262
6.04 Interest Rates   263
6.05 The Role of Government Policy   265
6.06 Estimated Project Life   266
6.07 Degrees of Risk   267
6.08 Rate of Return   269
6.09 Estimation of the Minimum Acceptable
     Return Rate, i_m 276
6.10 Development Projects   277
6.11 Investment Recovery   279
     References   283
     Problems   283

    Chapter 7
OVERALL CONSIDERATIONS IN PROJECT ANALYSIS   289

7.00 New Projects   289
7.01 Existing Plants   303
7.02 Replacement of Plant Components   309
7.03 Time Efficiency   317
7.04 Optimum Material Balance   324
7.05 The Use of Computers in Project Analysis   337
     Summary   338

References	338
Problems	340

Chapter 8
PROCESS PLANT COMPONENTS   349

8.00 Introduction	349
8.01 The Optimum Design of Process Piping	350
8.02 The Optimum Design of Heat-Transfer Apparatus	357
8.03 The Optimum Design of Insulation	380
8.04 The Optimum Design of Distillation Columns	385
8.05 The Optimum Design of Gas Absorption Columns	396
8.06 The Optimum Design of Liquid-Liquid and Solid-Liquid Extraction Systems	407
8.07 The Optimum Design of Continuous Direct Contact Rotary Dryers	408
8.08 The Optimum Design of Chemical Reactors	417
References	428
Problems	429

Appendix A
SUMMATION OF TIME SERIES   435

Appendix B
INFLATION   445

Appendix C
RAPID APPROXIMATION METHODS FOR THE DESIGN OF CHEMICAL-PROCESS EQUIPMENT   451

INDEX   489

CHEMICAL PROCESS ECONOMICS

Chapter 1

THE CHEMICAL INDUSTRY:
ITS PRESENT POSITION AND FUTURE PROSPECTS

1.00 INTRODUCTION

    Chemistry is defined by Webster's Collegiate Dictionary as "the science that treats of the composition of substances and of the transformation which they undergo." Thus, every substance in the world is a subject suitable for chemical study, and a great many substances have been so studied. As the chemists say, everything in the world is a chemical and therefore all the world's problems are chemical problems. An amusing statement that contains much truth. Certainly the science of chemistry is enormous in the extent and depth of its knowledge about material things, and of the reactions that they can be made to undergo. This knowledge has now been advanced to the point where the application of chemical principles to manufacturing processes has enabled mankind to produce an incredible number of substances that are useful and beneficial. Many of the products of the chemical effort have become valuable articles of commerce. These products influence essentially every aspect of human life and are the focal point of the effort of the chemical industry.

    Although scientifically knowledgeable people understand that every substance is a chemical and that the processes used to manufacture them are "chemical" in nature, this generalized viewpoint is not widely accepted. As a consequence, it is difficult to define the "chemical" industry in a way that is accurate and generally useful.

Thus, the glass and steel industries are most certainly "chemical" but these industries are usually considered by themselves, and not as part of the chemical industry. The same is true of the textile, food, pharmaceutical, fertilizer, copper, and petroleum industries.

As a matter of common usage, the chemical industry is considered to be concentrated on the production and sale of single pure elements or compounds. Such substances are often not of any great usefulness in themselves, as, for instance, an aspirin tablet is, but are sold to other manufacturers for use in the treatment or production of still other materials that are then sold to the general public. For instance, sodium hydroxide is a material that is produced in enormous quantities in many parts of the world as a single pure "chemical" compound. By itself sodium hydroxide is not very useful; it cannot be eaten, worn, or lived in. But when used in manufacturing processes, such as the making of soap, it becomes so useful that it is one of the basic industrial raw materials. Its production and sale is an enormous industry. Much the same statements can be made about many "chemical" compounds: sulfuric acid, chlorine, butadiene, phenol, citric acid, and trisodium phosphate, among many others.

The confusion between what is a "chemical" company and what is not becomes greater when it is realized that many "chemical" companies manufacture and sell thousands of products, some of which are pure compounds or elements, such as chlorine or vinyl chloride, and some of which are complex products, such as polyester fibers, penicillin, automobile radiator fluids, and plastic garbage bags. These products are sold to other industries and to the public.

To summarize:

1. The popular image of a "chemical" company is of one which manufactures and sells "chemicals," meaning single pure compounds or elements which are used as the raw materials for subsequent processing. A "chemical" company makes, for example, sulfuric acid, sodium carbonate, phenol, phthalic anhydride, or styrene. It does not make steel or paper, refine petroleum, or synthesize drugs, even though all of these are "chemical" processes.

2. The actual fact is that "chemical" companies engage in a wide variety of manufacturing and selling activities. Some of these are purely "chemical" in nature (as in 1 above) and others are more closely related to the familiar aspects of human life: food, clothing, shelter, medical treatment, and entertainment.

The Chemical Industry

To illustrate, the American Cyanamid Company, once the most "chemical" of chemical companies, has systematically diversified its activities by the acquisition of the Lederle Laboratories (ethical pharmaceuticals), Breck (hair preparations), Formica (plastic table tops), and Davis and Geck (surgical sutures). This diversification, away from strictly "chemical" products, has been quite successful, and now a substantial part of Cyanamid's profits come from these consumer products.

In fact, several of the famous chemical companies have recently renamed themselves, and, in so doing, have dropped the word "chemical." Presumably, this is to identify their names more closely with the broadened aspect of their business activities. For instance, the Monsanto Chemical Company, one of the most famous names in industry, has changed its name to Monsanto Company.

For the purposes of this book, attention will be centered on "chemical" enterprises, but this must not be construed in a narrow way. The "chemical" industry is a dynamic, ever-changing activity which embraces all aspects of the business spectrum.

1.01 SOME CHARACTERISTICS OF THE CHEMICAL INDUSTRY

The chemical industry is an industry like any other. It must buy and sell, borrow money, pay debts, hire and fire, plan for the future, live in the community, and make a profit as any other industry must. Also, like any other industry, the chemical business has some special characteristics.

1. It is heavily dependent on scientific knowledge and experience; only the electrical products and aerospace industries are more technical. Many chemicals are complex and are produced by processes and facilities that are also complex. These require the intense application of scientific knowledge for their successful conception and operation. The chemical industry is founded on chemical science. It follows the scientific and engineering work of others with careful attention, and also creates knowledge by its own research work. A high percentage of the employees of the chemical industry are scientists and engineers. They are used in all phases of the industry's activities.

2. A high rate of innovation and change in the industry's products, and in the processes by which they are made exist.

The chemical industry strives ceaselessly to create new products with superior properties, to modify older products so that their properties are improved, and to change the processes which make the products so that the processes are more efficient.

The result of this activity has been a continuous stream of new and modified products of all types. These new products can render the existing ones obsolete, create entirely new industries, and cause large changes in both the chemical industry and the industries it serves.

Not only are new products, such as polyester tire cord, invented and commercialized, but new processes for making important (and long-established) chemicals are continually being introduced. Thus, the most important process for making phenol has changed three times in 30 years.

This ability to introduce a series of new products and processes in a continuing manner has been largely responsible for the good profitability, the high growth rate, and the large changes in the industry's products over the years. The industry is constantly changing.

Before starting the lengthy and difficult work of commercialization of a process or a product the chemical industry conducts extensive studies into the economic prospects of the new venture. These studies include chemical research and development work, market research on sales volumes and prices, and economic evaluation of new capital requirements, return on investment, payout time, depreciation allowances, and taxation.

These matters are discussed in detail in subsequent chapters. In particular, the concept of venture worth, an economic criterion for project desirability, is presented.

3. The chemical industry is quite competitive. This competition exixts between companies manufacturing and selling the same material, and competition also exists between different products for the same end use.

It is difficult for a company to develop a product or a process that will operate essentially without competition. If the commercial market for the material appears to be profitable, competition will arise. The greater the profits, the greater the competition.

International competition is extensive. Many large chemical companies, originally founded in and now directed from one country, have manufacturing and sales operations in other countries. Not only must a company worry about competition from its domestic colleagues, but also from those overseas. Thus, many U. S.

chemical companies have operations in Europe, Canada, and South America, while European companies maintain competing facilities in the United States.

4. The chemical industry is relatively easy to enter. This can be done by anyone from one person to a large corporation. There are so many products which can be made in simple equipment, by easily understood processes, that very small groups can produce substantial quantities of chemicals in a profitable manner. It is extremely easy for anyone to enter the chemical business in a modest manner. Very often such small endeavors can be quite profitable and provide serious competition to larger companies.

If a larger corporation wishes to enter the chemical business in a substantial way so as to compete with the established chemical companies, one or both of two factors makes the entry fairly simple.

a. A special situation will exist whereby the corporation enjoys a cost advantage with respect to raw materials, power, water, transportation, or sales.

b. The process knowledge and engineering work can be purchased for a modest fee, or by means of cross-licensing of patents with other corporations.

At the present time there are many chemical engineering firms that can be employed, for a comparatively small fee, to design and construct chemical plants. Through the use of such firms any company with the necessary capital can purchase a swift and easy entry to the chemical business. The knowledge and experience of these engineering firms extends over the entire field of industrial chemistry. They are aggressive about selling their services, and in recent years have been successful in designing and building many chemical plants, of all types, all over the world. The existence of this powerful force devoted to expansion has been one of the principal reasons for the incredible expansion (perhaps overexpansion) of the chemical business in the last 20 years.

The most important example of corporations with special situations using the services of engineering contracting firms and thereby becoming powerful chemical manufacturers is that of the petroleum industry. These companies possess great strength in technical ability, raw materials, and money. Attracted by the large profits to be made in the chemical business, and knowing their own abilities, these companies have moved into the chemical manufacturing industry in a powerful way in the last 20 years. The result has been that 13 of the top 50 chemical companies in the United States are petroleum companies. Exxon is the fifth largest chemical company in the United States. Accompanying this expansion has been the creation of a large production capacity, the lowering of prices, and the intensification of competition in the chemical industry. This proceeded to the point where, in the 1960s, the large profits that attracted the petroleum companies in the first place disappeared, and almost all of the chemical industry experienced a substantial decline in earnings.

5. The chemical industry in the United States is made up of hundreds of firms. Some of these are quite large, and many others are much smaller. In 1973 the chemical and allied products industry sold products worth $84 billion. This was approximately 8.3% of the total supplied by all U.S. manufacturing industry. Of this $84 billion, 44% was sold by the 50 largest companies, 22% by the 10 largest companies, and 15% by the five largest. One company (the duPont company) supplied 5% of the total.

These numbers show that the industry is not dominated by the largest firms. The remainder of the industry is quite significant since it supplies 56% of the total sales. In fact, the chemical industry is dominated less by its largest firms than are several of the other manufacturing industries such as steel, aluminum, and automobiles. There are many small chemical companies that are quite competitive with the larger firms, are more profitable, and are growing at a faster rate.

However, it is true that there is a considerable degree of concentration with regard to many products. Thus, 80% of a certain chemical might be sold by three makers. However, such concentration, essentially a monopoly, is not highly profitable to the three makers. Other chemicals, made by other manufacturers, will compete strongly for the same markets.

6. The chemical industry is a large user of money for the construction of new and renovated facilities. In recent years, the chemical industry has used more than 15% of the new capital invested by all manufacturing industries, although the chemical industry produces only about 8% of all manufactured products.

Several factors contribute to the heavy use of capital by the chemical industry:

a. The chemical industry is growing at a rate that is faster than almost all other sectors of the manufacturing industry. Therefore, its demands for new capital are high.

b. Chemical equipment is complex and expensive. It is frequently made of special materials in order to resist high temperatures, high pressures, and the chemical action of corrosive substances. Furthermore, the equipment is frequently shaped into unusual forms that require special care and expense in fabrication.

c. Chemical plants require large amounts of utility services such as electrical generation or transmission, water cooling and distribution, and air and stream pollution control.

In the years prior to 1970 the U. S. chemical industry spent $2.5 to $3.5 billion a year on new plants and equipment. The amount spent increased by a substantial amount each year. In the period 1970-1971 the steady yearly increase in spending stopped, and there was a decline; but the amount spent was still quite large. In 1973 and 1974 the capital expenditures increased to record levels.

The necessary funds were obtained almost entirely from company surplus or from borrowing.

7. The chemical industry is not a large employer of labor. In many instances, the processes are continuous, automatically controlled, and the need for any kind of manual work is not great. Furthermore, it is obvious that the nature of the materials treated – frequently liquids and gases – make it impossible to use large amounts of manpower. While the chemical industry is well aware of the cost of labor, it is not as sensitive to labor costs as is the automobile industry, say.

The chemical industry needs skilled labor, or people it can train in various skills, and is willing to pay good wages to attract them. However, it does not need a great many of them.

8. The chemical industry has had an almost explosive growth since 1940. Much of this was started by World War II when large industries such as the synthetic rubber industry were created from essentially nothing. Since 1946 the rapid growth of the industry has continued, largely because of the incredible expansion in synthetic fibers, films, and thermoplastics.

At present, the value of shipments of chemicals is almost 20 times as great as in 1939, while the gross national product is only about 10 times as large. Obviously, the chemical industry grew at a rate that was much greater than that of the economy as a whole.

In the last few years this rapid expansion has slowed because of the serious deterioration in both the sales prices for the industry's products and the industry's profit margins. Figures 1.1 and 1.2 show these declines. However, despite these recent business difficulties the chemical industry is still expanding at a rate that is greater than that of almost all manufacturing industries.

In the years 1965-1971 this rapid expansion was slowed because of a serious deterioration in both the sales prices and the profit margins of the industry's activities. Figures 1.1 and 1.2 show the extent of the prosperous period of 1955-1965 and how extensive and how fast the declines from these high levels were. The period 1965-1971 was one of somber reflection on past mistakes, a great deal of pessimism about the industry's future, and intense activity directed towards a restoration of high prices and profit margins.

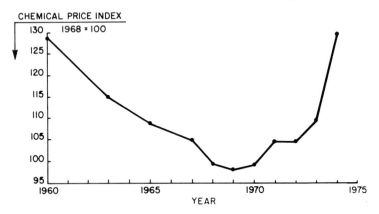

Fig. 1.1. Chemical price index-time record. Sources: Chemical and Engineering News, May 24, 1971, p. 19, and June 3, 1974, p. 45; Monthly Labor Review, U.S. Department of Labor, November, 1974, p. 113.

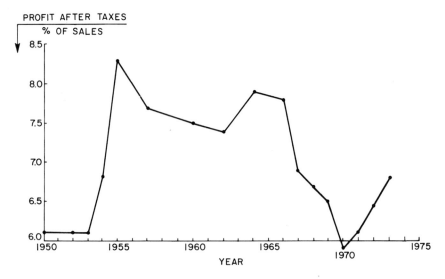

Fig. 1.2. Profit-time record, chemical and allied products industry. Source: Chemical and Engineering News, Aug. 27, 1962, p. 79; Sept. 6, 1971, p. 38A.

The Chemical Industry

However, despite the difficulties the chemical industry continued to expand its activities at a rate greater than that of almost all manufacturing industries.

In 1971 a dramatic reversal in the fortunes of the industry took place. As Figs. 1.1 and 1.2 show, there has been a sharp upturn in both prices and profitability since 1971, and this rise is continuing to this date.

9. Scientific research and development play a large part in the activities of the chemical industry, more so than in almost all other industries. Some companies spend appreciably more for research than do others, but almost all chemical companies spend something on research. It is estimated that the expenditures for research and development work in the United States by the basic chemical industry are now greater than $\$1 \times 10^9/\text{yr}$, and that the amount is still increasing. This is a very large effort indeed and leads to many discoveries of both theoretical and practical interest. Unfortunately, it is much more difficult to commercialize a discovery than it is to make the discovery in the first place. In order for research and development work to be useful to a company the results of the work must be commercialized.

There have been many spectacular advances achieved through industrial chemical research work. Entirely new industries have been founded on the discoveries of the laboratory, and many companies claim that a large percentage of their profits comes from discoveries made less than 10 years ago. Industrial chemical research work continues at a hard pace in the hope that the work will provide new products, new processes, and greater profits.

10. Chemical companies usually manufacture many different products. The larger companies may produce as many as 1000 substances, some quite different from others. The internal organization of the company will reflect this complexity. There may be as many as ten separate divisions, each essentially autonomous and concerned with one group of compounds, and each striving for maximum profitability.

These companies exhibit a high degree of vertical integration. Some divisions of the company will start with the basic raw materials and produce chemicals that are the raw materials for the work of another division. This arrangement is common and illustrates another factor of the chemical business: a large amount of internal (within one company) production and consumption. The intermediate compound will never be sold on the public market. In fact, it may never leave the chemical plant.

1.02 RAW MATERIALS, ENERGY, AND MANPOWER

Raw Materials

The raw materials of the chemical industry can be considered as belonging to one of two types.

1. Basic. These are substances which are found in nature. They are taken from the air, or mined from the ground, or extracted from the oceans. There are only 12 different materials required to produce the 50 most important industrial chemicals: Air, which supplies both oxygen and nitrogen. Water, which supplies hydrogen ion, hydroxyl ion, hydrogen, and oxygen. Sulfur, which is primarily used for sulfuric acid, the most important single chemical. Natural gas, which supplies carbon and hydrogen; its ethane, propane, and butane content may be treated to give ethylene, propylene, and butenes. Petroleum fractions, such as naphtha and gas oil, which are cracked to give ethylene and propylene, or may be differently processed to aromatic and cyclic hydrocarbons. Calcium carbonate, which is calcined to give calcium oxide which is in turn used for cement, calcium carbide, and an alkali. Bauxite, the most widely used source of alumina, which is in turn the raw material for alum and aluminum metal. Silica, or sand, which is processed into sodium silicate and silicones. Ilmenite (titanium dioxide), which is converted into white titanium dioxide pigment and into titanium metal. Sodium chloride, which is processed into sodium, chlorine, sodium hydroxide, and sodium carbonate. Coal, which is burned to give energy and heat or coked to give carbon and by-products such as benzene and naphthalene.

The annual United States production of the 50 most important chemicals is in excess of 325 billion pounds. For comparative purposes it should be noted that steel production in the United States is approximately 265 billion pounds a year. It is apparent that the supplying of the U. S. chemical industry with basic raw materials is an enormous business.

All of these raw materials are rather cheap, although by no means as cheap as only a few years ago. The year 1974 saw a strong increase in the prices for almost all raw materials. The most spectacular, and best known, was that of crude petroleum, but the costs of all other basic raw materials rose steeply. A list is given below of the December 1974 prices of some of the basic raw materials. It must be remembered that these prices will change in the near future and that such changes will force changes in the prices of almost all chemicals.

The Chemical Industry

Material	Price
Air and water	low
Salt, limestone, and phosphate rock	1 to 2 ¢/lb
Sulfur and petroleum naphtha	2 to 3 ¢/lb
Bauxite and ilmenite	2.5 to 3.5 ¢/lb

The cost of the raw material is frequently the major part of the cost of the final chemical. Chapter 5, especially Section 5.07 and Fig. 5.5, describes in detail the effect of raw material costs on the manufactured cost of the chemical.

The chemical industry pays careful attention to all of the factors that influence the cost of its raw materials.

a. The chemical company will own the raw material source and operate it as a part of the larger company. Thus, several chemical companies own large areas of central Florida, which is the source of phosphate rock. Mergers and purchases have taken place between chemical companies and oil producers so that the chemical companies will have an assured source of cheap hydrocarbons.

Business investments of this type are called "vertical integration," or "backward integration," and the terms clearly describe what has taken place. In order to be sure of their source of basic raw materials the chemical companies have bought control of companies producing the materials, or they have gone into the material-producing business themselves.

b. Chemical companies sign long-term contracts with producing companies for the delivery of specified quantities of the raw materials at special prices. Such contracts are usually good business for both parties since the customer is assured of a supply of the necessary raw material for a reasonable length of time into the future, and the producing company is assured of a steady market at a known price. These contracts can be complicated documents because both parties wish to be protected against unexpected changes in business conditions, technology, prices, and rates of use.

c. Transportation costs are an important part of the cost of the raw material. In an effort to minimize transportation costs chemical companies do one or all of several things.

(1) The producing plant will be built within a short distance of the source of its principal raw materials. Thus, petrochemical plants are frequently built within a large petroleum-refining com-

plex. An interesting example is the Midland, Michigan plant of Dow Chemical Company which was originally built there in order to exploit the natural salt wells that occur in that region.

(2) The producing plant will be built on the shores of deep water harbors, or on a large river, so that tankers, freighters, and barges can dock directly against the plant and discharge their cargoes straight into the manufacturing area.

(3) Chemical plants tend to concentrate in the centers of railroad and trucking transportation networks. In recent years both the railroads and the trucking companies have made intensive efforts to increase the speed and efficiency with which they transport large quantities of basic materials.

d. Many chemical companies keep up an unceasing technical effort to improve their processes so that more product is made from less raw material. Also since raw materials vary in quality, and the lower quality materials are usually cheaper, much effort is expended to utilize lower quality materials.

2. Intermediates. A second kind of raw materials much used by the chemical industry are what are known as intermediate materials. These are not the basic raw materials of Section 1, but neither are they articles used by the general public. Intermediates are usually single pure chemical compounds, such as sulfuric acid or propylene, which are produced from the basic raw materials in chemical plants specifically designed for that purpose (and for no other) and are sold to other chemical companies (or to other divisions of the same company) and there used as raw materials for the manufacture of more complex and more practically useful substances.

These compounds are what the general public commonly calls "chemicals." Their production and sale constitute a very large part of the chemical business. Almost all of the sales of these chemicals are to the chemical industry. A considerable portion of the production never appears in the public market, but is consumed within the producing company. The chemical industry is its own best customer.

The production rates for these chemicals can vary from a few million to many billions of pounds a year. Usually the prices for these chemicals range from 2 or 4¢/lb to 30 or 40¢/lb. In general, the greater the production rate, the lower the price, so that the general form of Fig. 1.3 is followed. This curve, known as an "Exclusion Chart" [1], shows roughly that chemicals with sales prices above about 8¢/lb cannot sell in volumes of above 7 billion pounds/year. The area to the upper right is "excluded." The plot of Fig. 1.3 is only for chemicals selling at rates in excess of

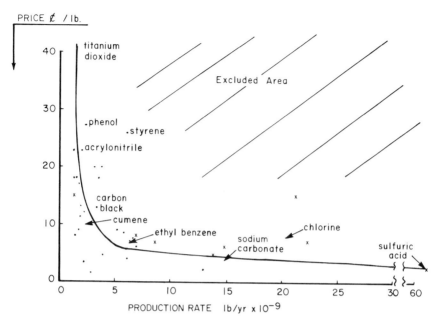

Fig. 1.3. Production rate-sales price relationship for heavy chemicals. Sources: Chemical and Engineering News, June 3, 1974, p. 24 (for production rates) and Chemical Marketing Reporter, December 2, 1974 (for prices).

1 billion pounds/year, but similar charts can be made for other chemicals selling at much lower rates; see Ref. 1.

The cost-production rate data were taken from tabulations [2,3] of the prices and production rates for the 50 chemicals having the largest sales volumes. It is interesting to note that of the 50 chemicals, 20 were only laboratory curiosities or were sold in quite small volume, as short a time ago as 25 years. Chemicals such as cyclohexane, cumene, styrene, terephthalic acid, and butadiene have all leaped from essentially zero sales into the large-volume class within the working careers of many active chemical engineers (including the authors).

By far the heaviest production is of sulfuric acid, at about 63 billion pounds/year. Other basic inorganic substances such as ammonia, sodium hydroxide, chlorine, and sodium carbonate are also produced in large quantities, ranging from 20 to 30 billion lb/yr.

Some organic chemicals have production rates of 6-10 billion lb/yr, and ethylene is the most heavily produced organic chemical, at a rate of over 22 billion lb/yr.

Such chemicals are usually called "heavy" or "basic" chemicals because they are the most fundamental materials produced by the chemical industry. They are used in many ways and are purchased by many manufacturers. They are produced in large plants, some of which are now as large as 1 billion lb/yr, they sell for low unit prices (in many cases less than 10¢/lb) and the profit margin on sales is low. However, the quantities produced continue to increase at a long-term rate of about 7.5% per year.

These chemicals are produced against rigid specifications as to assay, color, and kind and amount of impurities. The products from the various manufacturers are indistinguishable from one another. There is no competition on the basis of quality, and company name plays only a small part in the sales effort. The manufacturers do compete on the basis of price, delivery and the terms of the sales contracts. Transportation costs are an important part of the delivered product cost picture, and bulk transport in box cars, tank cars, pipe lines, and even specially designed ocean going tankers is now an integral part of the bulk chemical sales effort.

The costs which are printed in the public press for the various raw materials and intermediate chemicals must be regarded as approximations. The sale of these products is conducted by negotiation, and variations, especially price reductions away from the "list" price are frequently obtained. Competition in all phases of the chemical industry is quite severe, and manufacturers will frequently offer substantial discounts in order to sell their products.

In addition, it is often true that it is more convenient and cheaper for a prospective purchaser of a chemical to manufacture it than to buy it. This is true of several reactive gases such as acetylene, hydrogen cyanide, and carbon monoxide, and it is also true for nitric and hydrochloric acids.

When considering the present prices for chemicals it must be remembered that prices have risen rapidly in 1974 and are expected to continue to rise for some time to come.

Energy

The chemical industry is the largest consumer of energy of any manufacturing industry in the United States. Almost all of this energy is in the form of electricity generated by public utilities – although some is generated in the plants. In the period 1960-1970 the chemical process industries consumed an average of 30% of the electrical power generated in the United States.

At the present time the United States is facing a shortage of electrical power, and the chemical industry is certain to be seriously affected. The availability of fossil fuels – coal, petroleum, and natural gas – is being severely curtailed, and it is not at all certain that these shortages can be quickly alleviated by increased production or by the expansion of the production of electrical power from atomic energy.

It appears that several things will happen in the immediate future.

1. The chemical industry will continue to expand and will continue to need large amounts of energy.

2. It will become increasingly difficult for the electrical utility industry to supply the required energy, even at higher prices.

3. The chemical industry will be forced to generate more of its needed electrical power than it now does. This will raise many severe problems concerning greater capital investment, plant location, energy management in a plant, cooperation with the local utility company, and an increased involvement with air and stream pollution regulations.

4. The chemical industry will be forced to become more deeply involved with atomic energy plants.

5. Industry costs for energy will increase, production costs will go up, and profits may drop. Some of this took place in 1974, but profits did not drop because sale prices were increased.

6. New branches of the chemical industry will develop: synthetic natural gas from coal and petroleum, chemical manufacturing complexes centering around a nuclear reactor, and the production of a host of new and old chemicals as by-products from the coal and oil gasification plants.

The December 1974 fuel prices were approximately as follows:

Coal	$1.20/$10^6$ Btu
Natural Gas	$0.80/$10^6$ Btu
Fuel Oil	$1.30/$10^6$ Btu

Sources: The Wall Street Journal, November 1, 1974 and December 5, 1974; Sierra Pacific Power Company, Reno, Nevada, private communication, December 13, 1974.

The prices of the various forms of energy have risen dramatically in 1973-1974, and are still rising. The numbers given above are approximations and should be used only for estimation purposes, and for student problems. The future will probably see further increases which will be disturbing to the chemical industry and will cause many important changes.

Manpower

The U.S. chemical industry is not a large employer of labor, but it does employ somewhat over a million people, which is about 5% of all those employed in manufacturing in the United States. A substantial percentage of the chemical industry employees, about 40%, are not directly connected with chemical production, but work in administrative, sales, clerical, and research positions. Thus, only about 600,000 people are actively working in the chemical plants producing chemicals.

The number of people employed by the chemical industry has increased about 300% since 1939, but the ways in which these people work has changed substantially. The recent increase in centralized automatic control and automated materials handling has caused the industry to shift many employees from plant operation to instrument and equipment maintenance. In addition, there has been a large increase in employees engaged in research, engineering, sales, and consumer service. Thus, there has been a sharp upgrading of the skills of the industry's work force, and a shift away from simple equipment tending and materials handling.

The hourly wages and benefits paid to workers in the chemical industry have increased about three times in the last 25 years. However, there have been large increases in productivity in the same time due to sharply expanding volume, heavy capital investment, and mechanization. It is interesting to note that the ratio of the value of the industry's shipments to the value of the total payrolls has increased from 4.75 in 1947 to 6.25 in 1967 to about 10 in 1973.

1.03 THE PRICES OF CHEMICAL PRODUCTS

The sales prices of chemical products are dependent on several things:

1. The manufactured cost of the product

2. The quantity of that particular product currently being produced by the industry

3. The return on the capital investment that the company managers are willing to accept (see Chapter 6)

It is perfectly reasonable that the manufactured cost of the product has a strong influence on the sales price. No one will produce a chemical and offer it for sale at a price which is lower than the cost of making it. At least, not for long.

The manufactured cost of a chemical can be shown to be the sum of many smaller costs; see Fig. 5.5. Some of these costs contribute little to the final value, and others contribute much. Because there are many different kinds of chemicals, the various kinds of costs can contribute various amounts to the final cost. Thus, for some chemicals, made in modern plants at high production rates, the cost of the raw materials will be 60 to 80% of the final cost. The production of some other chemicals will consume large quantities of electrical power. Still others, made in small amounts by batch processes and having several processing steps, may have substantial labor charges.

Generally speaking, chemicals are produced in one of two ways:

1. Continuously, under automatic control, in well-maintained plants that are fairly expensive to build, and with small charges for manpower. Plants of this type usually produce only one chemical and the plant capacity is usually large—nearly always larger than 25 million lb/yr. In this case raw material and utility costs are most important.

2. Non-continuously, in small quantities to high quality control standards, and using a considerable amount of skilled manpower. Under these circumstances labor charges are high, and raw material and utility charges are not so important.

In recent years the first method has been driven to extreme lengths. It is now commonplace for chemical plants which produce only one pure substance to be made very large. Plants to produce styrene and ethylene at rates near 1 billion lb/yr have been built. Other basic chemicals, both organic and inorganic, are now being made in plants that have capacities over 500 million lb/yr.

The object in building plants of these large capacities is to take advantage of what is termed "the economies of scale." Several benefits accrue to the operators of such large plants:

1. The capital investment-related charges per pound of product are substantially reduced.

2. The large size of the plant enables the operators to take full advantage of modern advances in automatic control, materials handling and equipment design.

3. Economies can be achieved in raw material purchase, utilities supply, and manpower utilization.

There are, of course, difficulties. The total capital investment can be quite large, the plant must operate at close to capacity for long periods of time, any shutdown for emergencies or equipment repair is expensive, and the large quantity of product must be sold even if a smaller price has to be accepted.

To date, the economies effected have more than counterbalanced the difficulties. By building and operating such large plants the operators have been able to offset the general increase in the costs of equipment, labor, transportation, and construction. However, there are signs that it may not be justifiable to build still larger plants, or even more of the same. At least not for some years.

As a general rule, the greater the quantity of the substance produced, the lower the price; see Fig. 1.3. An entirely new product, originally made in small amounts, might command high prices and show large profits. However, as the production rate increases the sales price will decrease. If the initial profits are large, it is essentially certain that the production rate will increase. This effect seems to be a general law.

Figure 1.4 shows a plot of prices versus production rate for a spectrum of chemical products [4]. This graph is not exact, but it does show the price—production rate trend which has been observed.

Chemicals which have been produced for many years in large quantities tend to have more stable prices, but lower profits.

Some companies are willing to produce chemicals which have a fairly low price, a fairly low profit margin, and a low return on investment, but a high rate of production, so that the dollar volume of sales is large. Such companies are willing to start the production

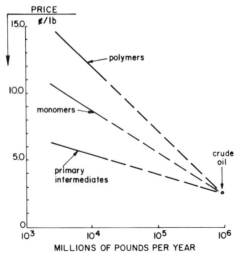

Fig. 1.4. General price trend for a spectrum of chemical products (Crude oil: approximate 1975 price, now increasing) [after Waggoner (4)].

The Chemical Industry

of a low price-low profit chemical, or to expand the existing production of such a product. They feel that it is preferable to be in a large-scale low-profit business which is fairly certain, than to risk failures in a high-profit essentially uncertain enterprise.

Other companies will produce only those chemicals that should have a high price and a high profit margin with a high rate of return on investment, even though the dollar volume of sales is not large. In order to be successful in this way it is necessary to introduce new products to the chemical business essentially continuously, because the profit margins of new products tend to decrease with time, often quite rapidly. Introducing new products is fairly risky; many do not succeed. However, if they do succeed the profit margins can be high, and the return on investment will be large.

A business philosophy such as that above requires intensive research and development work, clever marketing and sales efforts, and an intelligent, far-sighted, and flexible management. Several of the larger and more profitable chemical companies have introduced new products by the thousands over the years.

It is apparent from the above discussion that the two different approaches to business will strongly affect the prices of chemicals.

One graph will illustrate these points quite well. Figure 1.5 shows the production rate–sales price–time relationship for acrylonitrile (propene nitrile) produced in the United States during

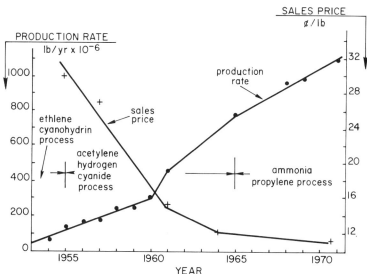

Fig. 1.5. Acrylonitrile production rate–sales price–time relationship for U. S. production only.

the period 1953-1971. This chemical is most widely used for the production of acrylic fibers, but it is also the raw material for plastic molding powders, Buna N rubber, and the very successful acrylonitrile-butadiene-styrene (ABS) resins.

Figure 1.5 shows two lines. One line is for the rate of production for each year between 1953 and 1971. The other line shows the sales price—time relationship over the same time period. Also included are three dimension lines showing which manufacturing process was most important during a period of time.

There are two principal observations to be made from studying this graph:

1. The production rate and the sales price both experienced unusually large changes. The production rate increased by 20 times and the sales price decreased 3 times in the 18-year period. It is apparent that there is a good correlation between the rising production rate and the falling sales price. The exact nature of the causal relationship is complex, but it is true that the sales price fell as the quantity produced increased.

In the beginning acrylonitrile was something of a specialty chemical and sold for a rather high price. The technology and the willingness to produce more were present, but at the high price the sales potential was not great. In an effort to increase demand the principal manufacturer deliberately lowered its price. This effort succeeded and led into the spiral of greater sales, better technology, low prices, and low profits.

2. Within less than 20 years the manufacturers adopted three different processes for making the product. Each process was more efficient than the last, used cheaper and more readily available raw materials, and produced a lower cost product. Thus, although the sales price fell, the cost of manufacturing also fell, and the companies were better able to combat the falling sales price.

Several of the organic compounds now sold in very large quantities—phenol, styrene, and vinyl chloride—have had similar histories. The business is now very large, but the prices have dropped until the profit margin is thin, and the business is unattractive.

Stobaugh [5] has analyzed the chemical price situation and drawn the following conclusions:

1. The three most important factors determining the manufacturing costs are raw materials, plant size, and technology.

2. As time passes, the size of the market for a given chemical increases, therefore the sizes of new plants can be increased because of the larger volumes involved. Increasing chemical plant size results in economies in operating and capital costs and therefore a lower cost product.

3. Technology improves with time (see Fig. 1.5), and better technology results in lowered raw material costs and reduced capital. A newer plant can have two advantages: larger size and better technology.

4. Raw material costs also decline in price with time because they are needed in larger quantities, and the economies of scale apply to raw materials just as they do to finished products.

5. As a consequence of all this, a new producer with new technology and a larger plant can produce a lower cost product and undercut the prices of the competition. This may be attractive business for a while, and as a result new producers will be attracted to the business. Older producers then suffer as the prices decline.

Waggoner [4] has also commented on this point and offers a view of the future. Waggoner believes that the trends of the past 20 years cannot continue much longer. He cites several reasons for this:

1. The steadily rising cost of new construction and new machinery. There is no reason to believe that this trend will be slowed, much less reversed, and future chemical plants will cost quite a bit more than the ones of today. This high cost of building new plants will restrict the amount of new manufacturing capacity which might be built in the future; see also McCurdy [6].

2. Competition from foreign producers. These companies now have technologies equal to those of the United States, and they are also well situated with respect to raw materials. Their plants cost less to build and their labor costs are lower. They can compete in the U.S. market on better than even terms and probably will continue to do so. Furthermore, several of the foreign firms—particularly English and German--have bought control of several outstanding U. S. firms and have become important factors in the domestic chemical business. In fact, the Imperial Chemical Industries of England has now become the largest chemical firm in the world because of its purchase of the Atlas Chemical Industries of Wilmington, Delaware.

3. The prices of raw materials will probably rise. U. S. prices for natural gas and petroleum derived feed stocks are now rising and the quantity is declining. The large supply of cheap raw materials

that has formed the base of the organic chemical industry in the United States is now not as available as before. More expensive raw materials mean more expensive products.

Precisely this happened in 1973-1974. The cost of raw materials increased greatly and the prices of the pure chemicals also increased. In the case of acrylonitrile the sales price rose from 11¢/lb in 1971 to 23¢/lb in 1974. Similar increases were experienced for phenol, styrene, and vinyl chloride.

1.04 RESEARCH AND DEVELOPMENT IN THE CHEMICAL INDUSTRY

The chemical industry in the United States does a large amount of research and development work. At present, the basic chemical industry spends almost $1 billion annually on such work. More than 20,000 scientists and engineers are employed to work in this field. The amount spent on research and development work by the basic chemical industry in the United States is approximately 2.2% of the value of the industry's annual sales. This is a very large effort, equivalent to the activities of a giant corporation. To place this in a business perspective, it should be realized that the after-tax earnings of the basic chemical industry in the United States in 1973 was 6.3% of sales. Thus, for various reasons, chemical corporations in the United States feel compelled to spend on research and development work a sum of money that is approximately 35% of its total profits. If it did not spend the money at all its profits might be higher—perhaps. On the other hand—perhaps not. The relationship of the chemical industry to its research and development effort is a complex one. Discussions concerning this are voluminous, heated, and never ending. Suffice it to say that no discussion of the chemical industry can be complete without an examination of the research and development effort, and the reasons for its existence.

The chemical industry has enjoyed a close relationship to chemical science since the beginning of chemical science. Early in the history of the chemical industry it was discovered that chemical research work in the laboratory could be exploited commercially so that profitable sales resulted and business activity increased. Over a period of years chemical manufacturers organized formal research groups, working in special laboratories, that did chemical research work on the products and processes of interest to the company. In addition, close ties were maintained with academic workers so that the technical level of the industry's work was always high, and new ideas were welcomed.

The Chemical Industry

In the chemical industry the effect of the research effort has been profound. New products and processes have regularly accounted for a large share of industrial profits, and many companies have grown many times larger in size by the investment of their surplus funds into large units for the manufacture of internally discovered products. Growth by internally generated knowledge powered by internally generated money has been a characteristic of the chemical industry.

The chemical industry's research effort has affected many other things beside the industry itself. No manufacturing company has been able to escape the influence of the chemical research work. Many firms have organized similar laboratories of their own, and have discovered many useful facts applicable to their particular business.

There are three forces behind the drive to do research and development work in industry.

1. Because of discoveries in the laboratory it becomes possible to introduce to other industries, or to the public, a new product never before known, that has superior properties. These may be exploited by both industry and the public to increase the volume of profitable business, or to raise the standard of living. The outstanding example of such a new product is nylon. There are many others, among them, tetraethyllead, polyethylene, and melamine.

2. Companies have a desire for an increase in knowledge concerning their own products or processes. The more knowledge a company has about its own processes or products, the stronger it becomes in respect to production and sales.

3. If a firm does adequate research work, and its competitors do not, the first will learn much more about the business and will shortly be able to take all the business for itself. Competing firms will be unable to match its internal efficiency and its intelligent and flexible sales effort. For this reason, almost all chemical companies should do some research and development work or find themselves badly beaten by their competitors.

As might be expected, the largest chemical firms do almost all of the research work done in the industry. Fifteen chemical makers account for 70% of the industry's sales and do 75% of the research and development work. Usually, smaller firms spend less on research work than the national average, although there are notable exceptions.

In the chemical industry two kinds of research and development work are done:

1. Investigations on the processes by which a product is made. Here there are two kinds of researches: (a) The search for a new process to make a new chemical. This is pioneering research and

can be quite risky because of the danger of failure from both the technical and marketing aspects. Work of this kind is done almost exclusively by the larger firms because of the long time and the heavy expense involved. On the other hand, success in this field can lead to years of profitable sales. The Dow Chemical Company's activities in the field of styrene production is a good example of this work. (b) The search for a new process for an old chemical. No process can be regarded as unchangeable. Improvements are always possible and may well provide substantial savings in the manufacturing process. Furthermore, the development of a new process for an old chemical is a low-risk project because it is not necessary to develop a market for the chemical. Thus, the most economical process for the production of phenol has changed three times in 30 years. In this connection, work with cheaper raw materials is important.

2. Investigations of new products and modifications of old ones. The chemical industry's research effort creates thousands of new products every year. Some of these are really quite new, for instance stereospecific rubbers, and some are simply modifications of older products, for instance, a change in the physical treatment of a catalyst particle.

Work of this kind is quite detailed and specialized, and if not carefully managed, can be more expensive than it is worth. Such work requires close contact with potential customers, a considerable amount of empirical experimentation, and a detailed knowledge of the needs and workings of the industry to be served. To give one example, several of the larger companies maintain fairly large groups which specialize in the synthesis, testing, and application of insecticides. It must be obvious that a good all-purpose insecticide would be a product worth having.

For the reasons outlined above the chemical industry believes that research and development work is a useful activity to support, and it intends to continue such support. However, it must be realized that the chemical industry is a profit-making business like any other. It must balance its expenditures against the possible benefits to be gained therefrom. Money spent on research and development is not excepted from this rule. During the 1950s and the early 1960s requests for money from research departments were approved almost without question. Research was regarded as the key to a more profitable future, and research efforts were strongly supported. However, the last decade has not been a good one for the chemical industry, and its research effort has not been as productive as had been hoped. As a consequence, all industry expenses are now being carefully investigated from the accountant's viewpoint, and the research departments are not exempted from this.

The venture worth concept, introduced and discussed in Chapter 3, is the best method yet devised for evaluating the economic

The Chemical Industry

attractiveness of a new venture. Section 6.09 discusses the economics of development work, and uses the venture worth criterion to determine the advisability of doing such work.

There have been several results of this more intense and detailed examination of research projects:

1. The amount of money in contemporary dollars (not corrected for inflation) that the chemical industry spends on research continues to grow and was 110% greater in 1973 than in 1960.

2. As a percentage of industry sales this outlay for research has decreased from 4.2% in 1960 to 2.2% in 1974. The rate of decrease has been almost linear since 1970.

3. If the industry's expenditures for research are corrected to 1958 dollars, the sums spent annually are no greater in 1974 than in 1960, and are significantly smaller than in 1958.

4. The number of persons employed in the research effort is now less than the number in 1960 and almost 16% less than the number in 1968.

It is apparent that the chemical industry has found it necessary to reduce the level of its spending on research work. Several reasons have been advanced for this reduction:

1. The cost of doing research increased substantially in the period 1960-1974. In fact, the cost of research increased more than any other aspect of the industry's costs, except for the cost of construction.

2. It became increasingly difficult to commercialize the results of the research work because of the high cost of construction for new plants, higher interest rates on borrowed money, increasingly severe competition, and the diversion of industry funds to air and stream pollution abatement construction.

3. Chemical technology became increasingly sophisticated and mature. The "easy" things had been done, and new chemical developments of commercial promise were few and far between. There was a great deal of expensive searching, but not much discovery.

4. The poor level of business for the chemical industry in the 1960s led many companies to seek increases in sales and profits by the acquisition of other firms in other industries. Growth by commercialization of research work did not appear to be a promising path. It was accorded a lower priority with respect to other investments. This, in turn, led to a reduction in spending on research work.

The present situation in industrial chemical research work seems to be as follows:

1. Industry is continuing to do research work. However, the extent of the industry's commitment to this activity is substantially less than it has been in the past. The gross number of dollars spent continues to rise, but when measured as a fraction of the industry's sales, the industry's support of research work has declined 50% in ten years.

2. The work now being done in the chemical industry's research laboratories is more strongly business-oriented than in the past. Research projects that appear to have a fairly long time to commercialization, or whose business goals are not clear, are not strongly supported. Every project is periodically subjected to a thorough technical and economic evaluation. Research work is not allowed to drift along; it is either commercialized or terminated.

3. It is apparent that this attitude is not sympathetic to pioneering research. Concentration on short-term goals may not produce new developments of importance. Many scientists argue that the present attitude is self-defeating, and in the long run the industry will suffer, not profit, from its present approach. On the other hand, the business managers say that in the long run we are all dead, but that today we must make money.

1.05 THE FINANCIAL ASPECT OF THE CHEMICAL BUSINESS

Chemical companies in the United States, and in a considerable portion of the rest of the world, are publicly owned, publicly financed, stock corporations operating in a profit-oriented economy, under a considerable degree of government control. This means that chemical companies are a part of what is known as profit-making private industry.

In order to obtain money with which to start operations at the company's formation, or at other times during its life, a company will borrow from banks and insurance companies, sell bonds, or preferred and common stock to both private and institutional investors. Such companies are owned by the common stock holders and are operated by a board of directors who are elected by and are responsible to the stockholders for the profitable operation of the company. The board of directors appoints the principal officers of the company and, in turn, holds them responsible for profitable operation.

The incentive for any institution or private person to invest in a company is to receive a monetary reward. This reward can take one of several forms: interest on loans or bonds, with eventual repayment of the principal; dividend payments on preferred or common stock; and, for common stockholders, participation in the growth of the company, meaning that the dollar value of the shares of stock increases with time. Many chemical companies have thousands of stockholders and the shares of these companies are traded on the securities markets of the world in large quantities every day. Of the 132 large, medium, and small chemical and allied products companies listed in Chemical and Engineering News for September 6, 1971, every one was a publicly owned stock corporation.

Chemical companies can have financial structures that are made up of bonds, bank loans, and preferred and common stock in all possible combinations. There are usually excellent reasons for a given company having a given arrangement of stocks, bonds, and bank loans. Also, as business and financial conditions change these arrangements may change.

Table 1-1 shows the financial structure of 15 typical chemical companies in 1970. Included in the table are six large firms—the industry leaders—and five medium size and four small ones. Also listed are the chemical sales, the after-tax profits (as percent of sales), and the financial structure, described in terms of the percentage of the total capital contained in each category. Several points are noteworthy:

1. There are great differences in size between the various groups of companies.

2. The financial arrangements of the companies are similar despite the great differences in size.

   a. Twelve of the 15 companies have some funded debt (bonds or loans). For some this is fairly large (about 25% or more of the total capital); for others it is much smaller (about 11%).

   b. Three of the companies have no, or almost no, debt.

   c. The financial structure of a company is independent of its size.

3. There is a surprisingly large variation in the profit record between the companies. Also, size apparently has no effect on profitability. Small companies can be just as profitable as the large ones.

4. A general average of the company financial structure would be 20% funded debt, 0% preferred stock, 80% common stock.

There are any number of statistics by which the success or failure of a business firm may be measured. Three of these numbers, of great interest to common stockholders, easy to understand, and available in widely distributed financial publications are the stock market price of the common stock of the company, the earnings

Table 1-1 Financial Structure of Chemical Companies—1970[a]

Company	Chemical sales ($ × 10^{-6})	After-tax profits (% of sales)	Funded debt (% of total capital)	Pfd stock (% of total capital)	Common stock (% of total capital)
duPont	3224	9.1	2.6	2.4	95.0
Union Carbide	1876	5.2	23.5	0	76.5
Monsanto	1716	3.9	26.6	6.0	67.4
Allied Chemical	837	3.4	32.4	0	67.6
Am. Cyanamid	602	7.5	11.4	0	88.6
Dow Chemical	1643	6.9	26.0	0	74.0
Five medium size companies					
Stauffer Chem.	483	5.4	26.4	18.7	70.3
Rohm & Haas	445	5.7	11.0	0	89.0
Air Products	250	5.7	30.0	0	70.0
Lubrizol	184	11.9	0	0	100.0
Reichold	178	1.8	34.0	0	66.0
Four small companies					
Ansul	40	3.3	36.8	0	63.2
Emery Indust.	84	8.2	23.0	0	77.0
Great Lakes Chem.	12	11.4	11.5	0	88.5
Lawter Chemical	16	16.4	0	0	100.0

[a]In Section 6.08 rates of return are discussed. The rate of return used above is presented only for comparison purposes. For design purposes a rate of return based on net capital investment is preferred. Such rates are discussed more fully in Section 6.08.

per share of the common stock, and the dividends paid by the company. Of particular interest are the historical records of these numbers. Figure 1.6 shows a schematic diagram of the time record of the three numbers mentioned above for the Dow Chemical Company for 1946-1970. The numbers have been rounded off slightly so the graph is not an accurate plot. True accuracy is not important; the important thing is to study the trend.

The Chemical Industry

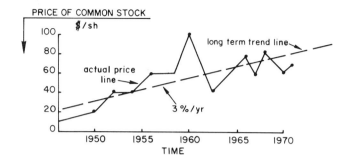

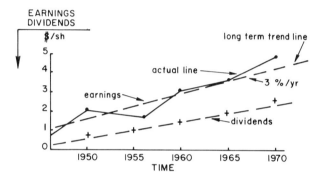

Fig. 1.6. Business record: Dow Chemical Company 1946-1971.

There are several interesting conclusions that can be drawn from the lines in Fig. 1.6:

1. The Dow Chemical Company has been quite successful. Its record of earnings has shown a steady rise over the last 25 years, and the dividends paid to the common stockholders have risen almost in step with the earnings. This is always a sign of a strong and successful business.

2. The record of the price of the common stock is somewhat erratic, but an average long-term line drawn through the actual price line shows that the price of Dow common stock has been rising steadily for 25 years and shows no signs of falling. From the standpoint of business operations, a steady long-term rise in the price of the common stock represents success.

In the years 1971-1973 the price of Dow Chemical Company stock had an unprecedented rise. The price increased by more than

Table 1-2 Sources and Applications of Company Funds

	1966	1970
	% of total	
Sources		
Net income (business profits)	45.1	35.4
Depreciation	29.4	32.5
Other corporate sources	6.1	13.1
Long-term debt	15.4	17.3
Sale of stock	4.0	1.7
Applications		
Dividends	24.8	22.7
Additions to working capital	5.5	6.7
Capital expenditures	60.3	53.6
Reduction of long-term debt	2.5	8.4
Other applications	6.9	8.6

100% in that time. The earnings of the company also rose dramatically—a stunning achievement.

Many other chemical companies have similar records. None of the larger companies have better records, and some have records that are substantially worse. Several smaller companies, which are fairly specialized in their business operations, have shown more profitable activity and higher rates of growth.

It is interesting to note the sources of company funds and the application (disbursement) of these. Table 1-2 shows the record for the major U. S. chemical companies for the years 1966 and 1970 [7]. A study of these numbers reveals several interesting facts about the business operations of chemical firms.

1. Substantially less than one-half of the companies' operating income is derived from the sale of chemical products. In fact, in 1970 this income was only slightly greater than one-third of the total income.

2. The cash flow from depreciation allowances is quite large and in 1970 was almost equal to that from chemical sales.

3. Long-term debt (borrowing and bond sales) is an important source of funds and is becoming more important despite high interest rates.

4. The sale of common stock is not a significant contributor of funds. In fact, few large chemical firms have sold any common stock in the last 20 years. The very large number of common shares now outstanding from each of the large chemical corporations has resulted from numerous distributions of shares (stock "splits") to the existing stockholders.

5. The greater part of the income to the company is spent for new capital investment. This number is in accord with the knowledge that the chemical industry is a large consumer of capital and that the greater portion of the new capital comes from within the company.

For a good portion of the period 1962-1972 the chemical industry did not add new equipment at a rate sufficient to meet the expected demands of the future. However, the profitable years 1973 and 1974 saw a substantial increase in new capital investment.

6. The chemical industry has been rather generous about paying dividends on common stock. It is improbable that the percentage of income paid out in dividends will increase, because of the heavy demands for the addition of new equipment and the replacement of old.

7. The percentage of income devoted to retiring long-term debt has increased to a substantial value. This reflects two trends: (a) an increase in debt financing so there is more debt to retire and (b) a desire to pay off the debt and return to a more conservative financial posture.

1.06 FUTURE PROSPECTS

Almost all responsible predictions of the future of the chemical industry in the United States indicate that the annual volume of production of basic industrial chemicals will increase by a factor of between 2 and 3 during the 1970s. This is not a startling prediction since a somewhat larger increase took place during the 1960s.

The chemical industry is growing at a faster rate than almost all other manufacturing industry. The overall average annual growth rate during the 1970s should be about 7% per year. This is substantially higher than the growth rate of the gross national product or of the growth rate of all manufacturing industry.

Some sectors of the industry will grow more rapidly than others. Organic chemicals, especially thermoplastics, will have larger than average gains while others, such as inorganics, will have smaller ones. It is almost certain that several new products will be introduced and will have rapid sales gains and high profits. It is also fairly certain that there will be a continual rearrangement and restructuring of the industry as companies seek to increase their profit margin by diversification, acquisition, and merger.

Despite the near certainty that the industry's sales will increase very much, it is by no means certain that the industry's profits will increase similarly. Furthermore, it is almost certain that the industry's problems will become more severe. Several of the future problems of the chemical industry can be listed:

1. The steady increase in the cost of doing business. The nationwide monetary inflation (see Appendix B) is causing all industry's costs—for raw materials, energy, manpower, new facilities, and transportation—to increase steadily. In order to offset such increases the industry must become more efficient, increase its prices, or shift its activities to more profitable fields.

2. The decrease in the amount of energy available, accompanied by an increase in its costs. As discussed in Section 1.02, a worldwide energy shortage is imminent, and this will surely affect the chemical industry. Over the short term, the effect is almost certain to be bad.

3. The decrease in both the quality and quantity of certain raw materials. Among these are natural gas, petroleum fractions, water (both process and cooling), and certain ores. Shortages can be overcome by imports, by changing to other raw materials, or by using lower grade substances. Whatever happens, some important raw materials are certain to become scarcer, more expensive, and less pure. This will have, and is now having, an adverse effect on the industry.

4. A greater concern for the abatement of air and stream pollution. The chemical industry is now quite aware that the public and the government will not allow any significant amount of air and stream pollution. This awareness will take several forms: (a) the installation of greater amounts of equipment to remove any pollutants, (b) the modification and development of processes so that process-related pollution is greatly reduced, (c) the abandonment of older processes and plants that cannot be modified to reduce pollution, and (d) the non-commercialization of new processes that are otherwise excellent, but that produce unacceptable amounts of pollutants.

The Chemical Industry

5. The decreasing effectiveness of the industry's research and development effort. For a variety of reasons, such as discussed in Section 1.04, the chemical industry has been forced to reduce the size of the effort and to change the focus of its research work. There has been a definite reduction in the rate of introduction of new products and processes, and a greater emphasis on the solution of short-term, more easily solved problems.

6. The difficulty of obtaining new financing. At the present time, interest rates are near historically high levels, so that borrowing is expensive. The industry's recent business record has not been good, so that it may be difficult to sell common stock, and, because of a lower profit level, industry's retained earnings may not be sufficient to finance a substantial expansion. Yet it appears that there will be a large expansion, and the money must be found to pay for it.

The difficulties listed above are certainly serious; by themselves they are a formidable group of problems. There are, of course, many other problems, and new ones are sure to arise. The industry will battle with them as it has battled other problems in the past. Some problems will be overcome, some will not, and some will change with time.

It is also interesting to discuss briefly the strong, or good, points about the industry, and to see how these will influence the future.

1. Probably the strongest, most positive property that the industry possesses is that it is so fundamental to modern corporate and private life. The chemical industry, its products, its knowledge, and its procedures are all pervasive in modern life. This high degree of involvement derives from the basic nature of chemical science and the systematic exploitation of this that has characterized the chemical industry. There is no way to escape from the fact that human beings and human life are inextricably involved with chemical science. Any industry that is based on chemical science is in a strong position to influence human life and work. As the world population grows and as mankind strives for a more secure, comfortable, and rewarding life, the chemical industry is certain to grow. There is apparently no end to the demand for the industry's products.

2. The chemical industry is highly technical, possessing great resources in chemical, engineering, and managerial skills. These resources enable the industry to create new businesses, change old ones, shift resources, and exploit any scientific discovery. The industry is flexible, it does not feel committed to any particular line

of business, and it is confident of its ability to succeed in new ventures because of its great strength in science, engineering and marketing.

3. The research and development staff and facilities of the chemical industry are superb. They can be counted on to produce a workable solution to almost any technical problem. These research staffs have in the past, and will in the future, produce many new products that will be useful, and they will also find better and cheaper ways to produce the basic chemicals now in use.

4. The chemical industry is intensely competitive. Chemical companies know a great deal about hard work, creative research, economical operations, and imaginative marketing. In the last 10 years they have passed through a difficult period that has taught them much. Their efforts to increase profits in the teeth of the difficulties of the last 10 years has given them valuable experience and knowledge which will help them in the difficult years ahead.

5. The chemical industry is in good financial condition. While there has been a decrease in profits and some increase in funded debt, there have been no reduced or passed dividends, no defaulting on bond issues, the common stocks still command reasonably good prices, and there has been a general increase in the industry's operational efficiency.

Since the above was written many of the companies in the industry have experienced a substantial improvement in their business operations. The years 1972 and 1973 were quite good and 1974 should be as good or better. The industry is now spending heavily on new plants and equipment for future operations.

REFERENCES

1. H.W. Zabel and M. Marchitto, Chemical Engineering, October 19, 1959.
2. Chemical and Engineering News, June 3, 1974, p. 24.
3. Chemical Marketing Reporter, December 2, 1974.
4. J.V. Waggoner, Chemical Engineering Progress, 67, No. 7, 17 (1971).
5. R.B. Stobaugh, Chemical Engineering Progress, 60, No. 12, 13 (1964).
6. R.C. McCurdy, Chemical Engineering Progress, 65, No. 5, 19 (1969).
7. Chemical and Engineering News, September 6, 1971, p. 39A.

Chapter 2

PRINCIPLES OF ECONOMIC EVALUATION

2.00 BASIC CONCEPTS

In the chemical industry no money will be spent by a manufacturing company unless a good argument can be advanced that the spending will assist the company in maximizing its financial return. Chemical companies spend large sums of money for research, maintenance, replacement of old or obsolete equipment, the construction of new facilities, and numerous other activities designed to improve the company's operating efficiency. Long experience tells the company's officers that such spending can result in greater profits within a rather short time following the spending of the money.

The criterion by which the worth of the spending is judged is its estimated contribution to the economic well-being of the company. Every project is subjected to a technical and economic evaluation [1]. In the economic evaluation, the same general principles apply whether the problem is the choice between various designs for complete plants, the design of the individual parts of the plant so that the different parts meet the same economic criteria, or the technique of economical operation of existing plants; see Section 2.15.

The general considerations that form the framework on which sound decisions can be made are simple. Often, the calculations based on these principles can be made with a considerable degree of exactness, but just as often they cannot. In the development of a commercial enterprise there may be too many intangibles to allow an accurate analysis. In these cases the judgment must be intuitive, but based on much experience and detailed analysis.

2.01 GROSS INCOME

Commercial operation involves selling a product to customers for a price, s. Therefore the annual cash income to the producer (seller) is equal to sQ, where the s is the selling price in $/lb and Q, the lb of product sold/year. The producer has a manufacturing cost c which reflects the cost to the producer of the raw materials, the utilities (steam, cooling water, electrical power, and others), the maintenance of plant equipment and the labor and supervision. (See Fig. 5.5 for a list of items to be included in the manufacturing cost.) The cost c does not include any charges for depreciation or taxes. Then,

$$R = sQ - cQ,$$

where R = gross income to the seller - $/yr (does not include deductions for depreciation or income taxes).

2.02 CAPITAL INVESTED IN FACILITIES AND WORKING CAPITAL

The producing company must invest a sum of money I in order to purchase and install the facilities needed to manufacture the product. Since chemical plants can be built quickly this sum of money will be spent within two years or less. This is usually a small fraction of the useful life of the project. In addition to I, there is needed an amount of capital I_w, called the working capital. This is the quantity of money invested in raw materials, intermediate and finished products, accounts receivable, and cash required to operate the project. The capital invested in nonproducing or service facilities such as utilities may or may not be included as part of the facility capital I, depending on accounting procedure. In this treatment such auxiliaries are not included in I.

Principles of Economic Evaluation

2.03 INCOME TAXES LEVIED BY GOVERNMENTAL AGENCIES

The gross income of corporations, minus an allowance for equipment depreciation (see Section 2.04), is taxed quite heavily by the U. S. government and rather lightly by the governments of the states and communities where the corporation owns property or does business.

The Federal tax rates are high, they are dependent on the economic policies of the current administration, on the domestic and world political situation, and on the nature of the company's business. The exact computation of Federal taxes is so detailed and complex that almost all corporations employ tax specialists who spend all their time on the company's tax problems. It is apparent that for a company with many millions of dollars of sales, a small fractional saving in taxes may amount to hundreds of thousands of dollars. This is large enough to have a strong influence on the company's business practices and policies. Consequently, tax problems, with all their ramifications, are now a special field of great sensitivity and importance.

In a preliminary analysis of a new venture it would be foolish and impractical to go into a detailed calculation of the expected taxes. However, a reasonably good approximation of the expected tax charge must be made.

In 1970 Federal income-tax rates for corporations were as shown in Table 2.1.

Table 2-1   Federal Income-Tax Rates

Taxes	%
Normal tax (on gross income - depreciation allowance)	22
Surtax (on gross income - $25,000 - depreciation allowance) (a tax on all gross earnings above $25,000)	26

It is apparent from this that if $25,000 is small in comparison to the tax base, and it is for all except the smallest companies, the Federal income-tax rate may be taken as 48% of the gross income, R, minus the depreciation allowance.

State and local taxes may be taken as 3% of the tax base Thus, all taxes together may be approximated as 51% of the tax base. For convenience in the approximation calculations used in economic evaluation work it is adquately precise to use a tax rate of 50%, or half of the difference between the gross income and the depreciation allowance.

2.04 ALLOWABLE DEPRECIATION FOR ACCOUNTING AND INCOME TAX

As discussed in Section 2.03, the Federal government allows corporations to make deductions from their gross incomes before levying a tax on the corporation's earnings. This allowance is made in recognition of the fact that plant and equipment are subjected to constant physical wear, and are threatened by technical obsolescence. By allowing corporations to deduct a certain amount of money each year from their gross income, so that the sum of money on which taxes are levied is reduced, the government assists corporations in meeting the expenses which are incurred when worn or obsolete equipment is replaced.

As mentioned in Section 2.03, the taxation laws and depreciation allowance rules are complex [2, 3], and are not the concern of this book. Suffice it to say that it is in the interest of the corporations to make the deductions large, but the U. S. Treasury Department regulations restrict these deductions within narrow limits [2, 3].

On some occasions, the government may authorize special fast depreciation rates in order to stimulate investment in new plants and processes. These accelerated depreciation rates are now subjects of political controversy. When these rates are employed in an economic evaluation a clear understanding should be reached with the tax authorities.

The Internal Revenue Service allows a corporation to calculate the yearly depreciation deduction in one of several ways. These methods are discussed in some detail in Chapter 3, especially Section 3.7. An illustrative example of the calculation is given in Example 3.2.

Here, it can be said that there are two types of procedures for calculating depreciation: (1) the straight-line methods, and (2) nonlinear, or accelerated, methods. The first method has the virtue of simplicity and is widely used in preliminary evaluations. The second methods are somewhat more complex but recover a greater portion of the investment in the first years of a project's life. This factor is regarded as quite important by industrial corporations.

The straight-line method assumes that the value of the property decreases linearly with time, and each year a deduction equal to

$$\frac{\text{Original value - salvage value}}{\text{Number of years of property life}}$$

is subtracted from the gross income, R. In many instances in the process industries the salvage value is very small (say zero) and the number of years of property life allowed by the Internal Revenue Service is essentially 10. Therefore the yearly depreciation deduction used is 0.1 (I). Special permission must be obtained from the Internal Revenue Service before a greater depreciation rate may be used. A discussion of accelerated depreciation methods is deferred to Section 3.06.

In addition to the charge against income which is allowed, and regulated, by the Internal Revenue Service for the purpose of computing the Federal income taxes, there is another charge made for the purpose of recovering, for the investor(s), the original capital investment. This is called "the depreciation charge for accounting." As is pointed out in Section 6.01, all investors wish to recover the money which they gave to the enterprise at the beginning. In order to make sure that such a recovery does take place, the operators of the project make a yearly charge against the income from operations, after an allowance for taxes, so that after a number of years the sum of all these charges will be equal to the original investment.

This situation is made clearer by a reference to an investment in U. S. government bonds. Here an investor gives $1,000, for one bond, to the U. S. Government in return for a contract which says: (1) The investor will receive, say, $40 each year for the life of the contract, (2) at the end of the life of the contract the U. S. Government will return to the investor the original investment of $1,000. [The authors are well aware of the vagaries of bond prices. They put forward the above illustration in its simplest form.] The investor will not give his money to anyone unless he is reasonably certain that he will receive a yearly interest payment and will also have his originial investment returned after a known number of years. The similarity between a bond investor, a depositor in a savings bank, and an investor in a chemical company, or the chemical company itself, should be clear. No one, and no corporation, will invest in an enterprise unless the financial structure is such that the return of the original investment is programmed into the operations. The effect of monetary inflation is discussed in Appendix B.

From the above it can be seen that there are two depreciation charges made by the corporation.

1. A charge (actually an allowance, not a charge) which accounts for deterioration and obsolescence of buildings and equipment. This charge is used in the computation of the taxation of the corporation's profits, and is controlled by the U. S. Treasury Department. The deduction is always made because it lowers the corporation's taxes, but the amount of the charge and the procedure for computing it are essentially out of the control of the corporation. This charge against gross profits lowers the corporation's tax liability and so increases the number of dollars the corporation can retain.

2. A charge is made by the corporation against its own income, after an allowance for taxation, so as to ensure that the original investment in the process, buildings, and equipment will be completely recovered at the end of some number of years. The magnitude of this charge is under the control of the company. Thus, the charge can be made quite high and the investment recovered quickly; or vice versa. Also, the fraction of the investment recovered each year can be changed during the life of the process.

If the actual life of a plant is known, the exact rate of capital recovery can be established using a series of uniform payments whose sum at compound interest just equals the investment. The interest rate to be employed in economic studies is generally somewhat higher than that corresponding to the current rate for borrowing capital.

From all of this it can be seen that the net profit returned to the corporation may be pictured by the flow diagram in Fig. 2.1. It is important to note that all of the income after taxes, $R - (R - dI)t$, is returned to the company. Customarily it is split into two parts:

1. The portion of the company's profits after taxes that is assigned to the recovery of capital is eI, where e is the fraction of capital to be recovered each year. This number is under the corporation's control and is usually based on an estimated project life of n years.

2. The remainder, the net profit, is seen to be $R - (R - dI)t - eI$, and this sum is also returned to the company. The net profit is used by the company's directors to pay dividends on preferred and common stock and to provide an excess (retained earnings) which may be used for further capital investment. In nearly every instance these flows of cash are not separated, but are lumped together in the pool of corporate funds.

Principles of Economic Evaluation

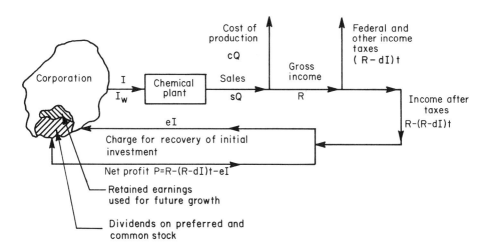

Fig. 2.1. Simple money flow diagram for a chemical plant.

It is important to note that in Eq. (2.1) and Fig. 2.1 the gross income R is computed before any charge for depreciation is included in the cost of the product. In many texts, and in many examples of these economic calculations, a charge for depreciation is included in the product cost. It is customary to include in the list of individual cost items, which add to the total manufacturing cost, an item for depreciation. (See Fig. 5.5.) Thus, when the product cost is used in calculations, the depreciation charge is already included. This method is straightforward and simple, and it establishes a basic selling price for the material. Since this charge is made before the taxes are calculated, the depreciation charge must be made in accordance with the U.S. Treasury Department regulations.

When the depreciation charge is included in the product cost, the money flow diagram of Fig. 2.1 changes to the money flow diagram of Fig. 2.2.

The equivalence of the two methods and the significance of the stream dI may be seen by an application of some simple algebra. From the definitions given above

$$R = sQ - cQ \tag{2.1}$$

$$R' = sQ - c'Q \tag{2.2}$$

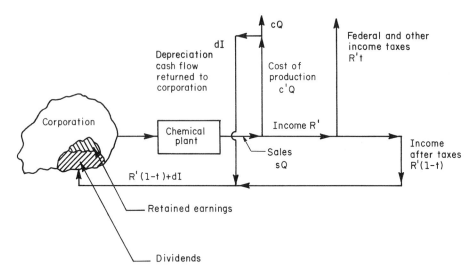

Fig. 2.2. Simple money flow diagram for a chemical plant.

$$c' = c + \frac{dI}{Q} \qquad (2.3)$$

$$R' = sQ - cQ - dI \qquad (2.4)$$

Income after taxes

1st case: $R - (R - dI)t = R - Rt + dIt$ (2.5)

2nd case: $R'(1 - t) = (R - dI)(1 - t) = R - Rt + dIt - dI$ (2.6)

Obviously, in the second case, the income after taxes is equal to the term "net profit" of the first case if $e = d$. That is, if the company's charges for depreciation d and capital recovery e are the same in both cases.

The flow of money returned to the company in the first case is larger than that in the second case by the quantity dI. In modern corporate accounting it is customary to subtract the term dI from the cost stream $c'Q$ and add it to the net profit stream $R'(1 - t)$ to make the total cash flow stream returning to the company. When this is done, and $d = e$, the two methods become equal.

Principles of Economic Evaluation 43

It would appear to be preferable to use the first method, where e and d are kept in separate streams, so that their differing roles can be clearly seen. However, it is common practice, especially in evaluation work, to make e = d, use a straight-line depreciation evaluation method, and add the depreciation charge to the basic operating cost. This method has the merit of establishing a minimum value for the material selling price.

The streams dI or eI are known as "the depreciation cash flow." In modern financial work the magnitude of these streams is considered quite important because when they are sufficiently large business is good, company financial practice is conservative, and the company's investment in its physical plant is being recovered rapidly. In some instances the depreciation cash flow can exceed the net profit; see Table 1-2, Section 1.05.

It should not be forgotten that the money income which causes these cash flows of net profit, taxation, and capital recovery comes from sales to customers. The selling price s must be set sufficiently high so that all these demands on the project may be met. When setting a sales price management takes notice of c and I, and of reasonable values for d and eI, and works backward from these to what s must be. Hopefully it will prove possible to blend all together to provide a profitable business operation.

2.05 MINIMUM ACCEPTABLE RETURN RATE

Modern industrial corporations, which have a fairly long history of profitable operation, know from experience that the funds that they invest in ventures of one kind or another can be counted on to return to the corporation a certain fraction of the investment each year. These investments may or may not be in the main activities of the company. If attractive investment opportunities are not available in new chemical plants, investments can always be made in real estate, the securities of other corporations (in the chemical industry or not), or in the bonds of Federal and state governments. There have been several instances of chemical companies that owned many millions of dollars worth of the stocks and bonds of other corporations. The financial officers of companies insist that no new funds be committed to a venture unless the estimated rate of return from that investment be equal to or better than that which they are certain that they can secure by making the same investment in many of the activities of the economy.

The rate of return that can be secured from investments in general (including the company's own work) is known as the company's present rate of return on invested capital, i. New projects, which have been shown to be technically feasible, must show a predicted rate of return i_m, which is greater than i; i_m is known as the minimum acceptable rate of return for new investments in plants and processes. Sections 6.07, 6.08, and Table 6-1 give an extended discussion of this matter.

The company's financial officers will specify what value of i_m they consider desirable. It must be equal to or higher than i, but not so much higher that an attractive investment would be eliminated.

2.06 METHODS FOR EVALUATING INVESTMENT OPPORTUNITIES

A good method for evaluating the desirability of an investment should meet the following criteria:

1. It should summarize all the information that is relevant to the decision and none that is irrelevant.

2. It should be applicable to all types of proposals.

3. The result of the evaluation should consist of a single figure: a percentage, or a number of years, or a sum of money.

4. The result should be relatively simple to compute.

5. The evaluation method should show the project's worth simply and without need for elaborate explanation.

Such an ideal is difficult to realize, and there are some differences of opinion among specialists and corporate managers about the best method. In actual practice there are a number of methods used, and since most of the indices are easy to compute, no single one is used to the exclusion of the others. Usually all are calculated, and all are studied.

When evaluating a new project there are many questions that managers ask, but some are more important and penetrating than others. Managers will require simple answers to the following questions:

Principles of Economic Evaluation

1. What will be the total amount of new capital invested at the time of the start of operations? In other words, how deeply will the company go into debt before repayment of the debt will start?

2. How much time will elapse between the start of the project and the time the plant will be in full operation? In other words, how long will the project be a drain on company funds before profitable sales begin?

3. What will be the total manufacturing cost of the product? If the factory cost of the product is known, an experienced salesman of chemicals can quickly predict the probable selling price of the product, and from this the probable market acceptability.

4. What will be the rate of return on the invested capital? A rate of return lower than i_m will probably not be regarded favorably.

5. What will be the length of time of profitable operation before the new capital invested will be recovered? This should be appreciably shorter than the allowable depreciable life of the equipment.

Some of the simpler methods employed in economic studies are developed below. Their principal virtues are simplicity of concept, ease of computation, and overall applicability.

The following assumptions are employed:

1. Project life is assumed as not known exactly. In this case the depreciation rates for accounting (e) and taxation (d) are assumed to be constant and equal.

2. The new capital investment I is spent at one instant of time and successful operation at a uniform rate of return starts at that time. Especially for larger projects, this would not really be true; see Ref. 1, p. 38.

3. The salvage value of the equipment is taken as zero. In actuality equipment has a low salvage value, but not zero (there is a large business in second-hand chemical process equipment), and there are some tax advantages to be obtained from taking a loss on a capital investment. In a later chapter it is shown how salvage value may be taken into account.

2.07 THE CAPITAL INVESTMENT

As mentioned above, the new capital I that will be required for the project is a number which is carefully investigated before approval is given to start the project. As is discussed in Section 2.19, no corporation has unlimited funds, and projects requiring large capital expenditures are regarded less favorably than those requiring smaller ones. Few corporations will invest in a project where the capital required is so high that the percent return and the payout time are beyond normally accepted limits, or where a possible failure of the project could endanger the company's financial integrity.

The size of the new capital investment influences the initial investment, I; the percent return on the new capital investment; the number of years that must elapse before the investment is recovered (the payout time); and, through the depreciation charge for taxes, d, the manufacturing cost of the product. All of these items are favorably influenced by a low new capital investment and unfavorably influenced by a large new capital investment.

When computing the manufacturing cost of the product, the depreciation charge for taxes is calculated from the expression

$(d \times I)/Q$

and if $d = 0.10$ and Q is large, say 5×10^7 lb/yr, then $(d \times I)/Q$ will not usually exceed 1¢/lb of finished product. This is not a large part of the final factory cost of the product. Process development engineers, with their attention firmly fixed on the technical aspects of the project, tend to neglect the rather small contribution of the new capital investment to the final factory cost. They tend to be rather uncritical of substantial increases in I. This is a mistake, because financial officers are quite sensitive to the amount of new capital required. And, financial officers say, Yes or No.

The amount of the new capital required is a most sensitive quantity that should be carefully studied by all concerned before any action is taken.

2.08 THE MANUFACTURED COST OF THE PRODUCT

The total factory cost of the product is the most important number examined by evaluators. A chemical made for a high price

Principles of Economic Evaluation

will have only a limited market, even if it has many desirable qualities, and it will have competition from several other chemicals. One having a low cost may sell in very heavy tonnages (see Fig. 1.3) and will be a strong competitor, even if its properties are not excellent. In general, chemicals sold in large quantities sell for less than 25¢/lb. More expensive chemicals, such as pharmaceuticals and specialty plastics, command smaller markets and sell for prices above 50¢/lb. When a chemical is made in large amounts the capital investment may be high and the production system complex. However, the factory costs will be low. When the production is small the capital investment is usually kept low and the processing methods simple. However, the factory cost may be high.

2.09 THE NET PROFIT

The net profit P is shown in Fig. 2.1 and is simply the gross income less taxes and depreciation. It is the money remaining after costs, depreciation, and taxes have been deducted from the sales revenue.

$$P = R - (R - dI)t - eI \qquad (2.7)$$

The higher the profit, the better the project.

2.10 CASH POSITION; THE CASH POSITION DIAGRAM

The cash position of an enterprise at any time (usually at the end of a year) is equal to the sum of all the previous years' cash flow back to the corporation (see Fig. 2.1) minus the capital investment. It is the amount of money owed to the corporation by the enterprise after the end of a number of years of operation.

$$\text{Cash position} = \sum_0^n (P + eI) - (I + I_w)$$

This concept is most easily visualized by means of a graph; see Fig. 2.3. In a real case the construction of the graph would be much more complex due to changing costs and prices, capital invested over a period of time, and more complex ways of calculating depreciation.

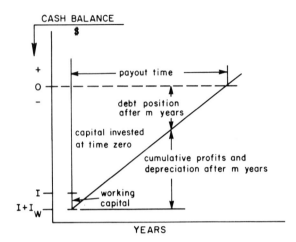

Fig. 2.3. An idealized cash balance position diagram.

At the point where the line crosses the zero position ordinate, the abscissa, n, is the payout time.

2.11 RATE OF RETURN ON INVESTMENT

The annual rate of return on original investment expressed as a fraction is

$$p = \frac{P}{I+I_w} = \frac{\text{net profit in a year}}{\text{total investment}}$$

This ratio of earnings to investment is the simplest and perhaps the most widely used index of the attractiveness of ventures. Sometimes return rate is broken into two factors which are also useful as indications of attractiveness

$$\text{Turnover} = \frac{\text{sales}}{\text{investment}}$$

and

$$\text{Margin} = \frac{\text{net profits}}{\text{sales}}$$

The product of margin × turnover is equal to rate of return. These additional indices are useful in estimating the importance of changes

Principles of Economic Evaluation

in sales volume and price in situations where the relative capital investment is small.

It is an interesting statistic from the chemical industry [4] that many companies have turnover ratios of approximately unity. That is, almost every year the company's sales are equal to its capital investment. However, this is by no means a universal rule. Some of the smaller companies, with lower capital investment, have higher turnover ratios.

When considering rate of return as a method for measuring the productivity of capital, it should be remembered that a rate of return is a fraction and that the fraction can be made larger by increasing the numerator or by decreasing the denominator, or both. Using a smaller $I + I_w$ will increase the rate of return and so increase the profitability rating of the project. It can happen that useful additions to a project may be eliminated because of the desire to maximize the rate of return. From the point of view of an acceptable return (which is not the maximum), or of a larger number of dollars income from the project, the higher I and a lower rate of return may be better.

All rate of return indices suffer from this disadvantage: if rate of return is maximized, I may be reduced so far that opportunities for capital investment at acceptable rates of return are eliminated.

2.12 PAYOUT TIME

As mentioned before, one of the criteria that evaluators examine is the length of time required for the profits of the project to recover the initial capital investment. This number is called the payout time. It may be computed in two ways:

1. Not accounting for taxes or depreciation

$$T = I/R \qquad (2.8\,a)$$

2. Accounting for taxes and depreciation

$$T = I/[R - (R - dI)t] \qquad (2.8\,b)$$

When using these criteria it is necessary that the definitions of payout time be clear and unequivocal. In this book the payout time will be taken as shown in Eq. (2.8a), not accounting for taxes or depreciation.

It is recognized that this definition yields a result that is too short, because, in the real world, taxes and depreciation must be taken into consideration. However, because there are several different methods for the calculation of taxes and depreciation, confusion and complication can easily arise when a definition such as (2.8b) is used. For the purpose of describing the economic attractiveness of a project it is simpler and less confusing to use definition (2.8a).

It must be remembered that economic criteria are only numbers calculated in a carefully defined way. These criteria are used to evaluate the merit of an investment by comparing the same number calculated in the same way for each of one or more investments, or by comparison with a number held in the mind. Provided that the numbers are all calculated in a similar way it is not necessary that the number be totally realistic.

The use of rate of return on investment and payout time as methods for evaluating investment opportunities will indicate that a small investment at a high rate of return is preferable to a large investment at a lower rate of return.

Some evaluators prefer to use a modification of the annual rate of return method. This modification recognizes that the original capital investment is being partially recovered each year and will be almost completely recovered at the end of the project's life. Since this is true, they argue, with reason, that I should be replaced in the numerators of Eqs. (2.8) by an average value of the investment. Average investment is calculated as the sum of the undepreciated value each year of the project's life divided by the total number of years. If straight-line depreciation is used, average value will equal the initial value plus the final value divided by 2. The working capital will remain at full value. This method yields answers that are considerably higher than those obtained by calculating rate of return on original investment, which is the method used for the most part in this book.

2.13 INCREMENTAL RATE OF RETURN

In almost all projects involving the investment of capital funds the relationship between the net profit P and the investment I may be depicted graphically by a curve similar to the line ODMN of Fig. 2.4.

Principles of Economic Evaluation

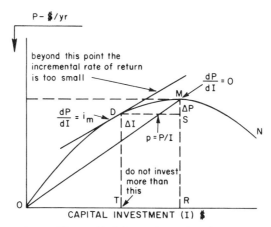

Fig. 2.4. The net profit-capital investment relationship.

In Section 2.11 the rate of return on an investment was defined as

$$p = \frac{P}{I + I_w}$$

The inclusion of the working capital I_w in the denominator is necessary because I_w is part of the plant total investment and, furthermore, is often as much as 10% of that investment. When the investment is less than plant size it is frequently convenient to neglect I_w because it will be small, and it is difficult to visualize the working capital concept as applying to a single piece of equipment or to one section of the plant. The working capital really is used only in connection with business operations of a plant—buying and selling raw materials, utilities, and products.

In this book, the economic evaluation of portions of a plant such as heat exchangers or extraction columns will not include a sum for the working capital. The economic evaluation of an entire plant will include I_w.

The term p is the same kind of function as (see Section 2.05) i, the company's usual rate of return from past and present business operations; i_m, the minimum acceptable rate of return from the project under review.

It is intuitively apparent that a small investment will result in a small profit and that as the investment becomes larger the profit will also become larger, but with a declining rate of increase.

Furthermore, it must also be apparent that as the investment becomes larger still, the numerical value of the profit must pass through a maximum and then decline. This is a result of the so-called "law of diminishing returns," a conclusion based on years of human experience [5].

It is also true that in many pieces of process equipment the value, or performance, that can be obtained from the apparatus is limited by thermal, phase, and chemical equilibrium conditions. The size of the apparatus, and also its expense, may be increased without bound. However, the extent of the physical or chemical change which can be produced in the apparatus is limited by the equilibrium conditions. Thus, the exit stream from a countercurrent heat exchanger cannot be heated above the entering temperature of the hot stream. Similar considerations apply to phase-contacting apparatus such as distillation columns and solvent extraction systems. Thus, the value obtainable from a system may be limited while its cost may increase indefinitely. Accordingly, the profit must decline as investment increases.

When considering an investment, the investors would certainly not wish to invest more capital than that corresponding to the maximum in the curve shown in Fig. 2.4. At the maximum in the curve $dP/dI = 0$. This is the so-called optimum point. The return on the investment, $p = P/I$, is equal to the slope of the chord, 0-M.

If the total capital investment at the optimum point, $I = 0R$, is divided into two parts, the first 0T and the second TR, the profits DT and SM can be associated with the two successive investments. The two rates of return on the two investments are DT/0T and SM/TR. It is apparent that DT is much larger than SM and that the rate of return on the investment (TR) is much smaller than the rate of return on the first investment (0T). The ratio SM/TR is called the incremental rate of return. It may well be that the incremental rate of return near the optimum point will be so low as to be unattractive to the investor. In the limit, at the optimum point, the incremental rate of return does become zero. Therefore, the investor will not only stop investing at the optimum point, but he will stop investing at the point where the rate of return on the incremental investment becomes smaller than minimum acceptable rate of return.

If the investment 0R is divided into many small amounts ΔI, each ΔI will have an incremental profit ΔP associated with it. In the limit of a continuous series of incremental investments it is apparent that the rate of return at any value of I will be dP/dI; the slope of the P-I curve at that value of I. When $dP/dI = i_m$ a point will have been reached where further investment is not desirable.

Principles of Economic Evaluation

The investor wishes the last increment of capital investment made to earn a return equal to i_m. The exact value of i_m must be chosen by financial management. A further discussion of i and i_m is deferred to Chapter 6.

The preceding discussion, which points out the importance of the incremental rate of return, also shifts the center of attraction from the classical optimum point (where $dP/dI = 0$) to the point where $dP/dI = i_m$. For the purposes of making economic-balance calculations it would be desirable to have a new method of plotting such that the new curve, using different coordinates, would have a slope of zero at the point where $dP/dI = i_m$. The new curve would go through a classical optimum at the point where the rate of return has the minimum acceptable value. Now, from Eq. (2.7)

$$P = R - Rt + dIt - eI$$

then

$$\frac{dP}{dI} = \frac{dR}{dI} - t\left(\frac{dR}{dI}\right) + dt - e = i_m$$

dP/dI will be zero when

$$\frac{dR}{dI} - t\left(\frac{dR}{dI}\right) + dt - e - i_m = 0$$

Introducing a new ordinate termed Y, then

$$\frac{dY}{dI} = 0 \quad \text{when} \quad \frac{dR}{dI} - t\left(\frac{dR}{dI}\right) + dt - e - i_m = 0$$

$$\text{or} \quad Y = R - Rt + dIt - dI - i_m I$$

$$\text{or} \quad Y = P - i_m I$$

The new coordinates become $P - i_m I$ as ordinate and I as abscissa. This curve will have a maximum at the point where $d(P - i_m I) = 0$.

This method of plotting is equivalent to charging the minimum acceptable rate of return as a cost against gross income. By using the new Y-I curve, calculations for optima automatically include an incremental rate of return based on that which is regarded as the minimum acceptable. Utilizing the results from this calculation an investor would not invest capital at a rate of return smaller than i_m.

An example of the application of the incremental rate of return is in the selection of a waste heat boiler for the recovery of heat from the stack gases of an industrial furnace. The smallest size available would operate at the largest temperature difference and would show the greatest recovery of heat per square foot of heat transfer area. This would correspond to the highest rate of return per unit of investment. That is, the steeper portion the lower left-hand part) of the curve in Fig. 2.5. However, at this point, although the rate of return is high, the gross amount of the investment is low and the dollar return is small. It might be attractive to install a larger and more expensive waste heat boiler because, even though it showed a lower rate of heat recovery per square foot, the recovery would represent a much larger dollar income and a good return on the investment. No investment would be made beyond the point where $dP/dI = i_m$ or where $dY/dI = 0$.

From the standpoint of not rejecting lower rate of return investments, the total profit P or cash position is significant. However, this index has the disadvantage that a large investment even at an inordinately low rate of return may show a substantial total profit. Thus a 20,000 bbl/day petroleum refinery in an isolated community might show an attractive rate of return. A unit several times as big in the same location might not be able to market its products advantageously. Profit on the latter plant, though larger than on the 20,000 bbl/day refinery, might represent such a small additional rate of return on invested capital that it would not be attractive.

When making economic-balance calculations it is frequently not necessary to employ total sales, total costs, and total capital investment. If the incremental profit or saving obtained by a unit investment in equipment can be computed, and if the incremental capital cost is known, the ratio of the two may be set equal to i_m. This equation may then be solved for some typical dimension or size of the equipment, which is the optimum size. Also, the earnings or profits above i_m, that is $P - i_m I$, may be used and a true maximum obtained, that at $dY/dI = 0$. In Chapter 5 the incremental costs of many items of equipment are given.

The use of the incremental rate of return and the incremental capital investment concepts are illustrated at several points later in this book.

2.14 THE VENTURE PROFIT CONCEPT

The situation that really exists in a going business is that the attractiveness of a venture is measured by the incremental total

Principles of Economic Evaluation   55

return above that obtainable from the rate of return i_m currently considered attractive. The incremental profit, which is the driving force attracting new capital, has been designated as venture profit. The following expression for venture profit is similar to that presented by Happel and Aries [6]:

$$V = P - i_m(I + I_w) = (1 - t)R - (e - dt + i_m)I - i_m I_w \qquad (2.9)$$

For approximate purposes, we may assume e and d to be constants and roughly equal. Then, there is obtained

$$V = (R - dI)(1 - t) - i_m(I + I_w) \qquad (2.10)$$

It will be noted that the venture profit V is identical to the new ordinate Y introduced in Section 2.13. In the evaluation of projects, the one showing the highest venture profit will be the most attractive. The chief drawback in the use of the venture profit is that it does require a minimum acceptable rate of return to be specified.

The following example illustrates how the venture profit method compares with the other commonly used indices.

Example 2.1

Consider the following group of mutually exclusive projects. Management has decided to increase production in a plant unit producing motor fuel by cracking. The following information is available regarding possible alternatives: (a) additional facilities added to the existing plant at an investment of $200,000 would return a gross annual profit of $250,000; (b) a new process which involves to some extent use of existing facilities at a new investment of $60,000 would provide a gross return of $550,000/yr; and (c) a complete new plant costing $1,000,000 would return a gross profit of $700,000/yr.

It is desired to compare net profit, rate of return, payout time, and venture profit.

The following assumptions are to be made. The minimum attractive return rate on invested capital after taxes and depreciation i_m is 15%. State and Federal income-tax rate will be 50%. Salvage value and working capital will be neglected. Depreciation for accounting and tax purposes will be taken constant, $e = d = 0.10$.

Solution. Results of calculations employing the formulas just developed appear in Table 2-2.

Table 2-2   Comparison of Indices of Attractiveness

	Alternatives		
	(a)	(b)	(c)
Investment, I	$200,000	$600,000	$1,000,000
Gross return, R ($/yr)	250,000	550,000	700,000
Depreciation, 0.1 I ($/yr)	20,000	60,000	100,000
Base for tax, (R - 0.1 I) ($/yr)	230,000	490,000	600,000
Tax, 0.5(R - 0.1 I) ($/yr)	115,000	245,000	300,000
Net profit, P ($/yr)	115,000	245,000	300,000
Rate of return, p (fraction/yr)	0.575	0.408	0.300
Payout time, T (years)	0.80	1.09	1.43
Minimum return, i_mI ($/pr)	$ 30,000	$ 90,000	$ 150,000
Venture profit, V ($/yr)	85,000	155,000	150,000

Note that, on the basis of total profit, alternative (c) appears most attractive, with an annual return of $300,000. On the basis of rate of return, however, alternative (a) appears best at 0.575. Alternative (a) naturally also shows the lowest payout time, 0.8 year. When the alternatives are compared on the basis of venture profit, alternative (b) is the most attractive. The same result would be obtained by comparing alternatives on the basis of rate of return on incremental investment. Thus the additional investment required for (c) over (b) is $400,000; this corresponds to an incremental net profit of $55,000/yr. The rate of return on the increment is 0.138, which is less than the minimum acceptable i_m = 0.15, and therefore this increment is not justified.

Acceptable return on incremental investment is a criterion that has been generally advocated as the proper method for comparing alternatives. This index is in agreement with the venture profit method, as shown by the following example quoted from a classic text in engineering economy by Grant [7; example 34, p. 215].

Principles of Economic Evaluation

Example 2.2

Assume that a number of different possible plans are under consideration for the construction of a building for commercial rentals on a piece of city property. Lines A, B, and C of Table 2-3 show the estimated total investment in land and building for each plan, and the estimates of annual receipts and disbursements. It is assumed that regardless of which plan is adopted the property will be held for 10 years, and at the end of that period will have a resale value that is exactly equal to the original investment in the land and building. The use here of an example that assumes 100% resale value is intended to simplify the calculation of prospective rate of return. This permits concentration of attention on the conclusions to be drawn from rate of return on total investment and rate of return on extra investment. The minimum attractive rate of return is considered to be 7%. That is, this is believed to be the prospective rate of return that could be obtained from other investments having the same risk.

Solution. Line D of Table 2-3 gives annual net receipts: that is, gross receipts minus disbursements. As the expected salvage value is 100%, no portion of the net receipts needs to be considered as a partial recovery of capital. Hence the figure in line E for rate of return on total investment is obtained by dividing the annual net receipts (line D) by the total investment (line A).

Lines F, G, and H relate to the calculation of the rate of return on the extra investment above the next lower investment under consideration. For example, consider the comparison of Plan IV with Plan III. Plan IV increases the investment from $152,000 to $182,000, an increment of $30,000. It increases prospective annual net receipts from $13,300 to $15,700, an increment of $2,400. This increment in annual net receipts constitutes an 8.0% return on the increment in investment.

If all the calculated rates of return in line E had been less than the minimum attractive return rate of 7%, it would have been evident that none of the plans was attractive. But when two or more plans show a return that meets this standard, the choice among them calls for a calculation of return on extra investment, such as shown in line H. This permits application of the principle that each avoidable increment of investment must pay its own way by meeting the standard of attractiveness. In this case each increment of investment to be attractive must show a prospective return of at least 7%.

Table 2-3 Comparison of Relative Economy, Example 2.2

		Plan					
		I	II	III	IV	V	VI
A	Total investment	$100,000	$130,000	$152,000	$182,000	$220,000	$260,000
B	Annual receipts	11,000	16,800	22,200	26,200	30,400	33,100
C	Annual disbursements	5,100	6,300	8,900	10,500	12,800	15,000
D	Annual net receipts	5,900	10,500	13,300	15,700	17,600	18,100
E	Rate of return on total investment (%)	5.9	8.1	8.8	8.6	8.0	7.0
F	Extra investment above next lower investment		$ 30,000	$ 22,000	$ 30,000	$ 38,000	$ 40,000
G	Increase in annual net receipts		4,600	2,800	2,400	1,900	500
H	Rate of return on extra investment (%)		15.3	12.7	8.0	5.0	1.2

Principles of Economic Evaluation 59

The application of this principle to the data of Table 2-3 leads to the selection of Plan IV. Even though the 8.6% return on total investment is below that of Plan III, the extra $30,000 investment over Plan III is still an attractive one. That is, the prospective 8.0% return on this extra investment is better than a 7% investment elsewhere that has equal risk.

On the other hand, Plan V is unattractive despite its prospective 8.0% return on total investment. It contains $38,000 of avoidable investment yielding a prospective return of only 5.0%. The 8.0% return on the total $220,000 investment may be thought of as resulting from a combination of an 8.6% return on a $182,000 investment and a 5.0% return on a $38,000 investment. It is good sense to avoid this unnecessary $38,000 investment that fails to earn the minimum attractive rate of return. Plan VI, with its 1.2% return on extra investment, is evidently even less attractive than Plan V.

If this example were to be solved by the venture profit method, the same result would be obtained. Assuming that the figures in line D of Table 2-3 refer to net profit P, the venture profit is obtained by subtracting the minimum acceptable profit from each investment. This will be 7% of the total investment figures listed in Table 2-3. Thus Table 2-4 may be constructed.

Table 2-4   Venture Profit Calculation for Example 2.2

	Plan					
	I	II	III	IV	V	VI
Annual net receipts	$5,900	$10,500	$13,300	$15,700	$17,600	$18,100
Minimum acceptable return	7,000	9,100	10,640	12,740	15,400	18,200
Venture profit	-1,100	+1,400	+2,660	+2,960	+2,200	-100

It is clear that Plan IV, which shows the highest venture profit, is the most attractive. Grant, in connection with the same example, shows that caution must be exercised in interpreting calculated rate of return on extra investment. The following modification of Example 2.2 is also taken from his book.

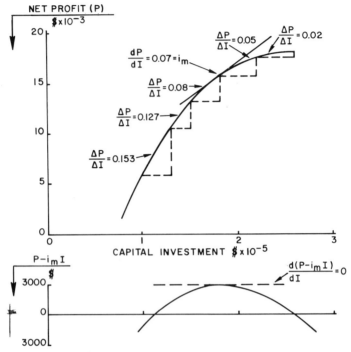

Fig. 2.5. Incremental rate of return.

The results of the calculations shown in Tables 2-3 and 2-4 can be shown graphically in a way similar to Fig. 2.4. In Fig. 2.5 the top curve shows a plot of the net profit as a function of the capital investment. It is obvious that the profitability of the investments is trending toward a maximum. The various incremental investments and incremental profits may be marked off and the incremental rates of return easily shown. The slope dP/dI of $0.07 = i_m$ is also shown, and it can be seen that this value is reached at Plan IV, or I = \$181,000. If $P - i_m I$ is calculated, as in Table 2-4, and this number plotted, a clearly defined maximum is obtained at Plan IV, or I = \$182,000. This is all in accordance with the discussion in Section 2.13.

Example 2.3

A presentation of prospective rates of return such as that shown in Table 2-3 may sometimes be misinterpreted, particularly

Principles of Economic Evaluation

where the rates of return on extra investment follow an irregular pattern. To illustrate this, assume an additional plan, Plan IV a, involving an investment of $200,000 and annual net receipts of $15,800. The section of the table relative to Plans IV, IV a, and V would then contain the somewhat misleading figures shown in Table 2-5.

Table 2-5 Comparison of Relative Economy, Example 2.3

		IV	IV a	V
A	Total investment	$182,000	$200,000	$220,000
B	Annual net receipts	15,700	15,800	17,600
E	Rate of return on total investment (%)	8.6	7.9	8.0
F	Incremental investment	$ 30,000	$ 18,000	$ 20,000
G	Increase in annual net receipts	2,400	100	1,800
H	Rate of return on extra investment (%)	8.0	0.6	9.0

Solution. At first glance it might appear that although Plan IV a is evidently not attractive, Plan V now meets the standard of attractiveness by showing 8.0% return on total investment and 9.0% return on extra investment.

On more critical examination, however, it is clear that the 9.0% return on the $20,000 extra investment on Plan V over Plan IV a has no significance as a guide to the choice among the alternatives. The showing of a 0.6% return on the $18,000 extra investment on Plan IV a should eliminate Plan IV a from consideration. Plan V should therefore be judged by comparison with the next lower investment that meets the required standard: namely, Plan IV.

Stated a little differently, one cannot make the $20,000 extra investment on Plan V over Plan IV a yielding 9.0% without also making an $18,000 investment yielding only 0.6%. The sum of these two is $38,000, which, as already pointed out, yields only 5.0% and is therefore unattractive.

This ambiguity is avoided by the use of the venture profit method. Thus, the venture profit on IV a is $15,800 - 14,000 = $1,800. Therefore the investment is unattractive because Plan IV shows a higher venture profit. Plan V is eliminated without any explicit comparison with IV a or IV, simply because it shows a lower venture profit than the best alternative, in this case Plan IV.

2.15 COMPONENTS OF A PROCESS UNIT

In the previous section the venture profit concept was introduced, and it was shown how this number could be used to compare the economic worth of alternative plans for the accomplishment of an entire project. The alternative having the maximum value of V was considered the best plan. It will be remembered that the heart of the venture profit concept is the profit over and above that considered the minimum acceptable.

When chemical plants are designed they are frequently designed section by section. That is, chemical reactors are designed as one section, distillation or extraction columns as another, and materials-handling equipment as still another. Frequently the groups responsible for these designs have little communication with one another (in fact some of the groups may not be employed by the same company), and each one may apply its own criteria concerning an optimum design. As a consequence, the final plant may have its various sections designed with various economic criteria in mind, and each may be operating at a different economic level. The sections of the plant will not be well proportioned economically and the plant will not operate at the optimum point. Some portions of the plant will have too much capital invested and some too little.

It is essential that the component parts of the plant be designed so that all are operating at maximum economy. In order for this to occur, the economic criteria which designs must meet should be the same for all portions of the plant and the same as that for the entire project; that is, the same value of incremental rate of return will be used for all designs. By basing the decisions on the maximum value of venture profit this criterion will be met. This problem is also the same as that faced in situations involving the replacement of part of an existing facility.

For computational purposes concerning plant components some simplifications are possible in many cases:

Principles of Economic Evaluation 63

1. The money income from the sale of the product, sQ, will be the same for all cases under consideration.

2. The working capital I_w will remain unchanged with new additions, and incremental working capital can be taken as equal to zero.

3. The tax rate will be the same on any portion of the plant as for all of it.

4. The depreciation charge d and the minimum acceptable rate of return i_m will be the same in each case.

Other charges such as those for raw materials, fuel, utilities, labor, and maintenance may change because the entire purpose of examining alternates is to find the combination of operating charges and capital investment-related charges which will maximize the venture profit.

Under these circumstances

$$R = S - C - mI \tag{2.11}$$

where S = the gross income from sales with no charge for maintenance in the product cost (\$/yr); see Section 2.01.

C = running operating expenses (raw materials, utilities, labor, and others) (\$/yr). This does not include a maintenance charge. Any one or all of these charges may change as the component design is changed.

mI = maintenance charge (\$/yr).

The reason for writing the gross profit in this way is to separate the operating charges of the unit from the capital investment-related charges.

When the maintenance charge is expressed in this fashion it is quite important that the proper value of m be used. Frequently, the purpose behind the installation of a more expensive unit (higher I) is to lower the maintenance charge, mI. This must be recognized in the calculation by the use of a lower value of m. In many instances m may be regarded as a constant, but in many others it is not, and care must be taken to account for this.

The venture profit V is defined in terms of a new function U.

$$V = (S - U)(1 - t) \tag{2.12}$$

then

$$U = C + \left(m + d + \frac{i_m}{1-t}\right)I \tag{2.13}$$

and

$$V = \left\{S - \left[C + \left(m + d + \frac{i_m}{1-t}\right)1\right]\right\}(1-t) \tag{2.13a}$$

Thus, U is evaluated as the total yearly cost before taxes, including a charge for minimum acceptable return on investment.

Minimizing U will be equivalent to maximizing V, since U and V are the only variables in Eq. (2.12). The item in parentheses in Eq. (2.13) may be considered as a "fixed charge" on investment, and may be given the symbol, r.

An alternative way of expressing the term $d + [i_m/(1-t)]$ is in terms of the maximum acceptable payout time. The point of maximum acceptable payout time corresponds to the marginal situation where $V = 0$. From formula (2.12) this is equivalent to $U = S$. If this equivalence is substituted into Eq. (2.13), noting that $C = S - R - mI$ and that $I = TR$, it is found that

$$T_m = \frac{1}{d + [i_m/(1-t)]} \tag{2.13b}$$

Then

$$U = C + [m + (1/T_m)]I \tag{2.13c}$$

and

$$V = [S - \{C + [m + (1/T_m)]I\}](1-t) \tag{2.13d}$$

Equation (2.13) may be used to choose between alternatives for the same task. The one showing the lowest value of U will be the alternative selected from a group of mutually exclusive design components or possibilities. In the case of different plants with the same throughput, I and C refer to total investment and operating cost. Where component parts are considered, it is sufficient to consider I and C as referring only to the investment and operating cost of the component to perform a given function. The remainder of the total I and C will be the same for all alternatives and will not affect the comparison.

In economic-balance problems both total costs and incremental costs are used. When the term U is computed from the expression

$$U = C + \left(m + d + \frac{i_m}{1-t}\right)I$$

Principles of Economic Evaluation 65

or from its more exact counterpart, Eq. (3.47),

$$U = C + \left(m + d + \frac{i_m - 0.35 i}{1 - t}\right) I$$

the values of all costs, both I and C, are the total costs.

The optimum point is found by one of two procedures:

1. U may be differentiated analytically and dU/dx set equal to zero, whereupon $-dC/dx$ and $r\, dI/dx$ may be equated and a solution found. In this case, the cost expressions originally substituted in the equations for U are the total costs. However, the analytical differentiation causes the constant term in the total cost expression to vanish, leaving only the incremental part. Therefore, when U may be differentiated analytically, the incremental cost may be used.

2. In many complex cases, U must be computed point by point and a graph drawn (see Fig. 8.4) before the optimum point is found. In such cases, the value of I used is the total installed cost. All of the detailed economic-balance problems discussed in Chapter 8 are treated by the total cost method.

Whenever the term dI/dx may be isolated from the expression for dU/dx, the incremental cost may be used. Whenever the expression for U may not be differentiated analytically, the total costs will have to be used.

It should be remembered that incremental costs are available only from total cost correlations; see Chapter 5.

Example 2.4

Determine whether a single-stage or a two-stage gas-engine-driven compressor should be installed to compress 600 cu ft/min of air from atmospheric pressure to 100 psig, based on the following information.

Fuel-gas consumption of the engines will be 10 cu ft/bhp-hr. Compressor efficiency will be 80% of that calculated on the basis of ideal adiabatic compression. Fuel-gas cost may be taken at $0.60/M cu ft. Installed cost of a single-stage unit is $100/bhp; a two-stage unit costs $150/bhp. Maintenance will be 6% of investment cost per year. Depreciation d will be taken at 0.10, total income tax t at 0.50, and minimum acceptable return rate i_m at 0.10. The compressor installation will have a 90% time efficiency.

Theoretical work for adiabatic compression may be calculated as follows:

$$W = \frac{kP_1v_1}{k-1}\left[\left(\frac{P_2}{P_1}\right)^{(k-1)/k} - 1\right]$$

where W = work (ft-lb)

P = pressure (lb/sq ft abs)

v = volume (cu ft)

k = specific heat ratio = 1.4 for air

Subscripts 1 and 2 refer to initial and final conditions for a single stage of compression.

Solution. For a two-stage compressor, the optimum operation involves the determination of the interstage pressure for which the total power requirement is a minimum [8]. For a fixed gas quantity this corresponds to the minimum work. If gas is cooled to the initial temperature between stages, the total work is given by

$$W = \frac{kP_1v_1}{k-1}\left[\left(\frac{P_2}{P_1}\right)^{(k-1)/k} - 2 + \left(\frac{P_3}{P_2}\right)^{(k-1)/k}\right]$$

where P_3 refers to the final pressure from the second stage. If this quantity is to be a minimum, then the derivative with respect to P_2 must be zero

$$\frac{dW}{dP_2} = \frac{kP_1v_1}{k-1}\left[\left(\frac{k-1}{k}\right)P_1^{\frac{1-k}{k}}P_2^{-\frac{1}{k}} - \left(\frac{k-1}{k}\right)P_3^{\frac{k-1}{k}}P_2^{\frac{1-2k}{k}}\right] = 0$$

Solving for the value of P_2 that satisfies this condition, we find

$$P_2^{(2k-2)/k} = P_3^{(k-1)/k}P_1^{(k-1)/k}$$

or

$$P_2 = (P_1 P_3)^{\frac{1}{2}}$$

Thus the optimum interstage pressure is the square root of the product of the initial and final pressures

Principles of Economic Evaluation

$$\text{Ideal horsepower (1 stage)} = \frac{1.4(14.7)(600)(144)}{0.4(33,000)} \left[\left(\frac{114.7}{14.7}\right)^{0.4/1.4} - 1 \right] = 107.8$$

$$P_2 \text{ (optimum)} = (114.7 \times 14.7)^{\frac{1}{2}} = 41.1 \text{ lb/sq in.}$$

$$\text{Ideal horsepower (2 stage)} = 2 \left\{ \frac{1.4(14.7)(600)(144)}{0.4(33,000)} \left[\left(\frac{41.1}{14.7}\right)^{0.4/1.4} - 1 \right] \right\} = 92.0$$

Cost of fuel, single-stage:

$$C_1 = \frac{107.8}{0.80} \times 10 \times 7884 \times \frac{0.60}{1000} = \$6374/\text{yr}$$

Cost of fuel, two-stage:

$$C_2 = \frac{92}{0.80} \times 10 \times 7884 \times \frac{0.60}{1000} = \$5440/\text{yr}$$

Investment dependent costs, single-stage:

$$\left(m + d + \frac{i_m}{1-t} \right) I_1 = 13{,}475 \left(0.06 + 0.1 + \frac{0.1}{0.5} \right) = \$4850$$

Investment dependent costs, two-stage:

$$\left(m + d + \frac{i_m}{1-t} \right) I_2 = 17{,}250(0.36) = \$6210$$

Equation (2.13) may be used to complete the evaluation.

$$U_1 = \$6374 + 4850 = \$11{,}224$$

$$U_2 = \$5440 + 6210 = \$11{,}650$$

In the example above, the difference between the two values of the compressor costs is small. In practical engineering work this difference would not be considered significant and more would be done to clarify the situation. Two points especially are susceptible to further refinement: (1) The maintenance charge for a two-stage compressor is probably higher than for a single-stage one; this would make U_2 higher yet and make the choice more clear-cut.

(2) The capital investment in each of the types of compressors could probably be made more precise. However, this lack of a clear difference is not important from the point of view of the illustration. What is important is to note that the greater fuel economy of the two-stage compressor is essentially offset by its greater investment-dependent costs, so that when the total costs, U_1 and U_2, are calculated the difference between the two is small.

This balance between a greater economy in operations and a greater cost due to increased capital investment is a constantly recurring problem in process economics.

In design problems it is often true that some parameters may be varied practically continuously over a considerable range. Only one value of the variable will be an optimum. Thus there are problems of optimum values for velocity of fluid flow, thickness of insulation, temperature to which to cool a stream from which heat is recovered, and others. If the variable in question is designated as x, then the desired value of x will correspond to $dU/dx = 0$. Thus after differentiation of Eq. (2.13) and setting $dU/dx = 0$ (assuming that C and I are the only dependent variables), there is obtained

$$-\frac{dC}{dx} = \left(m + d + \frac{i_m}{1-t}\right)\frac{dI}{dx} \qquad (2.14)$$

The capital investment-related charge factor, $m + d + [i_m/(1-t)]$ is often designated as r. Then Eq. (2.14) becomes $-dC/dx = r\,dI/dx$. The decrease in operating cost per unit increase in the variable must just balance the increased "fixed charge" per unit increase in the operating variable when the optimum is reached. Thus, in the case of a heat exchanger, additional surface is installed until fixed charges on the last square foot just equal the annual savings made by installing it.

Example 2.5

It is desired to determine the optimum temperature approach between flue gases from a furnace and a waste heat boiler recovering heat from these gases.

Assume that sufficient draft is available so that no blower or stack is required, that equipment will operate 8,400 hr/yr, and that steam will be generated at 100 lb/sq in. gage. Incremental waste heat boiler surface dI/dx is taken at $3.00/sq ft installed, and an overall coefficient of 5 Btu/(hr)(sq ft)(° F) is obtainable. The value of steam, including fixed charges on steam-generation equipment,

Principles of Economic Evaluation

is estimated to be $0.48/M lb. The heat required to produce 1 lb of steam at 100 lb/sq in. gage from water at 90° F is 1133 Btu.

Tax rate t = 0.50; depreciation e = d = 0.10; minimum acceptable profit i_m = 0.10. Incremental maintenance and other fixed charges m = 0.06.

Solution. Formula (2.14) may be applied,

$$-\frac{dC}{dx} = \left(0.06 + 0.1 + \frac{0.10}{0.50}\right)\frac{dI}{dx} = 0.36\frac{dI}{dx}$$

Therefore

$$-dC/dx = (0.36)(3.00) = \$1.08/(yr)(sq\ ft)$$

Since heat transferred per hour per square foot is

$$Q/\theta = (5)(1)(\Delta t)$$

the savings made per calendar year on the last square foot of heat transfer area will be

$$-\frac{dC}{dx} = \frac{(5)(1)(\Delta t)(8400)(0.48)}{(1000)(1133)} = 0.0178(\Delta t)\ \$/(yr)(sq\ ft)$$

Equating the values of $-dC/dx$ above gives the required Δt as follows:

$$1.08 = 0.0178(\Delta t)$$

$$\Delta t = 60.8°\ F$$

The calculation above, despite its neat simplicity, is a very good example of the difficulties the process engineer encounters when attempting to make a simple economic-balance calculation. There are a large number of assumptions made: i_m = 0.10, e = d, d = 0.10, t = 0.50, m = 0.06, U (the overall heat-transfer coefficient) = 5, dI/dx (the incremental installed cost of the waste heat boiler) = $3/sq ft, and that a suitable means of producing sufficient draft is available. Any one of these assumptions could be wrong and its use would cause the final answer to be incorrect. In fact, it may be impossible ever to get accurate numbers for all these variables. Accurate values for overall heat-transfer coefficients are very hard to get; suppose it were 4 or 6.

A sensitive item is dI/dx. In Chapter 5 it is shown that dI/dx may be written in the form

$$dI/dx = nK(x)^{n-1}$$

and that n is about 0.7. This means that the value of dI/dx to be used depends somewhat (to the -0.3 power) on the absolute value of x. In the present case of waste heat boilers, the value of dI/dx, where x is the steam capacity in thousands of pounds per hour, may range from \$1.50/sq ft for a large apparatus to \$5/sq ft for a much smaller one. A variance as great as this results in a difference of more than threefold in the value of the optimum Δt. The value used above, \$3/sq ft, appears reasonable. However, if the absolute value of x is not fairly well known in advance, an accurate value of dI/dx cannot be chosen. So the process engineer finds himself in the difficult position of needing to know the answer before he can compute that answer. A good deal of judgement and experience plus several trial calculations are necessary before a sound result is obtained.

It is important to note that the use of the function U has the advantage that the gross sales return R does not have to be calculated. It is fairly obvious that such a calculation would be difficult when the question involves one single piece of equipment in a group of many kinds of process equipment. As a consequence, it is a great simplification to be able to concentrate on the equipment costs and the equipment savings only.

However, for some kinds of process apparatus such as heat exchangers, it is somewhat difficult to assign operating expenses to the units. For instance, in the present case of a waste heat boiler, what are the operating costs? It is reasonably easy to determine the capital-related fixed charges because the capital investment I is already determined. However, the operating expenses, such as labor, electrical power consumption, and steam requirements are not easily calculated.

In the case of heat exchangers there are several operating expenses, such as pumping costs to force the fluids through the apparatus and labor requirements for operation and cleaning. However, these costs are quite low and for normal operations they do not influence matters strongly. The principal operating expense must be the heat which is not recovered. After all, the purpose of a heat recovery unit is to recover heat; if it does not do this it is reasonable to charge the unit with a cost proportional to the heat not recovered. An infinitely large exchanger can bring the exit cool stream to the same temperature as the entering hot stream, and all the heat available in the hot stream can be recovered. But this is an infinitely large exchanger with an infinitely high capital-related cost. Such an exchanger cannot be built. Consequently any real

Principles of Economic Evaluation

exchanger will have a charge against it amounting to the value of the heat which is not recovered. In Example 2.5 the term C was evaluated as the savings (negative cost) of the heat transferred.

Other commonly used pieces of process equipment such as distillation or absorption columns have more easily identified operating expenses: steam requirements for distillation column reboilers, cooling water requirements for condensers, electrical power for pumping gases through absorption columns, and the labor requirements for each. Absorption, extraction, and distillation columns also carry a charge for valuable material not recovered, but the charges for steam, cooling water, and electrical power are also important. In a heat exchanger the most important charge is the cost of the heat not recovered.

Figure 2.6 shows the relationship between the operating expenses (heat not recovered) and the total cost, U.

At the optimum point where $dU/dA = 0$ (and note that the minimum acceptable rate of return is included) the slopes, dC/dA, the change in the heat not recovered, and $(d/dA)[m + d + (i_m/1 - t)]I$, the change in the capital-related costs, must be equal and opposite.

By concentrating on the slopes dC/dA and $(d/dA)[m + d + (i_m/1 - t)]I$ when they are equal and opposite, it is possible to calculate dC/dA as $(U)(1)(\Delta t)$, the heat recovered by 1 sq ft of heat exchange area, and $(d/dA)[m + d + (i_m/1 - t)]$, the incremental capital-related cost for one additional square foot of heat exchange

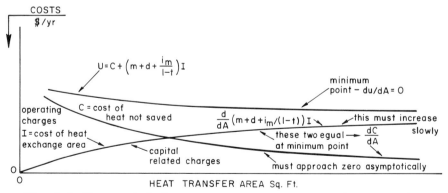

Fig. 2.6. Economic balance on a waste heat boiler (note long flat minimum).

area. These values of the slopes may then be used in Eq. (2.14) to calculate the optimum value of the terminal temperature difference.

From a practical viewpoint it is important to note the long flat minimum in the U vs A curve. The optimum point could be selected as being almost anywhere along the fairly flat portion of the U-A line. This means that the exit-gas temperature could be quite a bit higher than that at the optimum point. In conventional plant practice it is common for the exit-gas temperature to be above $300°$ F so that a better draft will be produced by a chimney.

Often problems of this type are more complicated, involving costs for fluid pumping and countercurrent flow of fluids which change in temperature. Various standard references consider these cases, but unfortunately they do not give details for the methods that should be employed to include minimum acceptable return and tax rate in the economic balance.

If more than one independent variable were involved, it would be necessary to express Eq. (2.13) in terms of these variables x, y, z, ... instead of just in terms of x. Differentiation with respect to each of these variables and setting $\partial U/\partial x$, $\partial U/\partial y$, $\partial U/\partial z$, ... equal to zero, respectively, would result in a series of simultaneous equations to be solved for the optimum values [9], subject to some limitations [10]. Often the relationships between independent variables will be discontinuous or not readily expressible analytically, in which case the use of graphs or plots is necessary. Optimums are then arrived at by trial-and-error procedures; see Example 4.3.

Techniques for solving economic-balance problems where more than one independent variable are involved will be discussed further in Chapter 4.

2.16 APPLICATIONS TO PLANT OPERATION

Thus far the process economics has been examined for a fixed plant throughput. For a new project, a number of optimum plants of varying size would be designed, and the one showing the greatest venture profit would be selected. For a plant already designed it is interesting to study the effect of throughput on the economics of the operation, because chemical plants are frequently required to operate at production rates both above and below the design rate. In this case the investment in physical facilities would be fixed.

Principles of Economic Evaluation

To investigate the effect of plant throughput on the process economics it is convenient to express R, the gross income, in terms of the plant production rate. Equation (2.1) was such an expression.

$$R = sQ - cQ \tag{2.1}$$

The total cost of production, c, may be divided into three groups of costs:

1. Costs which are independent of the production rate. In any plant there are certain charges which will be placed against the plant whether or not product is being made. These are charges for property taxes, protection, insurance, management, and maintenance (to some extent).

2. Costs which are directly proportional to the production rate. These are the charges incurred from the purchase of raw materials, utilities, and (to some extent) labor. Maintenance charges will be higher for a plant in operation than for an idle plant, but in the range of 20% to 90% of capacity the maintenance cost will be essentially constant.

3. Costs which increase at a rate which is more than proportional to the increase in production rate. These are called superproduction costs. They result from mechanical inefficiencies, higher maintenance, lower yields from chemical reactions, greater utilities costs, and a lower labor productivity. Such costs are a consequence of forcing the plant to work at a rate greater than that which is the optimum.

In terms of these three classifications the cost of a unit of product may be expressed as

$$c = F/Q + a + bQ^r \tag{2.15}$$

Here F is taken as the fixed charges (group 1), not including taxes or depreciation. It can be seen that if investment I does not change, then maximizing venture profit V, net profit P, or gross return R will all result in fixing Q at the same value. That is, $dV/dQ = dP/dQ = dR/dQ = 0$ all correspond to the same value of Q. The constants a, b, and r will depend on the particular problem involved. The superproduction cost is assumed to be a power function, bQ^r. Any other convenient expression could be used to fit specific circumstances. (See Example 2.6.)

Combining Eqs. (2.1) and (2.15) gives Eq. (2.16), which relates R, the gross income (before taxes and depreciation), the production cost of the product, and the yearly production rate

$$R = sQ - cQ = sQ - F - aQ - bQ^{(r+1)}$$
$$R = (s - a)Q - F - bQ^{(r+1)} \tag{2.16}$$

An interesting chart may be constructed by plotting the yearly dollar value of each term of Eq. (2.16) as ordinate against the production rate as abscissa. Figure 2.7 in Example 2.6 shows this graph. Several interesting answers may be obtained by examining this plot.

1. At production rate Q_1 the lines representing total annual sales return, sQ_1, and total annual cost, $aQ_1 + F + bQ^{(r+1)}$, will intersect. At production rates lower than Q_1 the plant costs will be greater than plant income, and operations will show a loss. At production rates above Q_1 (but below Q_4) the reverse situation will apply. Point Q_1 is called the break-even point.

2. By observing that the cost per unit of production c is

$$a + F/Q + bQ^r$$

the production rate Q_2 corresponding to the minimum cost per unit is found from

$$dc/dQ = -FQ^{-2} + brQ^{r-1} = 0$$
$$Q_2 = (F/br)^{1/(r+1)} \tag{2.17}$$

3. It is apparent from the curve that the profit per year goes through a maximum as production rate increases, because at higher rates the superproduction costs become more important. It is possible to find the point where the yearly profit is a maximum from

$$R = sQ - aQ - F - bQ^{(r+1)}$$
$$dR/dQ = s - a - (b)(r + 1)Q^r = 0$$
$$Q_3 = \left[\frac{s - a}{b(r + 1)}\right]^{1/r} \tag{2.18}$$

It should be noted that the production rate for minimum cost per unit is not the same as the production rate for maximum yearly profit. It is also noteworthy that the break-even point is not the throughput below which the plant will be shut down. The criterion

Principles of Economic Evaluation   75

for shutdown is discussed more fully in Chapter 3. Consideration of salvage value and tax credits are involved.

Example 2.6

A butadiene plant with a rated capacity of 5.56×10^7 lb/yr must operate at a reduced capacity, owing to a reduction in the price of butadiene product to 12¢/lb and an increase in the price of the feedstock, n-butane, to 10¢/gal. Fixed charges of $1284/day include direct operating labor, supervision, supplies, and their appropriate overhead charges. Production costs, proportional to butadiene production at 4.62¢/lb, include chemicals, catalyst, royalty, and utilities, as well as labor and material for maintenance (together with their proportion of overhead), but not the raw material, n-butane. Cost for butane feedstock is the only superproduction cost. The plant can be operated at higher yields if it is run at reduced capacity as detailed in the accompanying tabulation. The optimum rate of production is desired.

% of Rated capacity	$Q \times 10^{-7}$ (lb/yr)	Yield of butadiene on butane feed (%, wt)	Cost of feed in butadiene produced (¢/lb of butadiene)
100	5.56	50	6.17
80	4.45	55	5.63
60	3.34	60	5.13

Solution. In the discussion given above it was pointed out that the optimum rate of production is that which yields the maximum profit. This rate may be found by setting $dR/dQ = 0$ and solving for Q_3. For cases where the superproduction cost is given by a term of the form bQ^r, Eq. (2.18) may be solved for Q_3. When the cost-production relationship is tabulated, as in this problem, it is necessary to find an empirical equation which will express the relationship with a fair degree of precision. Here, by plotting on ordinary rectangular coordinates, it can be shown that the equation expressing cost as a function of production rate is a straight line

Cost of feed = $(4.68 \times 10^{-8})Q + 3.57$ ¢/lb

where Q is the production rate in millions of pounds per year. Then,

$$R = 0.01\,sQ - 0.01\,aQ - F - (4.68 \times 10^{-10})Q^2 - 0.0357\,Q$$

$$dR/dQ = 0.01\,s - 0.01\,a - (9.36 \times 10^{-10})Q - 0.0357 = 0$$

$$s = 12\cent/lb \quad \text{and} \quad a = 4.6\cent/lb$$

whence $Q = 40.9 \times 10^6$, or 74% of capacity.

It is also possible to plot the cost-production rate data on logarithmic coordinates (expressing the production rate as a percent of rated capacity), or use the method of least squares, and obtain another empirical expression which is just as good.

$$\text{Cost of feed} = 1.176(\% \text{ of rating})^{0.36}$$

$$= 1.176(100\,Q/Q_0)^{0.36} \quad \cent/lb \text{ butadiene}$$

where Q = production rate under discussion (lb/yr)

$Q_0 = 100\%$ capacity $= 5.56 \times 10^7$ lb/yr

When this expression for the superproduction cost is used Eq. (2.16) becomes

$$R = (s-a)Q - F - 0.01176(100Q/55.6 \times 10^6)^{0.36}(Q)$$

or

$$R = (s-a)Q - F - 0.01176(100/55.6 \times 10^6)^{0.36} Q^{1.36}$$

$$b = 0.01176(100/55.6 \times 10^6)^{0.36}$$

$$b = 1.$$

$$R = (s-a)Q - F - 1.01 \times 10^{-4} Q^{1.36}$$

From Eq. (2.16)

$$Q_3 = \left[\frac{s-a}{b(1+r)}\right]^{1/r} = \left[\frac{0.12 - 0.0462}{(1.01 \times 10^{-4})(1.36)}\right]^{2.78}$$

$Q_3 = 39.6 \times 10^6$ lb/yr, or 71% of rated capacity

These two answers are close enough for practical work. The lack of agreement arises from differences in goodness of fit of the equations to the data.

Principles of Economic Evaluation

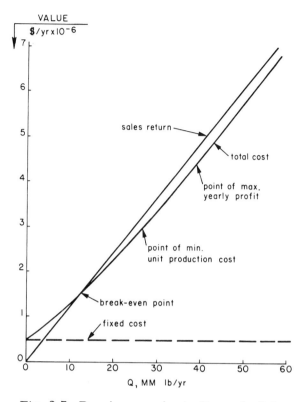

Fig. 2.7. Break-even chart. Example 2.6.

The conversion of tabulated or plotted data from an experimental unit or a plant test into a "good enough" empirical equation is a widely practiced device in economic-balance calculations. When digital computers are used this is necessary because computers cannot read graphs or tables. Sherwood [11] gives several examples.

It is instructive to draw a "break-even" chart, such as Fig. 2.7, using the information on butadiene production given in this problem. Two assumptions will be made:

1. The sales price will remain constant at 12¢/lb despite large changes in production rate. Thus the sales income to the company will be $0.12 Q$ \$/yr.

2. The superproduction costs will be given by the relationship $b(100Q/Q_o)^r$, which is the equation given above with percent rating expressed in terms of Q, the actual production rate, and Q_o, the rate at 100% capacity, in this case 55.6×10^6 lb/yr. Although the plant test data, tabulated previously, extend only over the range 60 to 100% of rated capacity it will be assumed that the relationship is valid over a more extensive range.

It is required that the terms

$$cQ = aQ + F + 0.01176(100Q/55.6 \times 10^6)^{0.36} Q, \text{ the total product cost in \$/yr}$$

and

sQ, the sales return in $/yr

be plotted against Q. Here a = 0.0462 $/lb and F = (1284)(365) = $469,000.

The results of this calculation and plotting are shown in Fig. 2.7. It is clearly seen that the "break-even" point is fairly precisely fixed at Q = 12,800,000 lb/yr or 23% of the rated capacity.

The optimum production point, or point of maximum yearly profit, has already been computed. It is shown on the graph.

It is also possible to compute the point of minimum unit cost production, that is, the point where 1 lb of product is made for the lowest cost. From Eq. (2.17)

$$Q_2 = (F/br)^{1/(1+r)}$$

F = \$469,000   r = 0.36   b = 1.01×10^{-4}

$$Q_2 = \left[\frac{469{,}000}{(1.01 \times 10^{-4})(0.36)}\right]^{0.735}$$

Q_2 = 26,920,000 lb/yr, or 48.4% of rated capacity

Based on this information, the gross return at 71% of capacity is determined by summing the three types of charges for 152,400 × 0.71 = 108,200 lb/day butadiene production.

A unit of the capacity employed in this example might involve an investment of approximately $10,000,000. A gross return of 836 × 365 = $305,140/yr would represent only a nominal return, so that to determine whether it was actually profitable to operate

	$/day	¢/lb
Fixed charges	1,284	1.19
Production charges	5,000	4.62
Superproduction charges	5,864	5.42
Total production charges	$12,148	11.23
Sales income at 12¢/lb	12,984	12.00
Gross return	$ 836	0.77

would depend on allowable depreciation for tax purposes and necessary working capital for the operation. However, if the plant were operated at all, these items would not enter into the calculation of the optimum at 71%.

2.17 THE LIMITED AVAILABILITY OF CAPITAL

The foregoing discussions concentrated on the selection of the alternative having the highest value of the venture profit. It was assumed that the investment capital necessary to produce the highest venture profit would be available. That is, the company would be willing to invest money up to the point where the incremental rate of return equalled the "minimum acceptable rate of return"; see Section 2.13 and Fig. 2.4. This is often not the case. An unlimited amount of capital is no more available to businesses than it is to individuals. Very frequently, planners find that investment must not be carried beyond a certain sum, even though venture profit calculations show that a larger investment would be more attractive.

It is probably true that there is an optimum range of investment magnitudes to which the concept of minimum acceptable rate of return applies. There can be projects for which the investment is so small—say, less than $5,000—that a venture profit analysis is not worthwhile. On the other hand, there may be an investment so large that in the event of failure the financial structure of the company would be seriously endangered.

The previous mathematical relationships are applicable, in principle, to situations in which there is a limit on the availability of capital. The optimum selection will be that which maximizes venture profit, but with the additional restriction of a fixed maximum $I + I_w$.

Perhaps the simplest situation that can be considered to illustrate the nature of this restriction is the case of a process plant in which optional investment is possible for only one of the unit components. Thus, suppose a plant is estimated to cost $900,000 without the optional unit, say a heat exchanger. The unit will operate satisfactorily without the exchanger, so that the size of the latter is purely a matter of the economics of heat recovery. If capital were unlimited in supply, the size of the exchanger would be justified on the basis that the last square foot of surface installed would have a rate of return equal to the minimum acceptable rate, as indicated by Eq. (2.14). If the exchanger cost $200,000 the total plant investment would then be $1,100,000. If now a capital limitation of $1,000,000 were imposed it is obvious that it would be desirable to install an exchanger costing only $100,000. The limitation of a definite ceiling on capital investment, if applied with strict mathematical logic, would rule out any optional investment over $1,000,000 no matter what the incremental rate of return.

It is seldom that definite ceilings on capital investment are applied with strict mathematical logic. For one reason, estimates of plant costs are not precise enough to allow the imposition of sharp upper limits; there is always an area of uncertainty. Also, small increments to a large investment will nearly always be permitted if the incremental rate of return is high. However, a large increment, say 25% or more, might be considered impossible. Financial officers of chemical companies often feel that a sharp upper limit on new investments must be set and observed in order to prevent the engineers from building an efficient, profitable plant that costs too much. However, the situation is frequently too complex for such a simple approach.

The actual situation is probably closer to one where the minimum acceptable return rate is a complicated function of both the size of the individual investment and the other company investments already approved or under consideration in the same risk category. The specification of minimum acceptable return rates under such circumstances might not be possible in any simple form, but the general principle of maximizing the total venture profit will still be the same.

If a combination of several projects that are not mutually exclusive is under consideration, the situation can become quite complex. Not only can there be alternatives among the various projects, but each project may also contain its own alternatives. Such complicated problems are frequently solvable by some of the more advanced methods discussed in Chapter 4. For those situations where the internal alternatives may be ignored, a trial-and-error procedure may be used to determine the maximum total venture profit, subject to the restriction of a maximum total $I + I_w$ for all projects approved.

SUMMARY

It has been shown how the variables of investment and plant throughput enter into economic-balance calculations. Simplified indices have been presented in which it was assumed that no change occurred from year to year in the alternatives under consideration.

Where a minimum acceptable return rate can be established, the venture-profit index is recommended. If minimum acceptable rate of return is not available from management, rate of return or payout time appear to be most useful. These indices in one form or another are widely employed in the process industries. Since all three of these indices are readily computed, perhaps the best procedure is to employ them all together, rather than placing reliance on any one index.

REFERENCES

1. D.G. Jordan, Chemical Process Development, Interscience Publishers, John Wiley and Sons, Inc., New York, 1968, p. 26.
2. Depreciation Guidelines and Rules, United States of America, Treasury Department, Internal Revenue Service, Publication No. 456, 1962.
3. M.S. Peters and K.D. Timmerhaus, Plant Design and Economics for Chemical Engineers, 2nd ed., McGraw-Hill Book Co., New York, 1968, pp. 215-217.
4. Chemical and Engineering News, September 2, 1968.
5. L. Abbott, Economics and the Modern World, Harcourt, Brace and World, Inc., New York, 1960, pp. 461-470.
6. J. Happel and R.S. Aries, Chemical Engineering Progress, 46, 115 (1950).

7. E.L. Grant, Principles of Engineering Economy, 3rd ed., Ronald Press, New York, 1950.
8. H.S. Mickley, T.K. Sherwood, and C.E. Reed, Applied Mathematics in Chemical Engineering, 2nd ed., McGraw-Hill Book Co., New York, 1957, p. 29.
9. Reference 8, p. 219.
10. R.G.D. Allen, Mathematical Analysis for Economists, Macmillan, London, 1947.
11. T.K. Sherwood, A Course in Process Design, The MIT Press, Cambridge, Mass., 1963.

PROBLEMS

2.1. A projected chemical process involves a new capital investment of $1,000,000 and the following estimated costs:

	$/yr
Raw materials	500,000
Manufacturing cost (includes labor, maintenance, utilities, and plant expense)	125,000

It is estimated that the gross income from sales, the term sQ of Eq. (2.1), will be $900,000/yr.

Assuming that the depreciation rates for accounting and for tax purposes will be the same at 10%, and that the total income-tax rate, after allowance for depreciation, will be 50%, compute

    a. The net profit after taxes and depreciation
    b. The payout time before taxes or depreciation
    c. The payout time after taxes but not including depreciation
    d. The cash position after 5 and after 10 years of operation
    e. The venture profit if $i_m = 0.15$

Principles of Economic Evaluation   83

2.2. The heat lost from 100 lineal ft of standard 4-in. steel pipe carrying steam is worth $35/yr if the pipe is lagged with 2-in. magnesia pipe covering and $357/yr if the pipe is bare. If the cost of installing this insulation is $2/lineal ft of pipe, calculate the annual rate of profit on the insulation installation after taxes and depreciation. Assume a depreciation rate of 10%/yr and maintenance charges of 5%/yr on investment. Tax rate on income will be taken at 55%.

2.3. A chemical is to be recovered from inert waste gases by absorption in an organic solvent. The solvent is circulated to an absorption tower where almost all of the chemical is absorbed, the chemical-solvent solution is then stripped of solvent in another tower, and the chemical recovered. The stripped solvent is then recirculated to the absorption tower.

The following relationship has been found to hold between the amount of the solvent circulated and the percentage recovery of the chemical:

$$L = 45/(100 - R)^{0.33}$$

where L = solvent circulating rate (gal/1000 cu ft of inert gas)

R = recovery of chemical (%)

The cost of solvent circulation, stripping, and other operating expenses is 10¢/1000 gal of solvent. The chemical recovered is worth $1/gal, and it is present in the gases being scrubbed to the extent of 1 gal/1000 cu ft of gas.

Assuming that the existing plant operation will not be changed except to increase or decrease the solvent circulation rate, what is the optimum percent recovery of the chemical?

2.4. A pump installation of 40 hp operates at 86% overall efficiency and costs $900 installed. A second installation of the same size costing $970 operates at 90% efficiency. Energy is worth 4¢/kw-hr, maintenance costs amount to 7% of the investment per year, and a maximum acceptable payout time of 2 years on incremental capital investment is required. How many hours per year must the installation operate in order to economically justify the second alternative?

2.5. A multiple-effect evaporator system is boiling 100,000 lb/day of water from an aqueous solution. The system operates for 300 days each calendar year. Each effect will evaporate 0.8 lb of water/lb of steam.

Cost of each effect $12,000

Maintenance and operating costs   10%/yr of initial investment

Cost of steam   0.60 $/1000 lb

Labor and other costs are independent of the number of effects. If the maximum acceptable payout time on incremental investment is 3 years, estimate the optimum number of effects.

2.6. A process involves a drying operation which, if not carefully controlled, will result in product damage. In the design of a plant involving this process it is possible to install a safety device which will shut off the power supply to the air blower if the flow of material being dried is interrupted. The installed cost of this device is $4,000 and the estimated life of the process plant in which it is proposed to install it is 10 years.

If the device is not installed, each stoppage of flow will result in a $1,500 loss, whereas if the device is installed the loss is reduced to $100. Interruptions occur to the extent of twice a year. Maintenance costs on the safety device are $100/yr.

(a) If the minimum annual return rate on optional investment is 15% and the overall tax rate is 55%, should the safety device be installed?

(b) If the safety device is installed what would be the annual rate of return on the extra investment? What would be the payout time? (These results should be calculated after both taxes and depreciation.)

2.7. In Example 2.6 the plant capacity had been reduced due to an increase in the price of the raw material. (a) Under the existing conditions of increased raw material cost, what sales price would have to be charged if the optimum production rate were to be 100%, that is, 5.56×10^7 lb/yr? (b) Using the new sales price and the production rate of 5.56×10^7 lb/yr, what would be the break-even point in terms of percent of rated capacity?

2.8. For part of a high-pressure process installation the decision has to be made as to whether to install two reactors in parallel or one single large one. Relative investment costs are $2,200,000 and $1,500,000. In the case of the larger reactor an annual inspection and maintenance period of 1 week will be necessary. In the case of the two-reactor system shutdown can be scheduled so that for the 1 week down time of each reactor, the other will be in operation and

Principles of Economic Evaluation 85

be able to handle 75% of the total throughput from the remainder of the system.

The value of the product after allowing for raw material costs and operating expenses is estimated at $70,000 for each day the operation proceeds at full capacity.

Assume the cost for inspection and maintenance of each reactor is $5,000. Assume a payout time before taxes and depreciation of 3 years. Which installation is preferable?

2.9. The following economic data are available for a proposed processing system to manufacture high-purity cyclohexane by hydrogenation of benzene:

Sales income	$1,900,000/yr
Manufacturing costs	$800,000/yr
Plant investment	$2,000,000
Working capital	$80,000
Plant life	10 yr
Expected earnings rate of firm	$i = 0.12$
Tax rate	$t = 0.50$
Minimum acceptable rate of return	$i_m = 0.20$

Would this be considered an attractive venture?

2.10. Which of the following two mutually exclusive projects would be considered the more attractive?

	A	B
New capital investment, I ($)	1,000,000	6,000,000
Gross profits, R ($/yr)	2,000,000	5,000,000
Years of operation, n	5	10

It may be assumed that the tax rate, the company's present earnings rate, and the minimum acceptable return rate are 0.50, 0.10, and 0.20. The differences in working capital and salvage value between the projects may be neglected.

2.11. A hot oil side stream is being drawn from a distillation column at 450° F at a rate of 300,000/lb hr. The stream is cooled and sent to storage. It is proposed that a heat exchanger be installed to use the hot oil to preheat crude oil, which is the feed to the distillation column, at a rate of 600,000 lb/hr and a temperature of 70° F. Determine whether it would be economical to preheat the crude oil to 250°F. (a) Is this the most economical preheat temperature? (b) What is the most economical preheat temperature? The cost involved in cooling the hot stream may be neglected.

The following data may be used in Problem 2.11:

Value of heat in process	0.50 $/10^6 Btu
Heat capacity of oil streams	0.50 Btu/(lb)(°F)
Heat-transfer coefficient for exchanger (defined by log mean temp diff)	60 Btu/(hr)(sq ft)(° F)
Maximum allowable payout time on invested capital	2 yr
Operating hours per year	8000
Installed cost of heat exchanger	$346A^{0.62}$ (see Chapter 5)
	A = heat exchanger area (sq ft)

Solution. (a) An overall heat balance states that

$$(300,000)(0.5)(450 - T) = 600,000(0.5)(t - 70)$$

$$\text{if } t = 250° F \quad T = 90°F$$

Then

Heat transferred = $(600,000)(0.5)(250 - 70)$

= 5.4×10^7 Btu/hr

Savings due to heat transferred = $(5.4 \times 10^7)(0.50)(10^{-6})(8000)$

= 216,000 $/yr

Heat exchanger area = $5.4 \times 10^7 /(60)(\Delta T_{lm})$

Principles of Economic Evaluation

$$\Delta T_{lm} = \frac{(450 - 250) - (90 - 70)}{\ln(200/20)} = 78.2$$

Heat exchanger area = $5.4 \times 10^7 /(60)(78.2)$ = 11,520 sq ft

Cost of heat exchanger area = $346(11,520)^{0.62}$ = \$114,000

From Eq. (2.13a), the venture proft is

V = {216,000 − [0 + (0.06 + 1/2)(114,000)]} (0.5)

= \$76,000

It is economical to preheat the feed to 250° F.

(b) In order to find out if this is the most economical temperature, and to find just what the most economical temperature is, it is best to assume a number of preheat temperatures, calculate the values of V, and find where the maximum lies. Thus,

T (°F)	t (°F)	ΔT_{lm} (°F)	A (sq ft)	$346A^{0.62}$ ($)	S ($/yr)	V ($/yr)
240	110	103	8,250	93,100	204,000	75,950
	100	91.2	9,800	103,800	214,000	77,950
250	90	78	11,520	114,000	216,000	76,000
255	80	63	14,700	132,000	222,000	74,050

Thus, the maximum is fairly sharp and is near 245°F. This is the most economical temperature.

Chapter 3

EXPANDED ECONOMIC EVALUATION EQUATIONS

3.00 THE TIME VALUE OF MONEY

Human beings have a decided preference for money existing in their possession now as opposed to the promise of the same amount of money some time in the future. It is a natural and understandable human act to satisfy present wants with presently owned money rather than to postpone such gratifications to some future date. People regard the usefulness of a present receipt of money as greater than the usefulness of the future receipt of the same amount.

Users of large amounts of money, such as governments, corporations, and entrepreneurs, find it necessary to offer a premium in the future payment of money in order to persuade people to exchange a certain sum of their money now for a sum of money some years from now. This premium is known as interest, growth, or capital appreciation. It is the price borrowers must pay for obtaining a large sum of money now, and it is the reward which the lender receives for waiting for his money. For a more complete discussion see Abbott [1].

These well-known facts establish a basis for an understanding of the "time value of money." Money which exists now is more valuable than money which may exist some time in the future. The simplest and best known example of the time value of money is personal saving at compound interest. Here a sum of money in existence now, P, is deposited in a savings bank at a rate of interest, i. If the principal P is allowed to remain on deposit, and if all interest earned is redeposited at the end of each year, it is easy to show that the sum of money on deposit at the end of n years, S, will be

$$S = P(1 + i)^n \tag{3.1}$$

and

$$P = S/(1 + i)^n \tag{3.2}$$

Equation (3.2) is a relationship between a sum of money believed to exist n years from now and the value that the sum is believed to have at present. Since i and n are both positive it can be seen that P will always be smaller than S. That is, a future sum of money is discounted by the factor $1/(1+i)^n$ when it is to be compared to a sum of money now existing.

In the chemical industry engineers and financial managers are constantly concerned with plans which call for the expenditure of a rather large sum of money now and the recovery of that sum over a period of years into the future. In the making of these plans it is necessary to compare sums of money which exist at widely separated periods of time. In order to make a valid comparison between the separate sums they must all be referred to the same point in time. That is, make an allowance for the time value of money. This is most clearly and conveniently done by reducing all sums to the present time.

Equations (3.1) and (3.2) are widely used for this purpose and there are extensive tables of $(1 + i)^n$ and $1/(1 + i)^n$ for many different values of i and n [2,3]. These equations are based on the payment of interest at the end of each year—termed annual compounding. Compounding can, and does, take place at other times—such as every 3 months. It can also take place continuously. There is merit to this assumption since, in practice, some financial payments flow into and out from a corporation, or an individual, continuously and not in lump sums at the end of each year. However, see below and Appendix A. Continuous compounding gives

$$dP/dn = iP$$

which can be integrated between 0 and n to give

$$S = P e^{in} \tag{3.3}$$

or

$$P = S e^{-in} \tag{3.4}$$

For times shorter than 10 years and interest rates lower than 15%

Expanded Economic Evaluation Equations 91

there is not a great deal of difference between the results computed from Eqs. (3.1) and (3.3).

Although continuous compounding has some theoretical appeal, industry and government almost always operate on an annual basis, with yearly appropriations and annual reports to citizens and stockholders. In this book only annual compounding will be used, although there is a discussion and an example of continuous compounding in Appendix A. The books of Taylor [3], Rudd and Watson [4], and Peters and Timmerhaus [5] may be consulted for further discussions on continuous compounding.

3.01 THE VENTURE WORTH METHOD

The economic evaluation of new projects is aided by the calculation of one or more indicators, or indices, which quantify the economic attractiveness of the projects by expressing that attractiveness by a single number. Thus, in Chapter 2, such indicators as payout time, rate of return, and venture profit were introduced.

These numbers are useful, but lacking in generality because they do not consider the time value of money. As discussed in Section 3.00, a proper analysis of a project with a projected life of more than 1 year must take into consideration the time value of money.

Grant's book [6] and an article by Weaver, Bauman, and Heneghan [7] give comparisons of some of the methods which take the time value of money into account. Monetary inflation, which also affects the future value of money, is discussed briefly in Appendix B. References [6] and [7] also describe some of the simpler procedures similar to those discussed in Chapter 2. The indices using compound interest formulations are designated as "present worth," "interest rate of return," and "capital recovery period."

Happel has devised the "venture worth" method. This is a modification of the present worth method of approach and is a logical extension of the venture profit concept. It is specifically designed to avoid the ambiguity associated with many calculations of this type.

The venture worth method judges the profitability of a project by the present value of the venture profit which is obtained during each year of the project's anticipated life. Cash flows in the future are discounted to the present at the company's projected average rate of earnings on invested capital i. The rate i is subject to some variations in interpretation, depending on whether it is to be identified with reinvestment of profits or with interest on borrowed funds. In fact, separate rates could be employed for each of these if desired. The attitude adopted in this book is that the company, in considering a new project, always is able, from other unspecified projects, to obtain a certain earnings rate which serves as a yardstick for new proposals. In Chapter 6 some of the problems of appropriate rate selection for individual project evaluation are discussed. In Appendix B the effect of monetary inflation on the value of i is discussed.

The method developed below is presented in the form of equations which include certain simplifying assumptions. The equations obtained are useful in deriving generalizations. In complicated cases, the use of tabulated year-by-year cash flows will be more convenient than using analytical expressions (see Example 3.2).

3.02 EXTENDED VENTURE PROFIT EQUATION

It will be recalled from Chapter 2 that the incremental return each year on an investment, above a rate i_m currently obtainable for the same type of project, was used as the criterion of attractiveness and designated as the venture profit, V. Equation (2.9) expressed venture profit in terms of the important variables involved.

A more complete expression for the venture profit may be obtained by considering several items in some detail:

1. The term "e" is a constant fraction of the investment I taken to represent the allowance for the recovery of capital. Occasionally, "e" may be set equal to "d," the depreciation allowance factor, as in Eq. (2.10). A more logical approach is to let e equal the sinking fund deposit factor, $i/[(1 + i)^n - 1]$. In this way a series of equal annual payments at compound interest results in the recovery of the initial investment at the end of the life of the venture.

2. When a process plant comes to the end of its useful life, or when it is prematurely shut down, the installed process equipment may be in fairly good condition and have some value for future use in another plant or for sale in the open market as used equip-

Expanded Economic Evaluation Equations

ment. This value is called the "salvage value," and it should be considered in any economic analysis.

In most process plants, the salvage value of the installed process equipment will be small because the savings made by using old equipment are balanced by dismantling costs. Also, the value of salvaged equipment is not as great as that of new equipment specifically designed for effective use in a new plant. However, if salvage value is recoverable at the end of the life of an enterprise, it constitutes a source of income, S_a. This is not tax-free, however, if in the course of tax computation the entire cost of the process equipment has been amortized. The net salvage value will amount to $S_a(1 - t)$, where t is the combined Federal and state income tax expressed as a fraction. The amount of income thus obtained is logically distributed uniformly over the life of the venture by means of the sinking fund deposit factor.

3. The working capital is, in theory, completely recoverable at any time, but the government allows no depreciation for its use. Also no sinking fund is required to recover it as is necessary with capital invested in nonsalvageable equipment. Usually the price of land is not considered as working capital, but, since it is treated the same in calculations of this type, it may be included with I_w. Land is usually only a small proportion of the investment in a chemical-process plant.

If these factors are taken into consideration, the following more complete expression for venture profit is obtained [9]

$$V = R - \frac{iI}{(1+i)^n - 1} - (R - dI)t$$

$$\begin{pmatrix}\text{Venture}\\\text{profit}\end{pmatrix} = \begin{pmatrix}\text{annual}\\\text{gross}\\\text{return}\end{pmatrix} - \begin{pmatrix}\text{payment to}\\\text{sinking}\\\text{fund}\end{pmatrix} - \begin{pmatrix}\text{income}\\\text{tax}\end{pmatrix}$$

$$- i_m I - i_m I_w + \frac{i(1-t)S_a}{(1+i)^n - 1}$$

$$- \begin{pmatrix}\text{minimum}\\\text{acceptable}\\\text{return}\end{pmatrix} - \begin{pmatrix}\text{cost of}\\\text{working}\\\text{capital}\end{pmatrix} + \begin{pmatrix}\text{annual credit}\\\text{for}\\\text{salvage}\end{pmatrix}$$

(3.5)

Here i is the rate of earnings of the company on its invested capital, and i_m is the minimum acceptable return rate on the proposed new project; i_m will usually be larger than i, depending on risk and other factors as discussed in Chapter 6. Terms in R and I may be combined to give the following alternative form:

$$V = (1-t)R - \left[\frac{i}{(1+i)^n - 1} - dt + i_m\right]I - i_m I_w + \frac{i(1-t)S_a}{(1+i)^n - 1} \quad (3.6)$$

3.03 THE VENTURE WORTH EQUATION

If an enterprise operates over a period of years, the attractiveness is judged by a summation of the series of annual venture profit values, represented by V_k for each year k. Any convenient reference time may be chosen, but the present is usually most convenient. From Eq. (A.3) there is obtained

$$W = \sum_{k=1}^{k=n} \frac{V_k}{(1+i)^k} \quad (3.7)$$

where W is the venture worth, and the summation is extended for n years of plant life. Each of the items in Eq. (3.6) can be similarly summed to obtain its contribution to W. In the case of rate of depreciation for tax purposes d, the subscript k varies from 1 to r years, where r represents the number of years allowed for depreciation by tax officials.

$$W = \sum_{k=1}^{k=n} \frac{(1-t)R_k}{(1+i)^k} + \sum_{k=1}^{k=r} \frac{d_k tI}{(1+i)^k}$$
$$- \left[\frac{i}{(1+i)^n - 1} + i_m\right] \sum_{k=1}^{k=n} \frac{I}{(1+i)^k}$$
$$- \sum_{k=1}^{k=n} \frac{i_m I_w}{(1+i)^k} + \left[\frac{i(1-t)}{(1+i)^n - 1}\right] \sum_{k=1}^{k=n} \frac{S_a}{(1+i)^k} \quad (3.8)$$

In the above relationship it is assumed that I, I_w, and S_a do not change with the year k. I is the initial investment, I_w is the

Expanded Economic Evaluation Equations

requirement each year for working capital, and S_a is the salvage value at the end of the nth year. The gross return rate and depreciation are assumed to change from year to year. If it were desired, more general formulations could be developed, allowing for variations in I and I_w.

In Appendix B it is shown that I does change with time. Monetary inflation causes the value of I to increase each year according to a compound interest relationship of the type $(1 + a)^k$, where a is an annual rate of inflation. This increase makes it harder for the project to recover the equivalent purchasing power of the original capital investment, and accordingly decreases the economic attractiveness of the project. The interest rate i is also affected by inflation, but the value of i "seen" by corporations is that resulting from their operations in an inflationary economic climate. Consequently, the value of i reported by corporations within the 2 or 3 years previous to the start of the venture worth analysis is the proper value of i to use. An allowance for inflation is automatically included when the corporation's recent earnings rate is used for i.

The venture worth analysis is used to compare the economic attractiveness of future projects. If all the projects are calculated on the same basis, which may include an allowance for inflation or not, then the comparisons between the projects will be valid and meaningful conclusions can be drawn. Future cash flows will all be inflated at the same rate.

It would appear to be reasonable to conclude that the venture worth concept can be applied in an inflationary economy even though the equations for W do not contain any specific allowance for inflation.

If an allowance for inflation is not made it must be remembered that at the end of the project's life there will not be a sufficient number of dollars available to repurchase the original investment, or to return to the owners the purchasing power of the original investment. Actually, this is not a serious matter. Inflation, at lower than present rates, has always been present (long before the venture worth concept was first published) so that it has never been possible to repurchase the original plant. This fact seems not to have bothered anyone unduly. Also, at the end of 10 or more years business opportunities and process technology will have changed so much that some entirely different investment will be under consideration, and there will be no question of buying back the original plant. Still further, payout times are customarily so short, 2 to 4 years, that inflation cannot proceed very far and the inflated value of I is not seriously different from the original value of I. However, if inflation should continue to increase at faster and faster rates, business planners will be forced to consider its effect much more carefully.

Since I, I_w, and S_a are presumed not to change, the summations for the terms involving them are obtained by the uniform annual series present worth factor, Eq. (A.5). The following result may be obtained, making use of the fact that $i + i/[(1+i)^n - 1] = i(1+i)^n/[(1+i)^n - 1]$, the capital recovery factor

$$W = \sum_{k=1}^{k=n} \frac{(1-t)R_k}{(1+i)^k} + \sum_{k=1}^{k=r} \frac{d_k tI}{(1+i)^k}$$

$$\begin{pmatrix} \text{Venture} \\ \text{worth} \end{pmatrix} = \begin{pmatrix} \text{summation of} \\ \text{gross untaxed} \\ \text{income} \end{pmatrix} + \begin{pmatrix} \text{summation of} \\ \text{discounted} \\ \text{tax credit} \end{pmatrix}$$

$$- \left[\frac{(1+i)^n - 1}{i(1+i)^n} \right] (i_m - i)(I + I_w) - I$$

$$- \begin{pmatrix} \text{present worth of} \\ \text{incremental return} \\ \text{on total investment} \end{pmatrix} - \begin{pmatrix} \text{initial} \\ \text{investment} \end{pmatrix}$$

$$- \left[\frac{(1+i)^n - 1}{(1+i)^n} \right] I_w + \frac{(1-t)}{(1+i)^n} S_a \qquad (3.9)$$

$$- \begin{pmatrix} \text{present worth} \\ \text{of cost of} \\ \text{working capital} \end{pmatrix} + \begin{pmatrix} \text{present worth} \\ \text{of salvage} \end{pmatrix}$$

The venture worth is thus the present worth of all cash flows above an acceptable minimum rate discounted to the present by means of the average rate of earnings on capital. It represents a sum of money which may be thought of as the "cash value" of an idea, to borrow a phrase from William James [10].

The numerical value of W is expressed in dollars, a sum of money; it may be positive, negative, or zero. When comparing one project, or alternative investment, with another, the more positive one is better.

The open structure of Eq. (3.9), with all important items making up W set apart from one another, allows the rapid inspection of the effects of the variables. Whatever the value obtained

Expanded Economic Evaluation Equations 97

for W in a first calculation, the changes required in the variables to produce a change in W can be obtained by inspection or by a small additional calculation.

In Happel's previous development [9] of this idea the terms i and i_m were set equal to one another. In that case the term involving $i_m - i$ vanishes, and a simpler equation results. This is convenient because it is often difficult to obtain precise values of i_m and i. As discussed later, i can be calculated from the company's financial records, but i_m is somewhat harder to evaluate. However, it is wise to make an effort to establish i_m with some precision rather than to set it equal to i. After all, the purpose of investing in a new project is to obtain a return that is higher than either i or i_m.

The summations involving R_k and d_k are sometimes difficult to evaluate. However, if they vary in some systematic fashion it is possible to develop formulations for the summations indicated in Eq. (3.9). Such summations are discussed later in this chapter. These algebraic expressions for the summations are quite useful because they reduce, by a fairly large factor, the time and effort involved in computing W.

It should be recognized that any calculation of W involves the evaluator in the prediction of the future. This is true whether R_k and d_k are held constant or assumed to vary as time passes. Long experience teaches that sales usually increase, costs decrease a small amount, and prices drop considerably. Such variations may be allowed for by making a year-by-year tabulation of the terms in Eq. (3.9) or by utilizing the algebraic equivalents of the sums for linearly or exponentially decreasing values of R_k. Section 3.07 has a discussion of the forms that the venture worth equation takes when R_k decreases steadily in a smooth manner.

Many workers in the field of process economics have discussed falling prices and changing costs along the following lines:

1. In an economic evaluation, the most important and the most uncertain assumptions are those of future sales volume, future sales price, and future production costs. It is obvious that a wrong estimate of sales price or sales volume will seriously distort the economic evaluation.

2. The assumption that prices and costs will remain essentially constant for a protracted period after plant start-up may lead to a serious overestimation of the project profitability.

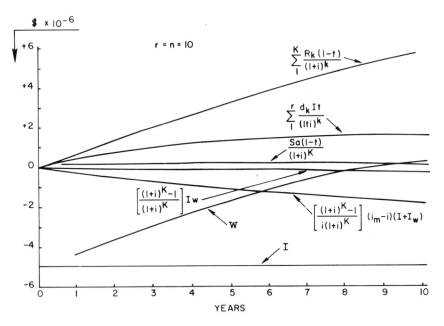

Fig. 3.1. Components of venture worth.

3. An accurate prediction of the future values of sales price, sales volume, and production costs is essentially impossible. However, a careful study of the price-cost-production rate history of similar products will serve as a useful guide. A study of this type will usually show that, for a new product, both the sales price and the production cost will decrease in an exponential manner. The sales price will usually decrease faster than the production cost. Consequently the profit margin will decrease steadily.

The results of Example 3.2 may be examined in some detail to obtain a clearer idea of the characteristics of the venture worth concept. Figure 3.1 is a plot of the components of the venture worth as a function of time, with the depreciation time being set at $r = n = 10$. Several items are worthy of close examination.

1. The venture worth, W, is the sum of six items, three positive and three negative. The final value for W is the sum of these six and a change in any of them changes the value of W. However, some of the items are more important than others.

2. The two most important items are (a) the summation of the discounted gross untaxed income and (b) the capital investment, I.

Expanded Economic Evaluation Equations

a. This item reflects directly the business success of the enterprise. As in any commercial operation the basic idea is to buy low and sell high. The larger the difference between buying (costs) and selling (prices), the larger the value of R, and the more successful the enterprise. A high value of R will contribute a strongly positive value to W and help make W increase rapidly. This is what is wanted. It is apparent that the process managers must exert every effort to make R large and to do it immediately upon the start of operations. The bad effects of a slow start-up, high initial costs, and low sales prices are obvious.

The effect of discounting future gross profits is also apparent. If the future cash flows were not discounted, the line for $\sum_{1}^{n} \frac{R_k(1-t)}{(1+i)^k}$ would rise essentially linearly, and the resulting value of W would be higher. The discounting has the effect of reducing the numerical values of the future cash flows, and causes the line to bend toward the horizontal. The gains from future profits thus become less important to the sum and to W. If the discounting rate is set quite high—say 10%—future cash flows will be very strongly discounted and contribute little to the sum and to W.

b. The initial capital investment I is considered to be a constant and a negative term. I is always the greatest penalty against the project. From an inspection of Fig. 3.1 it is easy to see why management regards a low I as desirable and a high I as bad. I also appears in two other terms, as discussed below. These two terms are of opposite sign and tend to counteract one another.

3. The term $\sum_{1}^{r} \frac{d_k It}{(1+i)^k}$ is the sum of all the money that is saved because the government allows the company to charge depreciation before paying taxes. This term is positive, a credit to the project, and is discounted so that it tends to become constant as time passes. If both accelerated depreciation techniques and compound interest discounting are used, as is the case here, the summed term will become constant a few years before the end of the project's life. In almost any worthwhile project the discounted tax credit term will not be as large as the gross profit term, but the tax credit term is not small and must be considered. Its value cannot vary widely because the government restricts the values of d_k and t to within rather narrow limits. This term is directly proportional to I.

In any calculation of the summation of the discounted tax credit (for a constant I), it must be realized that the numerical values for $d_k It/(1+i)^k$ depend on d_k, which in turn depends on the allowable depreciable life, r, and on the methods used to calculate d_k (straight line, declining balance, or sum-of-year's-digits). Therefore, the magnitude and the shape of the tax credit line depend on the value of r; if r were different the line would be different. Values of d_k and r are chosen in such a way that I will be recovered at the end of r years. If r were smaller, as in a special depreciation allowance from the government, the depreciation tax credit line would rise more steeply, and so would the line for W.

The highly negative values for W during the early years of the project reflect the large fraction of the capital investment that is still unrecovered.

4. Another term of importance is $\{[(1+\ )^n - 1]/i(1+i)^n\}$ $(i_m - i)(I + I_w)$. This is called the "present worth of the incremental return on total investment." This is a lump sum of money, counted in zero-time dollars, which represents the total amount of incremental interest payments on the total capital invested in the project for the years from 1 to k. $I + I_w$ is the total capital invested; $i_m - i$ is the increase in the interest rate earned by the project over that usually obtainable by investment in ordinary business activities when the project return is at least i_m. The term $[(1+i)^n - 1]/i(1+i)^n$ is the uniform series present worth factor. It is the multiplier that converts a uniform series of future payments, $(i_m - i)(I + I_w)$, into a single sum of money at the start of year 1.

In the venture worth equation this term is negative because the definition of venture worth is such that it is the excess over that earned by investment at the minimum acceptable rate, i_m; see Eq. (3.5). As time passes, this term becomes increasingly negative. The most important part of this term is the difference in interest rates: $i_m - i$. If $i_m = i$, the term vanishes, and no penalty is assessed against the project. If i_m is set fairly high, meaning that management will be satisfied only with a high rate of return, the term becomes strongly negative, and the project will be more heavily penalized. The economic position of any project can be distorted unfavorably by the selection of a value of i_m that is unrealistically high.

5. The remaining terms involving the working capital and the salvage value are small, and are opposed in sign, so that they tend to cancel one another. However, the discounted salvage value trends toward zero as time passes, while the working capital term becomes increasingly negative over the same period. Both working capital and salvage value are fairly difficult to establish with any reasonable degree of precision, and it is not especially rewarding to spend much time analyzing their contribution to W.

Expanded Economic Evaluation Equations

6. At time zero the value of W will be $-I + S_a/(1 - t)$ and as time passes it will become increasingly positive. Depending on the relative magnitudes of the items making up W, it will be positive, negative, or essentially zero at the end of n years. A positive value indicates that the best estimate available predicts that the revenue-creating powers of the project are sufficiently large to overcome the project deficits. A negative value would show the opposite effect. It would appear to be reasonable to conclude that a positive value of W indicates a project that may be profitable and a negative value, one which will not be profitable. Consequently, a logical action would be to use $W = 0$ as a cut-off point separating desirable projects from undesirable projects.

However, two factors come into consideration:

1. The numbers used in computing W are only estimates; they are not known with good precision. Rather small changes in selling price, product cost, interest rates i and i_m, and the capital investment can cause a large enough change in W to cause it to move from negative to positive, or the reverse. Consequently, slightly negative values of W should not be rejected, and slightly positive values of W should not be accepted without further study. The lack of precision in the calculation, and its subjective nature, should be recognized.

2. The Venture Worth, W, is normally used as a method for comparing the profitability of two or more alternative investments. The method involves calculating W for each investment and selecting the one with the highest value of W. Since all the investments will be calculated on the same basis, the one with the highest value of W is considered the best. For the purpose of the comparison it does not matter if some or all of the values of W are negative.

The words used to describe, or define, the venture worth would seem to exclude any investment with a negative value of W. (Who would accept a project where the money penalties are greater than the money income?) However, the situation is more complex than a simple yes or no, because a value of W is greater or smaller than zero. A thorough study of all the components of W is required before a decision is reached.

Example 3.1

A plant designed to recover propane from natural gas by a new process costs $3,000,000. Investment in land and other recoverable items amounts to $600,000. There is an assured supply of feed for 10 years on the basis of a long-time contract, but, after this contract expires, it is impossible to predict whether further operation will be economical. It is anticipated that, although the process is technically sound, it will probably take 2 years to complete construction and attain successful operation. During this period it is assumed that return will be balanced by operating costs so that the gross return will be zero. It is desired to compute what uniform return, R_k, will be the minimum required for the 8 years of successful operation, on the basis of the following assumptions.

A 15% return i_m on the entire investment after taxes and depreciation is the minimum acceptable to management. It is currently obtaining approximately 10% on present investments. The Federal and state authorities will allow the plant to be written off in a period of 5 years, assuming straight-line depreciation. Income-tax rate will be taken at 50%/yr. It will also be assumed that all profits and depreciation reserves will be reinvested at the current earning rate of the company or in other similar ventures (i.e., at 10% return after taxes and capital recovery). Also there will be no salvage value obtainable if the plant is abandoned.

Solution. Since the minimum return is required, W must be set equal to zero. The remaining terms in Eq. (3.9), excepting that involving S_a, which is zero, are evaluated as follows.

Summation of discounted gross untaxed income will be [see Appendix A, Eq. (A.8)]:

$$\sum_{k=3}^{k=10} \frac{1 - 0.50}{(1.1)^k} R = 0.5R \frac{(1.1)^{10-2}}{0.1(1.1)^{10}} = 2.20R$$

The return rate R_k is zero during the first 2 years and is taken constant at R for the remaining years, k = 3-10, inclusive.

Summation of discounted tax credit is

$$\sum_{k=1}^{k=5} \frac{(0.2)(0.50)(3,000,000)}{(1.1)^k} = 300,000 \frac{(1.1)^5 - 1}{0.1(1.1)^5} = 1,138,000$$

Present worth of incremental return on total investment is

$$\frac{(1.1)^{10} - 1}{0.1(1.1)^{10}} (0.05)(3,600,000) = 1,105,000$$

Expanded Economic Evaluation Equations

The investment I = 3,000,000, and the present worth of the working capital cost is

$$600{,}000 \; \frac{(1.1)^{10} - 1}{(1.1)^{10}} = 369{,}000$$

Therefore,

$$0 = 2.20R + 1{,}138{,}000 - 1{,}105{,}000 - 3{,}000{,}000 - 369{,}000$$

and R = $1,520,000/yr from the third through the tenth year.

Example 3.2

In Example 3.1 the propane recovery plant was analyzed using the assumption of a steady gross income R each year for the last 8 years of the life of the project. Furthermore, the depreciation rate d was assumed to be constant at 20%/yr. These rather simple conditions allowed direct substitution in Eq. (3.9), setting W = 0, and solving for R. In actual practice, the economic situation affecting the project will probably change continually, and a year-by-year tabulation may be required. Consider the following situation.

A corporation is investigating a process for the manufacture of a chemical that should have large sales in the synthetic fiber industry. The initial cost estimate shows that the new capital investment required will be $5,000,000; the working capital will be $400,000; the manufactured cost of the product (not including a depreciation charge) will be 15¢/lb, and the production capacity of the unit will be 3×10^7 lb/yr.

This set of numbers implies that the total new capital investment will be $5,000,000 and that it will all be spent, essentially at once, at the start of the project. In actual practice the total new capital investment may include charges for research, development, and engineering. Furthermore, these sums, plus the new plant investment, will be spent over a period of 3 or 4 years, rather than all at once. Such expenditures should be discounted, either forward or backward, to some convenient project starting time, such as the start of the project or the start of plant operations.

Adding research, development, and engineering charges to the new capital investment will increase I substantially. In the present case this was not done, but a more detailed analysis would do this; see Ref. [8], Table 1-2, p. 48.

The company's financial records show that its interest rate of return on total assets, i, is 12%. The financial officers are determined to increase this and have set the minimum acceptable rate of return on new chemical ventures at 18%, that is, $i_m = 0.18$.

With these numbers set, it is now necessary to compute a selling price s that will cause the project to meet the specifications given above. This number can be calculated from Eq. (2.9) by letting V = 0 and making some simple algebraic substitutions. For quick calculations let d = 0.10, e = 0.10, and t = 0.50. Then

$$s = \frac{i_m(I + I_w) - dIt + eI + Q(1 - t)c}{(1 - t)Q}$$

$$s = \frac{(0.18)(5,400,000) - (0.10)(0.50)(5,000,000) + (0.10)(5,000,000)}{(1 - 0.5)(3 \times 10^7)}$$

$$+ \frac{(3 \times 10^7)(1 - 0.5)(0.15)}{(1 - 0.5)(3 \times 10^7)}$$

s = 0.23 or 23¢/lb

Calculating a selling price in this manner, by setting the venture profit equal to zero, is equivalent to saying that the return on the investment will be exactly i_m, no more, and that the venture worth will be very close to zero. Actually, the price of 23¢/lb is a reasonable value and is approximately on the line of Fig. 1.3, which relates production rate and sales price.

This method of calculating a selling price assumes that the gross rate of return R_k will remain essentially constant during the lifetime of the project. Sometimes this is a valid assumption, and sometimes it is not. If the gross rate of return declines substantially as time passes (a rise in the return rate is a most unlikely occurrence), it is necessary to make the initial selling price much higher than that calculated from the method shown above, otherwise the declining rate of return will cause the project to be unprofitable; see Section 3.07 and Example 3.5.

A reasonable sequence of events is detailed below. As mentioned before, such a calculation involves a prediction of the future. But so does any other calculation, no matter how simple.

1. During year 1 the plant will operate at 50% of production capacity and the manufacturing cost will be 20¢/lb. The sales price will be 25¢/lb.

2. During year 2 the plant will operate at 80% of capacity, but the manufacturing cost will average 18¢/lb. The sales price will be 24¢/lb.

3. During years 3 through 8 the plant will operate at 100% of capacity, the manufacturing cost will be 15¢/lb, and the sales price will be 23¢/lb.

4. During years 9 and 10 the production rate will be increased to 40,000,000 lb/yr, and the manufacturing cost will be lowered to 14¢/lb. However, the sales price will be reduced to 20¢/lb.

5. The yearly depreciation charge will be calculated by the double declining balance method [see Section 3.06, Eq. (3.12)]. The unrecovered balance of the capital investment will be set equal to the salvage value.

Tables 3-1, 3-2, and 3-3 show the calculation of each of the individual components of the venture worth. In this case the venture worth at the end of 10 years is slightly positive. Figure 3.1 (on p. 98) is a plot of the results of the calculation against the number of years. It is interesting to note that the venture worth starts from a very low value and slowly climbs toward zero or above.

This example shows that the calculation of the venture worth is really a simple operation. Ordinarily, the venture worth is calculated only at the end of n years. In this case, only two terms have to be calculated each year; three terms need be calculated only once, and one term, I, is not calculated.

Chemical plants usually have a history of changing production rates, prices, and costs. Commonly this change moves in the direction of increasing production rates and decreasing prices and costs. For any new venture it would be foolish to expect that the plant will start immediately at time zero and run steadily under constant conditions until the project life is ended. Conditions will change, and this should be recognized at the beginning.

Strauss [11] has presented an ingenious diagram for graphically portraying the response of an economic indicator such as rate of return or gross profits, to changes in the process variables, such as capital investment or sales price, which affect the economics of a project. It is interesting to apply Strauss's diagram to the venture worth.

Briefly, Strauss plots on rectangular coordinates the percent change in the variables, from their most probable values as abscissa, against the numerical value of the economic indicator as ordinate. The intersection of the axes is placed at the most probable value of the economic indicator, found by calculations using

Table 3-1 Calculation of Terms in Venture Worth Equation—Example 3.2

Year of project life	Sales price, s ($/lb)	Mfg. cost, c ($/lb)	Production rate, Q (lb/yr × 10⁻⁶)	R Q(s − c) ($/yr)	$\dfrac{R_k(1-t)}{(1+i)^k}$	$\dfrac{d_k tI}{(1+i)^k}$
1	0.25	0.20	15	750,000	335,000	446,000
2	0.24	0.18	24	1,440,000	576,000	319,000
3	0.23	0.15	30	2,400,000	853,000	215,000
4	0.23	0.15	30	2,400,000	762,000	164,800
5	0.23	0.15	30	2,400,000	680,000	117,700
6	0.23	0.15	30	2,400,000	607,000	84,100
7	0.23	0.15	30	2,400,000	541,000	60,000
8	0.23	0.15	30	2,400,000	484,000	43,000
9	0.20	0.14	35	2,100,000	385,000	30,750
10	0.20	0.13	40	2,800,000	450,000	21,850

$$\sum_{1}^{10} \frac{R_k(1-t)}{(1+i)^k} = \$5{,}673{,}000$$

$$\sum_{1}^{10} \frac{d_k tI}{(1+i)^k} = \$1{,}502{,}200$$

Table 3-2 Calculation of Depreciation Charge ($d_k I$) Using Double Declining Balance[a]

End of year	D_{db}	$d_k I$ $0.2(5,000,000 - D_{db})$
1	0	1,000,000
2	1,000,000	800,000
3	1,800,000	604,000
4	2,404,000	519,200
5	2,923,200	415,360
6	3,338,560	332,288
7	3,670,848	265,830
8	3,936,678	212,664
9	4,149,342	170,132
10	4,319,474	136,105

Unrecovered balance = salvage value = S_a = $544,148

[a] I = $5,000,000; 10-year life; unrecovered balance = salvage value.

Table 3-3 Calculation of the Venture Worth[a]

Year	$\sum_1^k \dfrac{R_n(1-t)}{(1+i)^n}$ ($)	$\sum_1^k \dfrac{d_k tI}{(1+i)^k}$ ($)	$\left\{ \dfrac{(1+i)^k-1}{i(1+i)^k}(i_m-i)(I+I_w) \right\}$ ($)	$\left\{ \dfrac{(1+i)^k-1}{(1+i)^k}(I_w) \right\}$ ($)	$\dfrac{(1-t)S_a}{(1+i)^k}$ ($)	Venture worth, W ($)
1	335,000	446,000	288,000	42,800	241,000	−4,308,800
2	911,000	765,000	547,000	81,000	216,500	−3,735,500
3	1,764,000	980,000	776,000	115,200	193,600	−2,953,600
4	2,526,000	1,144,800	982,000	145,300	172,800	−2,302,500
5	3,206,000	1,262,500	1,166,000	168,200	154,000	−1,711,700
6	3,813,000	1,346,600	1,330,000	197,000	138,000	−1,229,400
7	4,354,000	1,406,600	1,477,000	219,000	123,000	− 812,400
8	4,838,000	1,449,600	1,609,000	238,500	109,800	− 450,100
9	5,223,000	1,480,350	1,723,000	256,000	98,100	− 177,570
10	5,673,000	1,502,200	1,830,000	271,500	87,500	+ 161,200

[a] $I = \$5,000,000$; $i_m = 18\%$; $Q = 3 \times 10^7$ lb/yr; $S_a = \$544,148$; $I_w = 400,000$; $i = 12\%$; $n = 10$ years.

Expanded Economic Evaluation Equations

Table 3-4 A Sensitivity Analysis for the Venture Worth[a]

Variable	% change from base	W ($ × 10⁻⁶)
$Q = 2 \times 10^7$ lb/yr	-33.3	-1.9
$Q = 4 \times 10^7$ lb/yr	+33.3	+1.9
$I = \$6 \times 10^6$	+20	-1.1
$I = \$4 \times 10^6$	-20	+1.1
$s = \$0.25$/lb	+17	+3.0
$s = \$0.18$/lb	-17	-3.0
$c = \$0.18$/lb	+20	-2.5
$c = \$0.12$/lb	-20	+2.5
$i_m = 0.15$	-17	+0.92
$i_m = 0.20$	+11	-0.61
$i = 0.08$	-33.3	-0.45
$i = 0.15$	+25	+0.18
Method of calculating depreciation allowance		
Straight-line; $d = 0.10$	0	0
Double declining balance		+0.07
Sum of year's digits		+0.21

[a] Basic conditions: $W = 0$; $Q = 3 \times 10^7$ lb/yr; $s = \$0.217$/lb; $c = \$0.15$/lb; $i = 0.12$; $i_m = 0.18$; $d_k = 0.10$ (straight line); $n = 10$ years; $I = \$5 \times 10^6$; $I_w = \$0.4 \times 10^6$.

the most probable values of the variables. Changes in the variables increase to the right and decrease to the left. A straight line is drawn from the intersection to the point located by the new value of the indicator calculated from the changed value of the variable.

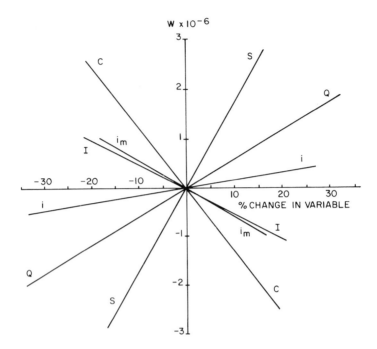

Fig. 3.2. A sensitivity diagram for the venture worth.

The slope of such a line is a measure of the sensitivity of the indicator to changes in the independent variable. A steep line indicates that the indicator is quite sensitive to changes in the variable; a shallow line shows the reverse.

In Example 3.2 the Venture Worth was calculated for one set of process variables. A sensitivity diagram is easily made (see Fig. 3.2). Table 3-4 shows a summary of the calculations. The basic, or most probable, values of the variables are those that cause the venture worth to be zero. Changes in the variables cause the venture worth to be positive or negative.

The sensitivity diagram is shown as Fig. 3.2. The results may be summarized:

1. W is quite sensitive to changes in the selling price and to the manufactured cost; see lines marked s and c. It must be perfectly obvious that any commercial enterprise is sensitive to the cost of its manufactured products and to the sales price of the same products. High prices and low costs are greatly to be desired.

Expanded Economic Evaluation Equations

2. W is moderately sensitive to Q, the production (or sales) rate. For a fixed capital investment and operating profit (s - c), W depends largely on Q.

3. W is moderately sensitive to the new capital investment, not so sensitive as to Q, but sensitive enough. As is obvious from Eq. (3.9) and Fig. 3.1, a high capital investment is a severe penalty to a project while a low capital investment is a benefit.

4. W is moderately sensitive to the value of i_m. This comes about because i_m enters the equation for W in the form $i_m - i$. Therefore, small changes in i_m have significant effects on W.

5. W is almost insensitive to the value of i. This results from the fact that i appears in every term but one, and the changes in the various terms, due to changes in i, tend to cancel one another.

The method used to calculate the depreciation allowance is also important to the numerical value of the venture worth. The accelerated allowances of double declining balance and sum-of-year's-digits cause a small improvement in the value of W.

3.04 RATE OF RETURN

A variation in the application of the present worth method is the use of a ratio of the incremental present worth, or venture worth, to the original investment. This may also be converted to a rate of return showing return per year on a uniform basis as a fraction of the original investment.

Rate of return has the advantage of measuring the productivity of capital in terms that are relatively familiar, and may be compared with investments in interest-bearing securities, where one obtains a given percent return on an initial investment and at the end of the period of investment recovers the original capital. It has the disadvantage, discussed in Chapter 2, of eliminating attractive investments which do not happen to show the absolute highest rates of return, though still showing excellent return rates with an opportunity for substantial capital investment.

The venture worth method may be applied to the determination of appropriate rates as follows.

In the cases where the venture profit formulation applies, an appropriate rate will be

$$v = V/(I + I_w) \tag{3.10}$$

Note that v, like p (Section 2.11), is defined as a rate of return on total investment. In the case of a nonuniform series, application of the interest and amortization factor enable one to obtain an equivalent fractional return per year.

$$v = \left[\frac{W}{I + I_w}\right]\left[\frac{i(1 + i)^n}{(1 + i)^n - 1}\right] \tag{3.11}$$

Reference to Eq. (3.9) will show that v is larger than the rate of return as usually computed (using the discount rate i) by the difference between i_m and i.

The use of the venture worth W in connection with a rate of return can be shown using the result of Example 3.2. Here W = $161,200 and $I + I_w$ = $5,400,000; then

$$v = \left[\frac{161,200}{5,400,000}\right]\left[\frac{(0.12)(1.12)^{10}}{(1.12)^{10} - 1}\right]$$

$$v = +0.00528$$

Since this rate of return is based on the venture worth, the actual rate of return is i_m + v, or 0.185. If W had been negative the rate of return would have been less than i_m, and if W = 0, the rate of return would have been i_m.

3.05 THE DISCOUNTED CASH FLOW METHOD

An inspection of Eq. (3.9) shows that values for i and i_m must be specified before a value for W can be computed. It is perfectly true that this is an unfortunate aspect of the venture worth concept, and there has been strong criticism directed against it because of the necessity of assuming i and i_m before calculating W.

Another method for evaluating projects, known as the discounted cash flow method (DCF), has been devised. This method avoids the necessity for assuming i and i_m before starting the calculation. The DCF method is carried out by letting i and i_m be equal, and equal to some new interest rate, i_d. The term W is set equal to zero, and a trial-and-error calculation is performed until a value for i_d is found that will make the right-hand side of Eq. (3.9) equal to zero. This procedure certainly avoids assuming i and i_m, and calculates an interest rate that causes all the future cash flows into the project to exactly equal the present value of the new capital investment, I. The new interest rate, i_d, is the rate of return on the company's investment from the proceeds of the project. Project desirability is gauged by the size of i_d; the larger the better.

The DCF method has gained a considerable amount of popularity in recent years, especially since the advent of the electronic computer, which makes short work of the trial-and-error calculation. The DCF method is a simpler variation of the venture worth method and it does have considerable appeal, but it suffers from several disadvantages.

1. The solution i_d, called the "solving rate," cannot be specified in advance, and it changes for each evaluation. Therefore, cash flows for different proposals are discounted at different rates and so cannot be compared on the same basis. It is difficult to establish a consistent value for the design of plant components.

2. Since a high value of the solving rate is desired, designers using the DCF method will tend to minimize I, the new capital investment, in order to maximize the solving rate. Thus, valuable investments, such as waste heat recovery, may be excluded when, in reality, such an investment may return a large number of dollars to the company at a satisfactory (but not the highest) rate of return. As discussed in Chapter 2, all rate of return indices suffer from this defect.

3. The most serious objection to the DCF method is that the answer is the interest rate which is used for the discounting of the future cash flows. If the project is a good one and the answer is high, the future cash flows will be discounted so severely that their contributions are almost negligible. Conversely, if the project is not a good one and the answer is low, the future cash flows will not be discounted strongly enough, and risk is not properly reflected. A distorted picture results. It would seem to be more realistic to assume an interest rate i which is close to the average value of the company's results over a period of years.

There is a further point that is not a disadvantage, but it is worthy of note. In the DCF method i and i_m need not be assumed in advance, but, once the calculation has been completed, the resulting rate i_d must be compared to some i_m which has been stored in the minds of the financial officers of the company. Thus, the

Table 3-5 Calculation of the Discounted Cash Flow Interest Rate[a]

Year	Assumed interest rate					
	10		15		20	
	$\dfrac{(1-t)R_k}{(1.10)^k}$ ($)	$\dfrac{d_k It}{(1.10)^k}$ ($)	$\dfrac{(1-t)R_k}{(1.15)^k}$ ($)	$\dfrac{d_k It}{(1.15)^k}$ ($)	$\dfrac{(1-t)R_k}{(1.20)^k}$ ($)	$\dfrac{d_k It}{(1.20)^k}$ ($)
1	341,500	455,000	326,000	434,000	313,000	417,000
2	595,000	330,500	543,000	302,500	500,000	277,500
3	901,000	226,500	780,000	198,200	694,000	174,800
4	819,000	177,000	685,000	148,200	579,000	125,000
5	745,000	128,900	597,000	103,300	482,500	83,400
6	677,000	93,700	519,000	71,800	402,000	55,600
7	615,000	68,300	451,000	50,000	335,000	37,100
8	559,000	49,600	392,000	34,750	278,200	24,750
9	445,000	36,050	298,500	24,150	203,000	16,480
10	540,000	26,200	346,500	16,800	226,500	11,000
$\sum$	6,237,000	1,581,750	4,948,000	1,383,700	4,013,200	1,222,630
$\dfrac{(1-t)S_a}{(1+i)^n}$	$105,000		$67,200		$43,900	
$\left[\dfrac{(1+i)^n-1}{(1+i)^n}\right]I_w$	$246,000		$308,500		$336,500	
W	$2,677,750		$1,090,400		-$56,670	

[a] Values of R_k and $d_k I$ are given in Tables 3-1 and 3-2.

Expanded Economic Evaluation Equations

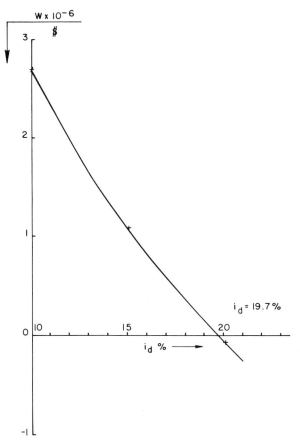

Fig. 3.3. Discounted cash flow method for the calculation of i_d.

selection of a value for i and i_m is not really avoided, but simply postponed. For a further discussion see Kapfer [11].

Example 3.3

The calculations for the discounted cash flow method may be illustrated using the problem of Example 3.2.

Letting $i_m = i$ and setting W = 0 there results

$$0 = \sum_{1}^{10} \frac{(1-t)R_k}{(1+i)^k} + \sum_{1}^{10} \frac{d_k It}{(1+i)^k} - I - \left[\frac{(1+i)^{10} - 1}{(1+i)^{10}}\right] I_w + \frac{(1-t)S_a}{(1+i)^{10}}$$

The procedure to be followed is to assume a value for i and then to calculate a value for the right-hand side of the equation. Three values for i, say 10%, 15%, and 20%, should bracket the most probable answers. A plot of i, as abscissa, is then made against the sum of the right-hand terms above as ordinate. The line should pass through an ordinate value of zero. The interest rate associated with this value will be the desired answer, the interest rate that makes W = 0.

Since only two terms need to be calculated for each year it is possible to set up a compact table and substitute numbers rather quickly. When done in this way, using the plot of W versus i, the calculation is really quite simple and straightforward—no trial and error at all. The most important thing to remember is that the discounting factors change for each interest rate.

Table 3-5 shows a general form for such calculations. The form includes spaces for the insertion of various production rates, sales prices, factory costs, and depreciation charges.

Figure 3.3 shows the plot of assumed interest rate against the corresponding value of W. The best curve through the three points crosses the W = 0 line at 19.7%. This value is the discounted cash flow rate. It is interesting to note that this value (19.7%) is close to the value for i_m of 18% set for the venture worth method.

3.06 INCOME-TAX CREDIT SUMMATIONS

In Section 2.04 the subjects of the Federal income tax and of the depreciation allowances permitted by the U.S. Internal Revenue Service were mentioned briefly. The importance, and the wide use, of the straight-line method of calculating depreciation was mentioned. In Examples 3.2 and 3.3 the double declining balance method was used, but no discussion was given. In this section three methods for calculating the depreciation allowance are presented in some detail. All are allowed by the U.S. Internal Revenue Service:

1. The straight-line method
2. The double-declining balance method (DDB)
3. The sum-of-the-year's-digits method (SYD)

It is also allowable to use any other consistent method that would not give an aggregate depreciation write-off at the end of the first two-thirds of the useful life any larger than that obtained by method 2.

The straight-line rate of depreciation is a constant equal to $1/r$, where r is the life of the facility for tax purposes. Thus, if the life of the plant is 10 years, the straight-line rate of depreciation is 0.1. This rate, applied for each of 10 years, will result in a depreciation reserve equal to the initial investment.

A declining balance rate is obtained by first computing the straight-line rate, and then applying some multiple of that rate to each year's unrecovered cost rather than to the original investment. Under the double declining balance method, twice the straight-line rate is applied to each year's remaining unrecovered cost. Thus, if the life of a facility is 10 years, the straight-line rate will be 0.1, and the first year's double declining balance rate will be 0.20. If the original investment is I, the depreciation allowance the first year will be 0.2 I. For the second year it will be 0.16 I (0.20 of the unrecovered investment of 0.8 I), and similarly until the tenth year has been completed. Since this method involves taking a fraction of an unrecovered cost each year, it will never result in the complete recovery of the investment. In order to overcome this objection the U.S. Internal Revenue Service allows the taxpayer to shift from the DDB depreciation method to the straight-line method any time after the start of the project. This shift is illustrated in Fig. 3.5 and discussed later.

The general, or k-th, term for the double declining balance factor is

$$d_k = (1 - 2/r)^{k-1}(2/r) \qquad (3.12)$$

The rate of depreciation for the sum-of-the-year's digits method is a fraction, the numerator of which is the remaining useful life of the property at the beginning of the tax year, and the denominator is the sum of the individual digits corresponding to the years of life at the start. Thus, with a life of 10 years, the sum of the year's digits will be $10 + 9 + 8 + 7 + 6 + 5 + 4 + 3 + 2 + 1 = 55$. The depreciation rate the first year will be $10/55 = 0.182$. If the initial cost of the facility is I, the depreciation for the first year will be 0.182 I, and for the second year 0.164 I, and so on until the last year. The SYD method will recover 100% of the investment at the end of r years. A shift from SYD to straight line cannot be made once the SYD method has been started.

Table 3-6 General Terms for Undepreciated Fraction of Original Investment

Depreciation method	Fraction of original investment undepreciated at end of year k
Straight line	$1 - k(1/r)$
Double declining balance	$(1 - 2/r)^k$
Sum-of-year's-digits	$1 - \left[\dfrac{2}{r(r+1)}\right](r - k + 1)$

Table 3-7 Undepreciated Fraction of Original Investment for 10 Years

Year	Straight line	Double declining balance	Sum-of-year's digits
1	0.9	0.8	0.818
2	0.8	0.64	0.655
3	0.7	0.512	0.510
4	0.6	0.41	0.383
5	0.5	0.328	0.274
6	0.4	0.262	0.183
7	0.3	0.21	0.11
8	0.2	0.168	0.056
9	0.1	0.134	0.018
10	0	0.108	0

Expanded Economic Evaluation Equations

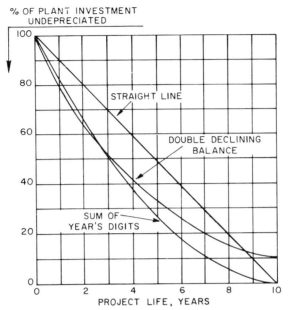

Fig. 3.4. Depreciation tax credit methods.

The general term for the sum-of-the-year's-depreciation rate is

$$d_k = 2(r - k + 1)/r(r + 1) \tag{3.13}$$

It is interesting to know not only the yearly rate of depreciation but the amount of the original investment, I, that remains undepreciated at the end of each year. General terms for this number can also be derived easily (Table 3-6).

Figure 3.4 and Table 3-7 show the values of the fraction of the original investment not depreciated as a function of time for a project with an estimated life of 10 years. Since 10 years is so often used as a project life, the graph and the table are useful.

From Fig. 3.4 it is apparent that the DDB and SYD methods allow a greater fraction of the capital investment to be retired in the first years of a project's life than does the straight-line method. For industrial firms embarking on a new project this is desirable. In the later years of a project's life the straight-line method allows a faster write-off than does either of the other two methods.

In Eq. (3.9), the venture worth equation, the depreciation tax credit term contains the depreciation factor, d_k. This factor may be expressed in any of the various ways discussed above; Eqs. (3.12) and (3.13). If d_k may be expressed in one of these ways it is possible to substitute one of these expressions in the summation term, perform the summation operation, and obtain an analytical expression for the summation term. This equation is usually much more convenient to use than a year-by-year tabulation.

When the straight-line method of calculating depreciation is used, $d_k = 1/r$. This may be substituted into Eq. (3.9) and the uniform annual series present worth factor employed to give

$$\sum_{k=1}^{k=r} \frac{d_k}{(1+i)^k} = \frac{1}{r} \sum_{k=1}^{k=r} \frac{1}{(1+i)^k} = \frac{1}{r}\left[\frac{(1+i)^r - 1}{i(1+r)^r}\right] \quad (3.14)$$

More generally, if it is desired to apply the straight-line method only to the last portion of the total tax period, Eq. (A.4) may be employed to give

$$\sum_{k=m}^{k=r} \frac{d_k}{(1+i)^k} = \frac{1}{r-m+1}\left[\frac{(1+i)^{r-m+1}}{i(1+i)^r}\right] \quad (3.15)$$

where m represents the first year on which tax is paid using the straight-line method. The factor is applicable to the undepreciated balance.

Designating m as the first year on which tax is paid using the straight-line method means that the change from the declining balance method to the straight-line method occurred at the end of year m - 1 and at the beginning of year m.

Because it is advantageous to change from the DDB method to the straight-line method at some point in the project's life, there arises the problem of selecting the point of change so that the discounted tax credit summation will be a maximum. This will be accomplished by changing when d_k by the straight-line method exceeds d_k by the declining balance method.

For the declining balance method, d_k is obtained from Eq. (3.12). If the switch is made at the start of year m, the unrecovered cost will be $(1 - 2/r)^{m-1}$. Then the depreciation factor for the straight-line method will be

Expanded Economic Evaluation Equations   121

$$d_m = (1 - 2/r)^{m-1} \left(\frac{1}{r - m + 1}\right) \quad (3.16)$$

$$\underset{\substack{\text{Unrecovered cost} \\ \text{at beginning of} \\ \text{year m}}}{} \quad \underset{\substack{\text{years remaining} \\ \text{from start of} \\ \text{year m to end of} \\ \text{year r}}}{}$$

If the annual credits for each method are set equal, there results

$$(1 - 2/r)^{m-1}(2/r) = (1 - 2/r)^{m-1}\left(\frac{1}{r - m + 1}\right) \quad (3.17)$$

$$m = 1 + r/2 \quad (3.18)$$

r	m
5	4
10	6
15	8

In the table above values of m have been given for several probable values of r. When m has fractional values, as when r is any odd number, the value of m must be set equal to the next larger integral value. An analytical expression for m, which always gives an integral value, is

$$m = (2r + \cos \pi r + 7)/4 \quad (3.19)$$

It is possible to combine the DDB values for d_k, the straight-line values for d_k, and the value for m so as to write an expression for the optimum total discounted tax credit.

$$\sum_{k=1}^{k=r} \frac{d_k}{(1 + i)^k} = \frac{2/r}{1 - 2/r} \sum_{k=1}^{k=m-1} \frac{(1 - 2/r)^k}{1 + i}$$

$$+ \frac{(1 - 2/r)^{m-1}}{r - m + 1} \sum_{k=m}^{k=r} \frac{1}{(1 + i)^k} \quad (3.20)$$

If the summations are performed as indicated, there results

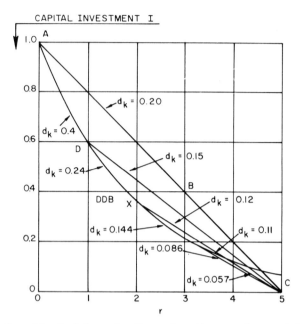

Fig. 3.5. Graphical determination of optimum depreciation rate.

$$\sum_{k=1}^{k=r} \frac{d_k}{(1+i)^k} = \left(\frac{2}{r}\right)\left\{\frac{1 - [(1 - 2/r)/(1+i)]^{m-1}}{i + 2/r}\right\}$$

$$+ \frac{(1 - 2/r)^{m-1}}{r - m + 1}\left[\frac{(1+i)^{r-m+1}}{i(1+i)^r}\right] \quad (3.21)$$

This is the maximum discounted tax credit factor that can be realized by combining the double declining balance method with the straight-line method of calculating depreciation.

The act of shifting from the double declining balance depreciation calculation to the straight-line method is best shown by a simple graph, Fig. 3.5.

If I is taken as $1.00 and the project life is assumed to be 5 years, the two lines ABC and AXC show the value of the unrecovered, or undepreciated, capital as a function of time. The line ABC shows the straight-line depreciation with a value for d_k of $1/r = 0.2$, or 20%/year. The line AXC shows the double declining balance

Expanded Economic Evaluation Equations

depreciation. The value of I declines more rapidly for the DDB method than for the straight-line method in the early years of the project and more slowly in the later years. Also, it is apparent that the DDB method will never recover 100% of I. After 5 years 7.8% of I remains unrecovered.

The consequences of making the switch from DDB to straight line can be seen by drawing straight lines such as DC or XC and noting the slope. Thus, if the switch is made after 2 years the depreciation rate is given by the slope of the line XC; $d_k = 0.12$. By doing this for each of the years 1, 2, 3, and 4 it is possible to get

Year	d_k (straight line)	d_k (DDB)
1	0.20	0.4
2	0.15	0.24
3	0.12	0.144
4	0.11	0.086

Thus, the answer obtained by this method is the same as that obtained from Eq. (3.19). It is only after about 3.5 years that it would be advantageous to switch from DDB to straight-line depreciation.

In the material given above attention was focused on the change from double declining balance depreciation calculation methods to those of the straight-line method so as to achieve maximum depreciation tax credits. In developing Eq. (3.21), a summation for the double declining balance term—the first term of Eq. (3.20)—was presented without discussion. This expression, and a similar one for the sum-of-the-year's-digits method, can be derived with the assistance of the geometric series summation methods of Appendix A.

The general, or k-th, term for the double declining balance factor is given by Eq. (3.12).

Then,

$$\sum_{k=1}^{k=r} \frac{d_k}{(1+i)^k} = \frac{2/r}{1 - 2/r} \sum_{k=1}^{k=r} \left[\frac{1 - 2/r}{1 + i} \right]^k \quad (3.22)$$

where the required summation is a geometrical series, which may be evaluated as follows:

$$\sum_{k=1}^{k=r} \frac{d_k}{(1+i)^k} = \frac{2/r}{1 - 2/r} \left[\frac{\left[\frac{1 - 2/r}{1 + i}\right]^{r+1} - 1}{\frac{1 - 2/r}{1 + i} - 1} \right]_1 \quad (3.23)$$

After simplification, the factor for double declining balance becomes

$$\sum_{k=1}^{k=r} \frac{d_k}{(1+i)^k} = \left(\frac{2}{r}\right) \frac{1 - [(1 - 2/r)/(1 + i)]^r}{i + 2/r} \quad (3.24)$$

and this expression is substituted into Eq. (3.20) to give Eq. (3.21).

When the discounted tax credit is computed using the sum-of-the-years'-digits method the k-th term is given by Eq. (3.13).

Here it may be noted that the sum of the digits in Eq. (3.13) is represented by $(r/2)(r + 1)$. This makes

$$\sum_{k=1}^{k=r} \frac{d_k}{(1+i)^k} = \frac{2}{r(r+1)} \sum_{k=1}^{k=r} \frac{r - k + 1}{(1+i)^k} \quad (3.25)$$

which may be split into the following two known summations:

$$\sum_{k=1}^{k=r} \frac{d_k}{(1+i)^k} = \frac{2}{r(r+1)} \left[(r+1) \sum_{k=1}^{k=r} \frac{1}{(1+i)^k} - \sum_{k=1}^{k=r} \frac{k}{(1+i)^k} \right] \quad (3.26)$$

The first summation in Eq. (3.26) is the uniform annual series present worth factor, Eq. (A.5). The second is Eq. (A.9).

Making substitutions for these summations gives

$$\sum_{k=1}^{k=r} \frac{d_k}{(1+i)^k} = \frac{2}{r} \left[\frac{(1+i)^r - 1}{i(1+i)^r} \right]$$

$$- \frac{2}{r(r+1)} \left\{ \frac{[(1+i)^r - 1](1+i)}{i^2(1+i)^r} - \frac{r}{i(1+i)^r} \right\} \quad (3.27)$$

which may be reduced by algebraic manipulation to

Expanded Economic Evaluation Equations 125

$$\sum_{k=1}^{k=r} \frac{d_k}{(1+i)^k} = \frac{2}{r(r+1)i} \left[r - \frac{(1+i)^r - 1}{i(1+i)^r} \right] \quad (3.28)$$

This term may then be used in the venture worth equation.

Example 3.4

It is desired to determine the most advantageous method for depreciation for income-tax purposes of a chemical plant investment. It may be assumed that the life of the plant may vary from 5 to 20 years and that the company's earning rate may vary from 5% to 15%.

Solution. This problem may be solved by the calculation of individual annual deductions or by the use of formulas (3.14), (3.24), (3.21), and (3.28). In order to illustrate the items involved, the detailed tabulation procedure for the case $r = 10$, $i = 0.10$ is shown in Table 3-8. As an illustration of the use of the summation formulas the totals for the above four schemes are computed using the appropriate formulas.

Equation (3.14) for the straight-line method gives

$$\frac{1}{r}\left[\frac{(1+i)-1}{i(1+i)^r}\right] = (0.1)\left[\frac{(1.1)^{10}-1}{0.1(1.1)^{10}}\right] = 0.1\left(\frac{1.59}{0.259}\right) = 0.614$$

Equation (3.24) for the double declining balance method gives

$$\left(\frac{2}{r}\right)\left\{\frac{1-[(1-2/r)/(1+i)]^r}{i+2/r}\right\} = 0.2\left[\frac{1-(0.8/1.1)^{10}}{0.1+0.2}\right] = 0.639$$

Equation (3.19) gives the value of m, the year (at the start of which) the switch from the double declining balance to the straight-line method should be made:

$$\frac{2r + \cos\pi r + 7}{4} = \frac{20 + 1 + 7}{4} = 7 = m$$

This value of $m = 7$ may now be employed in Eq. (3.21) to give the optimum for the combination method.

Table 3-8 Comparison of Tax Write-offs as Fraction of Investment[a]

Year	Straight line		Declining balance		Combination straight line and declining balance		Sum of the digits	
	Annual deduction	Present worth	Annual deduction	Present worth	Annual deduction	Present worth	Annual deduction	Present worth
1	0.10	0.0909	0.2000	0.1820	0.2000	0.1820	0.1820	0.1650
2	0.10	0.0826	0.1600	0.1320	0.1600	0.1320	0.1634	0.1350
3	0.10	0.0751	0.1280	0.0961	0.1280	0.0961	0.1454	0.1095
4	0.10	0.0683	0.1020	0.0696	0.1020	0.0696	0.1272	0.0870
5	0.10	0.0621	0.0820	0.0509	0.0820	0.0509	0.1090	0.0677
6	0.10	0.0565	0.0656	0.0370	0.0656	0.0370	0.0910	0.0513
7	0.10	0.0513	0.0524	0.0269	0.0656[b]	0.0336	0.0728	0.0372
8	0.10	0.0467	0.0420	0.0196	0.0656	0.0306	0.0546	0.0255
9	0.10	0.0424	0.0336	0.0142	0.0656	0.0278	0.0364	0.0154
10	0.10	0.0386	0.0268	0.0104	0.0656	0.0254	0.0182	0.0070
Total	1.00	0.6145	0.8924	0.6387	1.0000	0.6850	1.000	0.7006

Expanded Economic Evaluation Equations

Method of depreciation	Annual deduction d_k	Present worth of annual deduction $d_k/(1+i)^k$
Straight line	$1/r$	$\dfrac{1}{r(1+i)^k}$
Declining balance	$(1-2/r)^{k-1}(2/r)$	$\dfrac{(1-2/r)^{k-1}(2/r)}{(1+i)^k}$
Sum-of-years'-digits	$\dfrac{2(r-k+1)}{r(r+1)}$	$\dfrac{2(r-k+1)}{r(r+1)(1+i)^k}$

[a] $r = 10$; $i = 0.10$.
[b] Note shift to straight-line depreciation.

128   CHEMICAL PROCESS ECONOMICS

Table 3-9   Total Discounted Tax Credit Factor

Years of life	Straight line			Declining balance and straight line			Sum of the digits		
r	5%	10%	15%	5%	10%	15%	5%	10%	15%
5	0.866	0.758	0.672	0.898	0.808	0.712	0.895	0.806	0.733
10	0.772	0.614	0.502	0.818	0.685	0.588	0.829	0.699	0.605
15	0.693	0.508	0.390	0.747	0.589	0.493	0.770	0.615	0.509
20	0.623	0.425	0.314	0.689	0.515	0.410	0.718	0.546	0.430

$$\frac{2}{r}\left\{\frac{1 - [(1 - 2/r)/(1 + i)]^{m-1}}{i + 2/r}\right\} + \frac{(1 - 2/r)^{m-1}}{r - m + 1}\left[\frac{(1+i)^{r-m+1} - 1}{i(1 + i)^r}\right]$$

$$= 0.2\left[\frac{1 - (0.8/1.1)^6}{0.3}\right] + \frac{(0.8)^6}{4}\left[\frac{(1.1)^4 - 1}{0.1(1.1)^{10}}\right]$$

$$= \quad 0.567 \quad + \quad 0.118 \quad = 0.685$$

Finally, Eq. (3.28) gives the total write-off for the sum-of-the years'-digits method

$$\frac{2}{r(r + 1)i}\left[r - \frac{(1 + i)^r - 1}{i(1 + i)^r}\right] = \frac{2}{11}\left[10 - \frac{(1.1)^{10} - 1}{0.1(1.1)^{10}}\right] = 0.699$$

It is clear that the straight-line method and the double declining balance method would not normally be most advantageous to use. Since the straight-line method has been employed extensively in the past, it is included in the summary table (Table 3-9) of additional cases calculated by the summation formulas. The sum-of-the-years'-digits method appears to be most advantageous except in the case of $r < 5$ together with $i < 0.10$.

It must be remembered that each of the above factors must be multiplied by It to obtain the discounted tax credit term in the venture worth equation. The large discounting due to high interest rate and long times is apparent.

Expanded Economic Evaluation Equations 129

3.07 DECLINING GROSS RETURN

In Example 3.2 the sales price and the manufacturing costs of the product were arbitrarily changed from time to time because long experience in chemical plant operation teaches that such changes often occur. In many process plants sales prices decrease during the project's life. Furthermore, this is often accompanied by smaller reductions in the production costs and small increases in the production rates. In effect, the gross return rate R_k may decrease year by year.

Analytical expressions for a declining gross return, suitable for insertion in the venture worth equation, may be derived by assuming that the gross return decreases linearly or exponentially with time.

If it is assumed that the plant operates with a linearly declining gross return rate, there are three procedures for analyzing this situation:

1. It is assumed that the n-th, or last, year of operation will be satisfactory from the standpoint of profitable sales, but that the (n + 1)st year will be marginal. Because n is the assumed life of the plant, the capital investment will be recovered at the end of n years. Therefore all terms containing I will drop out of the venture worth equation. There will be no depreciation tax credit and no term for the recovery of I.

During the (n+1)st year, if the plant were operated, the cost of working capital would be $i_m I_w$, the cost of forgoing the return from salvage for 1 year would be $i(1 - t)S_a$, and the gross return for the last year after taxes would be $(1 - t)R_{n+1}$. To be marginal

$$R_{n+1} = \frac{i_m I_w}{1 - t} + i S_a \tag{3.29}$$

This reasoning provides a lower limit for the gross rate of return. From this value, and from R_1, the predicted gross return rate for the first year, a linearly declining rate of return, will give the following relationship for R_k

$$R_k = R_1 - (R_1 - R_{n+1})(k - 1)/n$$

where R_{n+1} is given by Eq. (3.29).

From Eq. (A.9), Appendix A,

$$\sum_{1}^{n} \frac{k}{(1+i)^k} = \frac{1+i}{i}\left[\frac{(1+i)^n - 1}{i(1+i)^n}\right] - \frac{n}{i(1+i)^n} \qquad (3.30)$$

and from Eq. (A.5),

$$\sum_{1}^{n} \frac{1}{(1+i)^k} = \frac{(1+i)^n - 1}{i(1+i)^n}$$

the following may be obtained

$$\sum_{k=1}^{k=n} \frac{k-1}{(1+i)^k} = \frac{1}{i}\left[\frac{(1+i)^n - 1}{i(1+i)^n} - \frac{n}{(1+i)^n}\right] \qquad (3.31)$$

It is now possible to evaluate the contribution of the term containing R_k, the summation of discounted gross untaxed income, in the venture worth equation, when the annual gross return decreases linearly with time.

$$\sum_{k=1}^{k=n} \frac{1-t}{(1+i)^k} R_k = (1-t)\sum_{k=1}^{k=n}\left[\frac{R_1}{(1+i)^k} - \frac{R_1 - R_{n+1}}{n}\frac{k-1}{(1+i)^k}\right] \qquad (3.32)$$

By employing Eqs. (3.29), (3.30), and (3.31) in Eq. (3.32) there is obtained

$$\sum_{k=1}^{k=n} \frac{1-t}{(1+i)^k} R_k = (1-t)\left\{\left[\frac{R_1(in-1) - \left(\frac{i_m I_w}{1-t} + S_a\right)}{in}\right]\frac{(1+i)^n - 1}{i(1+i)^n}\right.$$

$$\left. + \left[\frac{R_1 - \left(\frac{i_m I_w}{1-t} + iS_a\right)}{in}\right]\frac{n}{(1+i)^n}\right\} \qquad (3.33)$$

2. It may be that the rate of decline will not be to a marginal value of R. An assumption might be that R will decline a fixed amount each year, and that this amount will be a fraction, p, of R_1. Thus,

Expanded Economic Evaluation Equations

$$(R_1 - R_k)/(k - 1) = pR_1$$

and

$$R_k = R_1 - (k - 1)pR_1$$

If this occurs, then

$$\sum_1^n \frac{R_k(1 - t)}{(1 + i)^k} = \sum_1^n \frac{(R_1 - kpR_1 + pR_1)(1 - t)}{(1 + i)^k}$$

and

$$\sum_1^n \frac{R_k(1 - t)}{(1 + i)^k} = (1 - t)\left\{ R_1(1 + p)\left[\frac{(1 + i)^n - 1}{i(1 + i)^n}\right] \right.$$

$$\left. - pR_1\left[\left\{\frac{1 + i}{i}\right\}\left\{\frac{(1 + i)^n - 1}{i(1 + i)^n}\right\} - \frac{n}{i(1 + i)^n}\right]\right\}$$

(3.33a)

3. It is also possible that the rate of decline will be continuous at a constant fraction of the rate each year. Then,

$$dR_k/dk = -pP_k$$

and

$$R_k = R_o e^{-pk}$$

From Eq. (A.12),

$$\sum_1^n \frac{R_k(1 - t)}{(1 + i)^k} = \sum_1^n \frac{R_o e^{-pk}(1 - t)}{(1 + i)^k}$$

and

$$\sum_1^n \frac{R_k(1 - t)}{(1 + i)^k} = R_o(1 - t)\left\{\frac{(e^p + ie^p)^n - 1}{(e^p + ie^p - 1)(e^p + ie^p)^n}\right\} \quad (3.33b)$$

For values of p that are less than 0.15 per year, and for k that are 10 or less, the difference between the linear decline and the exponential decline is not great, and one treatment is as good as another; see Problem 3.6.

Example 3.5

An engineering economy study indicates that a certain gross annual return will be obtained from a proposed process plant. One way of applying this information to determine the attractiveness of the venture is to assume that this gross annual return will remain constant over the plant life. Essentially, this is what was done in Example 3.2, although some variations were considered.

How much greater would the initial calculated gross annual return have to be to make the venture look equally attractive, if, instead of assuming the return to remain constant, it were considered to decrease linearly during the plant life to a point of marginal operation?

Assume that plant life $n = 10$ years, minimum attractive return after depreciation and taxes $i_m = 0.15$, and cost of capital $i = 0.10$. For simplicity assume also that working capital requirement I_w and salvage value S_a will equal zero for this venture. This is not a bad assumption, since, in almost every case, they will not be far from zero.

Solution. For the case where R_k is assumed to decline linearly Eq. (3.33) will apply. Since $in = (0.1)(10) = 1$, the first term inside the braces in Eq. (3.33) will vanish and the equation reduces to the following for the summation at a declining rate:

$$(1-t) \sum_1^n \frac{R_k}{(1+i)^k} = (1-t) \frac{R_1 n}{(1+i)^n} = (1-t) \frac{R_1(10)}{(1.1)^{10}}$$

For the case where R_k is assumed to remain constant and equal to R_1, Eq. (A.5) applies.

$$(1-t) \sum_1^n \frac{R_1}{(1+i)^k} = (1-t)R_1 \frac{(1+i)^n - 1}{i(1+i)^n} = (1-t)R_1 \frac{(1.1)^{10} - 1}{0.1(1.1)^n}$$

Then, the ratio of the summation for the constant case to that for the declining return case will be

$$(1.1)^{10} - 1 = 1.59$$

For the declining return case, R_1 would therefore have to be increased 59% to make the two income sums equal. Note that since working capital is zero, i_m has no effect on the result.

Expanded Economic Evaluation Equations 133

In this particular problem it is a coincidence that the term "in" is 1 because i = 0.10 and n = 10. This coincidence simplifies the final relationship. A more general relationship between R_{1D} (the declining case) and R_{1C} (the constant case) —when the two venture worths are equal— is

$$R_{1D} = R_{1C} \frac{[(1 + i)^n - 1](in)}{[(1 + i)^n - 1](in - 1) + in}$$

3.08 TERMINATION OF OPERATIONS

The termination of operations in the plant may occur at the end of n years if the project follows the original estimates. However, there can be two departures from this idealization.

Operation for Fewer than r Years

If the plant fails and is shut down at the end of the year q (where q < r) there will remain some undepreciated capital. This may range from I, where the plant is a total failure and never produces anything (that is, q = 0), to a small fraction of I when the plant is shut down near the end of its depreciable life. In general, the undepreciated capital will be $I - \Sigma_1^q d_k I$.

In cases of this type the company may take one of several courses of action: (a) The plant may be dismantled and all the equipment sold for its salvage value, or (b) the plant may be allowed to stand idle for a number of years after shutdown, and then salvaged or, perhaps, put back into operation.

In case (a) where the failed plant is shut down, dismantled, and sold piecemeal (or discarded) the company will receive two credits:

1. The tax credit that comes from the loss of the undepreciated capital $(I - \Sigma_1^q d_k I)t$. The money loss $(I - \Sigma_1^q d_k I)$ can be carried on the records of the company for 5 years after the loss occurs and applied against the company's profits—from all its operations—at

any time in that 5-year period. The application of this loss against the company's profits reduces the taxes paid by the company by the amount $(I - \Sigma_1^q d_k I)t$. Thus, this quantity is called a tax credit and is a plus quantity in the economics of the individual project. This is the so-called "tax loss carried forward."

2. The after-taxes income from the sale of the dismantled plant is the salvage value: $(1 - t)S_a$. This term is the last one in the venture worth equation. Combining these two expressions the total credit to the company will be $(I - \Sigma_1^q d_k I)t + (1 - t)S_a$ if the plant is shut down at the end of the year q and quickly salvaged.

In case (b) the plant may be allowed to stand idle, not producing anything and consuming little or no maintenance, but accumulating depreciation tax credits, until the end of its depreciable life, or until it is salvaged or restarted. At the end of its depreciable life, salvage is also possible. If the plant is allowed to remain idle until the end of year r and then salvaged, the total credit accruing to the company is $\Sigma_{q+1}^r d_k It + (1 - t)S_a'$, where S_a' is a salvage value that might be different from S_a.

Which one of the two possible procedures company management may choose to follow will depend on the complex interaction between salvage values, depreciation methods, tax situations, company policies, and other matters. These are difficult problems, without good solutions, and best left to experienced financial officers.

It is important to recognize that in the case of a plant failure the entire invested capital is not lost. There will always be some money returned to the company from tax and salvage credits.

The term I appears in several places in the venture worth equation. This expression, Eq. (3.9), was derived for new projects, before any new capital was actually invested. For an existing project, where the capital has already been spent, the term I is no longer a variable in the venture worth equation. In some cases it is desirable to compute a venture worth for an existing project.

A situation which may be clarified by the concept of the venture worth for an existing project is that of developing a criterion for the possible termination of the project before the end of its depreciable life, $q < r$.

Expanded Economic Evaluation Equations

As shown above in connection with a project being terminated at the end of year q (q < r), there will be a tax-loss credit associated with the undepreciated capital investment $(I - \Sigma_1^q d_k I)t$. This sum may be entered as a credit in the venture worth equation for an existing project.

For a project being analyzed at the beginning of year p for a proposed shutdown at the end of the year q, the venture worth is

$$W_{pq} = \sum_{k=p}^{k=q} \frac{(1-t)R_k}{(1+i)^{k-p+1}} + \sum_{k=p}^{k=q} \frac{d_k tI}{(1+i)^{k-p+1}} + \frac{\left(I - \sum_{k=1}^{k=q} d_k I\right)t}{(1+i)^{q-p+1}}$$

$$- \left[\frac{(1+i)^{q-p+1}}{i(1+i)^{q-p+1}}\right] i_m I_w + \frac{1-t}{(1+i)^{q-p+1}} S_a \qquad (3.34)$$

This expression for the venture worth of a project already in operation is similar to that written before [Eq. (3.9)]. In Eq. (3.34), as previously, the credits to the project are balanced against the debits, and the difference is the value of the venture worth. This value may be positive, negative, or zero. As before, a positive W can be interpreted as good and a negative W as bad. If W from Eq. (3.34) should be zero or negative, serious thought should be given to terminating the project.

It should be noted that in Eq. (3.34) the only debit (negative) term is the interest on the working capital. This will not be large. Since the second, third, and fifth terms are always positive, it is apparent that R_k may become small, or negative, before W_{pq} will be zero. It is entirely possible that a plant operating at a loss ($R_k < 0$) can still be a profitable venture so long as the depreciation tax credits are applicable and are larger than the interest charges on the working capital.

Operation for More than r Years

If the plant operates for more than the allowable depreciation period, the terms $\Sigma_1^q d_k It/(1+i)^k$ and I disappear from the venture worth equation, Eq. (3.9). The plant is fully depreciated, and the profitability of the plant depends upon the comparative values of the yearly income R, the salvage (which probably decreases as the years pass), and the yearly cost of the working capital. The termination point is reached when, as in Eq. (3.29),

$$R = \frac{i_m I_w}{1 - t} + i S_a$$

Since both terms on the right-hand side of the equation will probably be fairly low, it can be seen that R need not be high in order that the plant stay in operation. Things which may work against continued operation are (1) high maintenance charges in an old plant and (2) increased competition from newer and more efficient plants and processes (this is by far the most serious). However, despite these it is apparent that a fully depreciated, or "written off," plant can be a considerable competitive advantage. New processes are often praised as being the answer to a company's problems, but, equally as often, the new processes cannot compete with those which do not have to carry the burden of an undepreciated plant.

3.09 MAXIMUM IN VENTURE WORTH AS A CRITERION OF PROJECT ATTRACTIVENESS

During the life of a plant the conditions which affect it will change, and after operations start it is necessary that periodic reviews be made to determine whether the plant should continue operating or be shut down. One method for characterizing the economic worth of an existing project is to compute a venture worth from the start of year p (when p > 1) to termination at the end of year q (when q < r). For example, at the start of the third year of operation (p = 3) the venture worth of the project might be computed through the end of the seventh year of operation (q = 7). Such a calculation would give the evaluators a fairly clear idea of the economic worth of operations from now (start of year p) to termination at the end of year q.

A good method for establishing the termination point of a project is to examine the situation existing when $\Delta W/\Delta k = 0$. That is, over a 1-year period there will be no increase in W. From Fig. 3.1 it will be seen that W increases steadily for a desirable project.

Expanded Economic Evaluation Equations

This relationship may be expressed mathematically by means of Eq. (3.34). At the end of year q

$$W_q = \frac{(1-t)R_q}{1+i} + \frac{d_q tI}{1+i} + \frac{(I - \sum_1^q d_k I)t}{1+i} - \frac{i_m I_w}{1+i} + \frac{(1-t)S_a}{1+i} \qquad (3.35)$$

This expression shows the venture worth of the project when calculated from the start of the year q to the end of the year q, 1 year. It is also possible to write an expression for W_{q-1}, the venture worth calculated for a single point in time, the start of year q.

$$W_{q-1} = 0 + 0 + (I - \sum_1^{q-1} d_k I)t - 0 + (1-t)S_a \qquad (3.36)$$

Several of the terms in W_{q-1} are set equal to zero. This comes about because W_{q-1} is being evaluated at the point in time q - 1, not for a number of years into the future as is usually the case for the venture worth. If the venture worth is evaluated at one single point in time there can be no contribution to W from R_k, $d_k I$, or $i_m I_w$ because all of these require a finite length of time in which they can accumulate value. W_{q-1} represents only the tax credit that a company can receive from the loss of the undepreciated capital and the salvage value of the abandoned plant.

It is desired that an expression for ΔW be found. This can be done by subtracting Eq. (3.36) from Eq. (3.35). Before the result can be simplified it is necessary that the undepreciated capital investment terms be put on the same basis. Note that

$$\sum_{k=1}^{k=q-1} d_k I = \sum_{k=1}^{k=q} d_k I - d_q I \qquad (3.37)$$

then

$$W_q = \frac{(1-t)R_q}{1+i} + \frac{(I - \sum_1^{q-1} d_k I)t}{1+i} - \frac{i_m I_w}{1+i} + \frac{(1-t)S_a}{1+i}$$

then

$$W_q - W_{q-1} = \frac{(1-t)R_q}{1+i} - \frac{i_m I_w}{1+i} - \frac{i}{1+i}\left(I - \sum_1^{q-1} d_k I\right)t - \frac{i(1-t)}{1+i}S_a \qquad (3.38)$$

Now if $W_q - W_{q-1}$ is set equal to zero

$$R_q = \frac{i_m I_w}{1-t} + iS_a + \frac{itI}{1-t}\left(1 - \sum_{1}^{q-1} d_k\right) \tag{3.39}$$

The three terms on the right-hand side of Eq. (3.39) are

$i_m I_w/(1-t)$ 1 year's interest on the working capital

iS_a 1 year's interest on salvage value

$\frac{itI}{1-t}\left(1 - \sum_{1}^{q-1} d_k\right)$ 1 year's interest on undepreciated capital investment

Equation (3.39) says that the gross income (before taxes) to the process during the year q must be greater than the sum of the various quantities of interest or the process should be shut down at the end of year q.

If $q > r$, that is, the plant is operated some years after full depreciation,

$$R_q = \frac{i_m I_w}{1-t} + iS_a$$

which is the same as Eq. (3.29)

3.10 COMPONENTS OF PROCESS UNITS

Very often the economics of a portion of a process unit may be simplified since a number of the variables involved will be fixed for all the alternatives being considered. Such situations are frequently met in the design of optional equipment in which size may be varied in small increments over a wide range as discussed in connection with the derivation of Eq. (2.13). Replacement of worn-out equipment in existing plants falls into the same category. In considering replacement, it is important to use a value for n, the life of the unit, that corresponds to the remaining life of the main plant or process for which the auxiliary being considered is only a part. In complicated cases it may be desirable to consider the venture worth of the entire proposal in order to assign correctly the life of the proposed replacement.

Expanded Economic Evaluation Equations

Sometimes the choice between alternative complete plants can be reduced to a simplified comparison like that described below, provided they are performing the same function or manufacturing the same product.

The following assumptions are thus often allowable. Sales realization is assumed to not be a function of the plant design (it may vary from year to year due to other circumstances, however). Working capital is assumed to be constant. Cost of production C is considered to be constant. Even though sales realization may vary, it is likely that the cost of production will remain more nearly constant.

The following expressions are then applicable:

$$R_k = S_k - C - mI \tag{3.40}$$

$$V_k = (S_k - U_k)(1 - t) - i_m I_w \tag{3.41}$$

In this discussion, as in Chapter 2, it is important to note that here R_k, V_k, and the others refer to a component of the process unit and not to the entire unit.

Minimizing the summation of U_k will be the equivalent of maximizing the summation of V_k values, with the assumptions made.

The summation of V_k, Eq. (3.7), is the fundamental definition of the venture worth. When Eq. (3.9) was derived, V_k was expressed in terms of gross income, depreciation tax credit, capital investment, interest rates, and salvage value. By means of Eqs. (3.40) and (3.41), V_k may be expressed in terms of U_k, C, m, and R_k. This expression may also be used in the definition of venture worth, and the two different equations for W set equal. After the algebraic manipulation has been completed, an expression for the summation of U_k, the total yearly cost before taxes, is obtained. It should be noted that the reason for using U rather than V is that when U is used sales do not have to be calculated. This might be difficult for the component parts of a unit.

$$\sum_{k=1}^{k=n} \frac{U_k}{(1+i)^k} = \sum_{k=1}^{k=n} \frac{C}{(1+i)^k} - \frac{S_a}{(1+i)^n}$$

$$+ \left[\sum_{k=1}^{k=n} \frac{m}{(1+i)^k} + \frac{1 + \frac{(1+i)^n - 1}{i(1+i)^n}(i_m - i) - \sum_{k=1}^{k=r} \frac{d_k t}{(1+i)^k}}{1 - t} \right] I \tag{3.42}$$

In this equation all terms are before taxes, therefore some terms are divided by (1 - t). In the venture worth equation (3.9), all terms are after taxes.

Multiplication of both sides of Eq. (3.42) by the capital recovery factor, $\dfrac{i(1+i)^n}{(1+i)^n - 1}$, gives

$$\left[\frac{i(1+i)^n}{(1+i)^n - 1}\right] \sum_{k=1}^{k=n} \frac{U_k}{(1+i)^k} = C - \frac{iS_a}{(1+i)^n - 1}$$

$$+ \left\{ m + \frac{\dfrac{i(1+i)^n}{(1+i)^n - 1}\left[1 + \dfrac{(1+i)^n - 1}{i(1+i)^n}(i_m - i) - \sum_{k=1}^{k=r}\dfrac{d_k t}{(1+i)^k}\right]}{1 - t} \right\} I$$

(3.43)

This formula is analogous to Eq. (2.13), which is $U = C + \{m + d_k + [i_m/(1-t)]\}I$. In both equations the expression in braces may be considered as a fixed charge on investment. However, Eq. (3.43) is superior because it allows a year-by-year calculation of U considering the time value of money, the incremental return on investment, the salvage value, and a nonlinear depreciation rate.

This equation may be used in the same way that the venture worth equation is used. By calculating $\sum_1^n U_k/(1+i)^k$ a number may be obtained which is characteristic of the costs peculiar to that alternative. For a number of alternatives the one having the minimum value of $\sum_1^n U_k/(1+i)^k$ will be the best one, just as the one having the maximum value of W is the best one.

It is possible to introduce a useful simplification to Eq. (3.43) by considering the maximum acceptable payout time on incremental investment (before taxes and depreciation). The maximum acceptable payout time corresponds to the situation where V, the venture profit, is equal to zero.

Letting V = 0 and noting that working capital and salvage value are considered unaffected when comparing alternatives to the same process, there is obtained

Expanded Economic Evaluation Equations

$$T_m = \frac{1 - t}{\frac{i(1 + i)^n}{(1 + i)^n - 1}\left[1 + \frac{(1 + i)^n - 1}{i(1 + i)^n}(i_m - i) - \sum_{k=1}^{k=r} \frac{d_k t}{(1 + i)^k}\right]} \quad (3.43a)$$

$$T_m = \frac{1 - t}{(i_m - i) + \frac{i(1 + i)^n}{(1 + i)^n - 1}\left[1 - \sum_{k=1}^{k=r} \frac{d_k t}{(1 + i)^k}\right]} \quad (3.43b)$$

Here payout time is defined as investment in plant facilities (excluding auxiliaries and working capital) divided by the annual rate of return before taxes and depreciation.

If straight-line depreciation can be employed and the tax period corresponds to the entire plant life: $d_k = d = 1/n$ (i.e., $r = n$)

$$T_m = \frac{1 - t}{\frac{i(1 + i)^n}{(1 + i)^n - 1} + (i_m - i) - \frac{t}{n}} \quad (3.44)$$

The approximation formula Eq. (A.52) of the first edition of this book

$$\frac{i(1 + i)^n}{(1 + i)^n - 1} \cong 0.65\,i + \frac{1}{n}$$

may be employed to give

$$T_m = \frac{1}{\frac{i_m - 0.35\,i}{1 - t} + \frac{1}{n}} \quad (3.45)$$

An inspection of Eq. (3.45) will show that the numerical value of T_m cannot have a very broad range. For any one company at any one time the numerical values of i, i_m, t, and n will all be essentially constant. Therefore, for project evaluations being carried out at nearly the same time the value of T_m will be a constant. For the ordinary values of i, i_m, t, and n frequently encountered in industrial work T_m will be between 2 and 3.5 years. A good first approximation, applicable to all projects, would be 2.5 years.

Starting with Eq. (3.43), noting the definition of T_m in Eq. (3.43a), and using the approximate formula, Eq. (3.45), for T_m, there is obtained a simple and useful form of Eq. (3.43).

$$\left[\frac{i(1+i)^n}{(1+i)^n - 1}\right] \sum_{k=1}^{k=n} \frac{U_k}{(1+i_m)^k} = C + \left(m + \frac{1}{n} + \frac{i_m - 0.35i}{1-t}\right) I \quad (3.46)$$

If U_k is constant, the left-hand side of Eq. (3.46) reduces to U, and Eq. (3.46) becomes

$$U = C + \left(m + \frac{1}{n} + \frac{i_m - 0.35i}{1-t}\right) I \quad (3.47)$$

This form of the expression for U may be compared to the form given by Eq. (2.13): $U = C + \{m + (1/n) + [i_m/(1-t)]\} I$. This expression was derived using the assumption that d and e were equal; see Eq. (2.9). Equation (3.47) corrects this defect by basing e on the compound interest capital recovery factor.

Example 3.6

A proposed ethylene manufacturing plant is expected to have an operating life of 12 years. However, because the product may be used in certain operations of military interest the government will allow the company to depreciate the capital investment for tax purposes in 5 years.

The company building the equipment is interested in investing money at a return of 15% or over, after taxes and depreciation; its current rate of return on invested capital is 10%. It is estimated that Federal plus state tax rate on income may be taken at 55%. What maximum acceptable payout time is required for justification of incremental expenditures on equipment such as heat exchangers and insulation?

Solution. Use Eq. (3.43a) and evaluate $\sum_{k=1}^{k=r} \frac{d_k t}{(1+i)^k}$ by the straight-line method, which is the only one allowable. For evaluation of the discounted tax credit factor, use Eq. (3.14):

$$\frac{1}{r}\left[\frac{(1+i)^r - 1}{i(1+i)^r}\right] = \frac{1}{5}(3.791) = 0.758$$

Expanded Economic Evaluation Equations   143

This result is also listed in Table 3.9. Using Eq. (3.43a),

$$T_m = \frac{1 - 0.55}{\dfrac{0.1(1.1)^{12}}{(1.1)^{12} - 1}\left[1 + \dfrac{(1.1)^{12} - 1}{0.1(1.1)^{12}}(0.15 - 0.10) - (0.758 \times 0.55)\right]}$$

$T_m = 3.32$ years

This is the maximum acceptable payout time, assuming that the optional facility does not involve any change in working capital or salvage. This is equivalent to a gross return of $100 \times 1/3.32 = 30.2\%$ before taxes and depreciation. It does not include maintenance. If maintenance or other charges on incremental investment amounted to 6%, an annual charge of $30.2 + 6 = 36.2\%$ on such investment would be required.

3.11 LIMITED AVAILABILITY OF CAPITAL

As in Chapter 2, the relationships developed in this chapter assume that unlimited funds for capital investment are available at the minimum acceptable rate of return. If capital is scarce, a higher value of i_m will be required before a project will be approved.

Where an arbitrary upper limit on available funds is set, the proper criterion for the selection of alternatives involves that which gives a maximum summation of venture worth for the selected projects, with the additional restriction of a maximum total $I + I_w$ for the projects chosen.

Where a number of design options, as well as basic alternative proposals, are available, the application of this criterion may become complicated.

SUMMARY

A general criterion based on the present worth concept has been developed for evaluation of the economic attractiveness of investment proposals. This yardstick, designated as "venture worth," determines the present cash value of any new proposal as judged from the improvement it can effect in an existing business structure. It is shown how variations in acceptable rate of return and tax may be taken into effect. The method is applicable not only to the evaluation of complete proposals but to the design of component parts of process plants as well. Alternative procedures which have attractive features may also be developed on the basis of present worth calculations.

The present worth methods are compared with other indices which evaluate profitability, taking the time value of money into consideration. One of these is of special interest when we do not have available a minimum acceptable return rate. It is designated as the "interest-rate-of-return" method and in effect determines the rate of return required to extinguish the incremental present worth or venture profit.

The methods described in this chapter have the advantage of enabling variations in future cash flows to be taken into account, provided they can be predicted in a reasonable way. If information on minimum acceptable return rate and current rate of return on invested capital is available, the venture worth method can be applied to advantage.

The venture worth method has the disadvantage of greater complexity and therefore is not as easily understood by those concerned. If adequate data are not available for predicting future cash flows, it is often more advantageous to employ simple indices like those developed in Chapter 2. In general, it is often desirable to express evaluations in several forms, depending on management preference, rather than basing all decisions on one method.

The general techniques discussed in Chapters 2 and 3 form the basis for subsequent detailed and practical treatment of economic-balance problems. It should be emphasized that, from the practical standpoint, all that is required in plant design problems is an appropriate "fixed annual charge" to assign to optional investments. Such a figure may be arrived at in a more or less arbitrary way rather than by detailed analysis of the types presented here. For those who prefer such a procedure, it will be entirely possible to make use of the results in later chapters without any acceptance of the methods outlined here.

The practical development of plant design procedures outlined in Chapters 7 and 8 is presented in a form so that it can be used independently of how one chooses to arrive at a minimum acceptable return rate or maximum allowable payout time before taxes.

REFERENCES

1. L. Abbott, Economics and the Modern World, Harcourt, Brace and World, Inc., New York, 1960, pp. 655-670.
2. Chemical Engineers' Handbook, 4th ed., McGraw-Hill Book Co., New York, 1963, pp. 1-32 through 1-38.
3. G.A. Taylor, Managerial and Engineering Economy, D. Van Nostrand Co., Princeton, New Jersey, 1964, pp. 439-479.
4. D.F. Rudd and C.C. Watson, Strategy of Process Engineering, John Wiley and Sons, Inc., New York, 1968, p. 85.
5. M.S. Peters and K.D. Timmerhaus, Plant Design and Economics for Chemical Engineers, McGraw-Hill Book Co., New York, 1968, p. 163.
6. E.L. Grant, Principles of Engineering Economy, 3rd ed., Ronald Press, New York, 1950.
7. J.B. Weaver, H.C. Bauman, and W.F. Heneghan, Chemical Engineers' Handbook, 4th ed., McGraw-Hill Book Co., New York, 1963, pp. 26-1 through 26-45.
8. D.G. Jordan, Chemical Process Development, Interscience Publishers, John Wiley and Sons, Inc., New York, 1968, pp. 38-43.
9. J. Happel, Chemical Engineering Progress, 51, 533 (1956).
10. W. James, Pragmatism, Longmans, Green, London, 1948.
11. W.H. Kapfer, Today Series, Process Economics, American Institute of Chemical Engineers, New York, 1967.
12. J.B. Weaver and R.J. Reilly, Chemical Engineering Progress, 52, 405 (1956).

PROBLEMS

3.1. (a) As a result of a $10,000 expenditure in connection with a research program, a company is able to obtain a process patent. In order to license rights it spends $1000/yr on process improvement and on the marketing of the process to prospective licensees. At the time the patent is issued there are no licensees, but, when it expires after 17 years, royalties amount to $50,000 a year. What is the present worth of the invention?

Taxes may be taken into consideration on the basis that Federal and state income tax is 55%/yr on net profit. The original expenditure and subsequent expenses are all taken as current normal costs of business. Assume that royalties increase at a uniform rate during the life of the patent. Minimum acceptable return rate on invested capital is 15%/yr, and current return rate on invested capital is 8%/yr.

(b) Assume that at the end of some number of years—say m— a competing firm obtains a new patent for the same process, and the value of the old patent drops to zero. What value should m have so that the firm will just recover its original expenditure of $10,000?

3.2. A projected chemical process plant involves a new capital expenditure of $1,000,000—$500,000 initially and $500,000 at the end of the first year. Construction and initial start-up runs will be completed after 2 years. From that time on for a period of 5 years the gross return is expected to be $350,000/yr. Thereafter, competition will probably reduce profits each succeeding year by $30,000/yr.

(a) How long will the plant probably be operated, and (b) what net average annual rate of return after taxes and depreciation will be obtained on the fixed capital investment during the useful life of the plant?

Assume that at the start of plant operation (after 2 years) there will be a working capital requirement of $200,000. The minimum attractive return rate to the company on new capital invested is 15%/yr, and actual current return rate on all capital is 10%/yr. Income tax rate is taken at 50%/yr. The company uses the sum-of-the-years'-digits method for depreciation for tax purposes, and the Internal Revenue Service will allow the plant to be depreciated over a period of 10 years from time of capital investment. It is assumed that there is no salvage value at the end of the useful plant life.

3.3. Using the data of Problem 3.2, calculate the rate of return by the discounted cash flow method.

3.4. A company is studying the investment of $5,000,000 in a new plant which can be depreciated over 10 years for tax purposes. The company wishes to consider the relative advantages of the following methods of tax write-off on the basis that present taxes are 55%/yr, but after 5 years may go up to 65%/yr: (a) sum-of-the-years'-digits method, (b) straight-line depreciation. Average rate of return is 12%/yr on invested capital. Which scheme is most attractive under the assumed conditions?

3.5. The accompanying table [from R.B. Smith and T. Dresser, Chemical Engineering Progress, 51, 544 (1955)] gives a comparison of costs for three types of heaters to supply heat to an oil stream in a process plant at a rate of 73,500,000 Btu/hr.

	Simple	Oil convection	Rotary air preheater
Liberation (MM Btu/hr)	127.0	114.0	96.5
Thermal efficiency (%)	57.9	64.5	76.1
Fuel cost at 33¢/(MM Btu)(yr)	$349,000	$313,000	$265,000
Power at 1.1¢/kw-hr ($/yr)	—	—	8,834
Capital (installed)($)	399,000	472,700	605,000

Fuel and power costs are based on 95% stream efficiency. Assume that the plant in which this equipment is installed will operate 10 years, that a tax rate of 55%/yr is applicable, and that straight-line depreciation will be employed over the entire 10-year period, that fixed charges including maintenance incurred by installation of this equipment will amount to 10%/yr of the investment, that a minimum acceptable return rate on invested capital after taxes and depreciation is 15%, and that the current rate on invested capital is 10%. Determine which of the three alternative installations should be selected.

3.6. A condenser for a process unit costs $20,000 and has an estimated life of 4 years. By employing special alloy tubes, the life can be extended to 6 years. The alloy tubes cost an additional $5,000 to purchase and install. It is assumed that the process has an additional life of at least 12 years. Average earnings rate for the company as a whole is 10%, but the minimum acceptable return rate for investments of this type is 25%. Should the alloy tubes be installed? A 53% total tax rate is applicable.

3.7. A chemical plant is to produce 50,000,000 lb/yr of a product. This product is to have a manufactured cost of 18¢/lb and will sell originally for 35¢/lb.

If the expected plant life is 10 years, the tax rate is 50%, and the discount rate is 10% (i = 0.10), calculate the $\sum_1^{10} R_k(1-t)/(1+i)^k$ term for each of the following conditions: (a) R_k is constant over the 10-year period. (b) R_k declines by 5% of R_1 each year. (c) R_k declines by 5% of R_k each year.

3.8. In Problem 3.7 it was shown that a declining rate of return would seriously reduce the value of the $\sum_1^n R_k(1-t)/(1+i)^k$ term. Such a reduction would cause the venture worth to become largely negative.

As in Example 3.5, the question arises as to what increases in R_1 would be necessary so that W would be zero despite the declining rate of return.

If, as in Problem 3.7, the rate of return declines exponentially at 5%/yr, and the venture worth for the constant rate of return situation is zero, what value should the initial sales price have in order that W = 0 for the declining rate of return situation?

3.9. Consider a project with the following economic parameters:
$I = \$5 \times 10^6$, $i = 0.10$, $t = 0.50$, $S_a = 0$
$I_w = \$5 \times 10^6$, $i_m = 0.15$, $d_k = 0.1$ (straight line)
Assume that R_k may be considered constant from the end of year 1 to the end of year 5. Calculate the value that R_k must have if W_{p-q} is to be zero when p = 2 and q = 5.

Chapter 4

SPECIAL MATHEMATICAL TECHNIQUES

4.00 INTRODUCTION

In previous chapters the problem of designing or operating a chemical-process plant has been developed in terms of a mathematical model, the venture profit or venture worth. The venture profit, or the venture worth, are functions of a number of independent process variables. These variables are to be controlled in such a way that the venture worth is maximized.

If the output of a given plant is assumed to be fixed, the variables may be considered to fall into two types: (1) those involving running expenses such as labor, utilities, and raw materials, and (2) those involving the investment of large sums of money in plant and equipment. The capital expenditures mentioned in (2) may be converted into equivalent running expenses by the application of suitable information on project life, tax rate, and minimum acceptable return rate.

The model employed is a considerable simplification of the broad subject of engineering economics. Corporate problems such as organization, plant location, transportation, and distribution are not considered. These fields are seldom the direct concern of the process engineer.

In this chapter attention is devoted to the problem of optimization and its formulation as applied to a variety of chemical-process problems, using the simplified model developed. In recent years the application of such models to problems in planning and control of manufacturing operations has been much facilitated by developments in theoretical methods for analysis and the availability of automatic high-speed computing equipment. A detailed discussion of the various types of computing equipment and their specific applications is not considered to be within the scope of this chapter. Sufficient details of the technique involved are presented so that the process engineer will be in a position to judge their applicability to his problems. Simple examples are given to illustrate the methods without recourse to the use of computing equipment. Anyone wishing to apply these methods to any considerable extent will find it desirable to study the references indicated in the text.

In the relatively short time between the two editions of this book the use of the computer has become widespread. At present it is much used to perform the complex computations associated with equipment design and has also found application in the solution of economic problems. Although these machines are quite expensive to buy or rent, the manufacturers of such machines provide many services to prospective customers in order to make the machines easy to use and available to all. There now exist computer programs for many different problems. These can be purchased, or rented, or they may be supplied free of charge by the manufacturer to the customer. The entire field of machine computation and its application to chemical-process engineering is now very large and will become larger.

Practically all chemical manufacturers, and all engineering firms, now use computers extensively. They have applied mathematics and computation groups that work on the application of special mathematical techniques to company problems. Even the individual process engineer, through the use of the "time-sharing" technique, may use a computer profitably. It is quite worthwhile for the process engineer to learn more than a little about special mathematical techniques and machine computation.

The application of special mathematical techniques to chemical-process optimization problems has received considerable academic attention in recent years and there is an extensive literature on the subject. Despite this intense activity and the rather broad knowledge that has been created, there is some resistance to the use of such techniques in industrial problems. It is argued that the accuracy of the data is not good, that costs and other economic information cannot be obtained with precision, that many design procedures are not precise, and that the cost of a detailed economic optimization study would be too high. For these reasons, advanced detailed optimization studies are often not done, and answers based

Special Mathematical Techniques 151

on empirical methods or past experience are used. Such methods are frequently incapable of analyzing a complex problem properly.

Arguments such as those listed above against optimization studies are really not valid. Good data can be obtained in the laboratory or pilot plant, design procedures using modern machine computations are quite precise enough, and economic data of good quality can be obtained from company records or vendor quotations. The time required to do the optimization calculations is greatly reduced by using special techniques and by machine computation.

The act of doing a detailed optimization study will force an expansion of the knowledge concerning the problem. It will delineate the relationships between the variables and their effect on the economics, and it will locate the optimum point of operation. The overall effect will be to produce a theoretical picture of the project and its economics that can be used to guide practical thinking about equipment design, operations, control, process changes, and future improvements.

Modern chemical-process economics is quite complex, involving many variables, and special mathematical techniques are often required in order to produce a useful answer to an optimization problem.

4.01 GENERAL FORMULATION OF OPTIMUM

In Chapter 2 the overall cost function U, equation (2.13), was shown to be a function of a single variable. It was also shown that an optimum value of the variable could be obtained by differentiating U with respect to the variable, setting the derivative equal to zero, and solving for the variable. Simple calculations of this type depend for their validity on the mathematical nature of the function. Some functions are of a type that allow such manipulations, and some are not. It is important to know the conditions that the function must meet in order for maxima or minima to exist.

Functions of One Variable: The Use of the Differential Calculus

1. If a function is a straight line, that is, a linear relationship exists between cost and independent variable, a maximum or

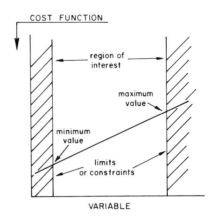

Fig. 4.1. A linear cost function

minimum can exist only at the limits of the variable (Fig. 4.1). This situation is sometimes described as saying that, for a linear function, the optimum always lies on a limit, an end value, a constraint. This principle is exploited in the technique known as linear programming.

2. If the function is nonlinear there are several conditions that must be met for the application of the ordinary differential calculus to find a maximum or a minimum.

a. The function must be continuous in the region considered.

b. The function must be differentiable in the region.

c. The first derivative must become zero at some point in the region. This would be the maximum or minimum point. For an illustration see Fig. 2.4; here a maximum occurs at point M, where $dP/dI = 0$.

d. If the first derivative becomes zero, according to c, at some point in the region some higher derivative must not become zero at the same point.

e. The first nonvanishing derivative must be an even one. That is, if the first nonvanishing derivative is $d^n c/dx^n$, n must be even.

f. If the first nonvanishing derivative is positive, the function has a minimum; if the first nonvanishing derivative is negative, the function has a maximum.

A good discussion of single-variable maxima and minima, with clear illustrations, is given by Boas [1].

Special Mathematical Techniques   153

Example 4.1

Consider the function for the theoretical work of adiabatic compression in two stages used in Example 2.4.

$$W = \frac{kP_1v_1}{k-1}\left[\left(\frac{P_2}{P_1}\right)^{\frac{k-1}{k}} - 2 + \left(\frac{P_3}{P_2}\right)^{\frac{k-1}{k}}\right]$$

It is desired to determine if this function has a maximum or a minimum with respect to P_2 by the use of the conditions mentioned above.

   a. Is the function continuous in the region $P_1 = 14.7$ to $P_3 = 114.7$? Yes.

   b. Is the function differentiable with respect to P_2 in the region 14.7 to 114.7? The derivative with respect to P_2 is easily taken and is

$$\frac{dW}{dP_2} = \frac{kP_1v_1}{k-1}\left[\left(\frac{k-1}{k}\right)P_1^{\frac{1-k}{k}} P_2^{-\frac{1}{k}} - \left(\frac{k-1}{k}\right)P_3^{\frac{k-1}{k}} P_2^{\frac{1-2k}{k}}\right]$$

as shown in Example 2.4.

   c. Does the first derivative become zero at some point in the region between P_1 and P_3? Example 2.4 shows that if dW/dP_2 is set equal to zero, $P_2 = (P_1 P_3)^{\frac{1}{2}}$ so that the first derivative does become zero at some point between P_1 and P_3.

   d. A higher derivative must not become zero at the same point. The second derivative at the point where $P_2 = (P_1 P_3)^{\frac{1}{2}}$ is

$$\frac{d^2W}{dP_2^2} = \left(\frac{kP_1v_1}{k-1}\right)\left(\frac{k-1}{k}\right)\left\{P_1^{\frac{1-3k}{2k}} P_3^{-\frac{1-k}{2k}}\left[\frac{2k-2}{k}\right]\right\}$$

and the second derivative cannot be zero.

   The derivative mentioned above is even (the second; n = 2) and must be positive. Therefore, by conditions e and f the function does have a minimum at some point in the region of interest. It is mathematically correct to differentiate W with respect to P_2, set the derivative equal to zero, and solve for P_2, thus calculating the pressure at which the compression work will be a minimum.

This was a fairly simple and straightforward problem, and an answer was obtained without much trouble. It might be said that the answer was obvious and the mathematical steps need not have been made. However, in process engineering complex equations can be, and frequently are, derived in order to express some function to be optimized [2,3]. Before any calculations are made it cannot be known if maxima or minima exist. To calculate many separate points to find the true shape of the curve might prove to be tedious, time-consuming, and expensive. Thus, performing the differentiations and checking against the conditions mentioned above — in order to prove or disprove the existence of maxima or minima — is well worth doing.

Functions of More Than One Variable:
The Use of Partial Differentiation

When more than one independent variable is involved in determining profitability, a more elaborate treatment is necessary than was employed in obtaining Eq. (2.13). Thus, for a system of $u, v, w, \ldots, z$ independent variables, the venture worth W will be a function of these variables.

$$W = f(u, v, w, \ldots, z) \tag{4.1}$$

Instead of venture worth W, profitability may be expressed in other convenient and equivalent forms (for example, U in Eq. (2.13) or the left side of Eq. (3.41). The problem is to specify $u, v, w, \ldots, z$ so that W will be a maximum. In general this will involve some type of trial-and-error procedure.

It is possible in some cases to set up an explicit function which will relate venture worth to the independent variables. Then differentiation will, as discussed in Chapter 2, result in a series of partial derivatives $\partial W/\partial u$, $\partial W/\partial v$, and others. Setting these partial derivatives equal to zero will result in the same number of simultaneous equations as the variables involved. These equations, subject to some limitations, can be solved for the corresponding optimum values of the variables. The necessary conditions [4] for a maximum at a point where the first derivatives become zero include restrictions on the second derivatives, continuity of the function involved throughout the range of the independent variables, and absence of optimum conditions at limiting values of one or more of the variables. These conditions often severely restrict the applicability of this system, as will be discussed later.

Special Mathematical Techniques   155

For the case of a two-variable function, $W = f(x,y)$, a point $x = a$, $y = b$ will have a maximum value $f(a,b)$ if the following conditions hold:

$$\frac{\partial W}{\partial x} = \frac{\partial W}{\partial y} = 0 \quad \text{or} \quad dW = 0 \tag{4.2}$$

$$\frac{\partial^2 W}{\partial x^2} < 0, \quad \frac{\partial^2 W}{\partial y^2} < 0 \tag{4.3}$$

$$\frac{\partial^2 W}{\partial x^2} \frac{\partial^2 W}{\partial y^2} > \left(\frac{\partial^2 W}{\partial x \partial y}\right)^2 \tag{4.4}$$

Similarly, if $\partial^2 W/\partial x^2$ and $\partial^2 W/\partial y^2$ are positive and Eq. (4.4) holds, there will be a minimum. But, if

$$\frac{\partial^2 W}{\partial x^2} \frac{\partial^2 W}{\partial y^2} < \left(\frac{\partial^2 W}{\partial x \partial y}\right)^2$$

then the point concerned will be a saddle point. The case where

$$\frac{\partial^2 W}{\partial x^2} \frac{\partial^2 W}{\partial y^2} = \left(\frac{\partial^2 W}{\partial x \partial y}\right)^2$$

is open and the value may be maximum, minimum, or neither.

Example 4.2

Consider the three-stage reversible adiabatic compression of a gas from an initial pressure P_1 to a final pressure P_4. If the fixed charges on the compressors are assumed to be essentially independent of the interstage pressure employed, then the optimum operation involves the determination of the interstage pressure for which the total power requirement is a minimum. (Note: Independent calculation indicates that, for optimum results, intercooling back to inlet temperature is desirable.) If the gas enters at a temperature T_1 and is cooled to T_1 between stages the total work is given by

$$E = NRT_1 \frac{k}{k-1} \left[\left(\frac{P_2}{P_1}\right)^{(k-1)/k} + \left(\frac{P_3}{P_2}\right)^{(k-1)/k} + \left(\frac{P_4}{P_3}\right)^{(k-1)/k} - 3 \right] \tag{4.5}$$

where N represents the pound-moles of gas compressed, R is the gas constant (1,544), k is the ratio of specific heat at constant pressure to that at constant volume of the gas compressed, and T_1 is the inlet gas temperature in degrees Fahrenheit absolute. E then represents the total work as foot-pounds.

Solution. If this quantity is to be a minimum, first $\partial E/\partial P_2$ and $\partial E/\partial P_3$ must be zero.

$$\frac{\partial E}{\partial P_2} = NRT_1 (P_1^{(1-k)/k} P_2^{-1/k} - P_3^{(k-1)/k} P_2^{(1-2k)/k}) \tag{4.6}$$

$$\frac{\partial E}{\partial P_3} = NRT_1 (P_2^{(1-k)/k} P_3^{-1/k} - P_4^{(k-1)/k} P_3^{(1-2k)/k}) \tag{4.7}$$

Setting $\partial E/\partial P_2$ and $\partial E/\partial P_3 = 0$,

$$P_2^2 = P_1 P_3 \quad \text{and} \quad P_3^2 = P_2 P_4$$

whence

$$P_2 = (P_1^2 P_4)^{1/3} \quad P_3 = (P_4^2 P_1)^{1/3} \tag{4.8}$$

If we now differentiate $\partial E/\partial P_2$ with respect to P_2 and substitute the values for P_2 and P_3 from Eq. (4.8), the following is obtained:

$$\left(\frac{\partial^2 E}{\partial P_2^2}\right)_0 = 2NRT_1 \frac{k-1}{k} (P_1^{(1-5k)/3k} P_4^{-(1+k)/3k}) \tag{4.9}$$

The subscript 0 indicates that the value of $\partial^2 E/\partial P_2^2$ is at the possible minimum which we are attempting to verify. Since the powers of P_1 and P_4 must be positive, and $k > 1$, it is clear that, from (4.9), $(\partial^2 E/\partial P_2^2)_0$ is greater than zero, a positive quantity. Similarly, by differentiation of (4.7) with respect to P_3 and substitution of values from (4.8),

$$\left(\frac{\partial^2 E}{\partial P_3^2}\right)_0 = 2NRT_1 \frac{k-1}{k} (P_1^{(1-3k)/3k} P_4^{-(1+3k)/3k}) \tag{4.10}$$

This by similar reasoning also is positive. The product of the second derivatives

Special Mathematical Techniques

$$\left(\frac{\partial^2 E}{\partial P_3^2}\right)_0 \left(\frac{\partial^2 E}{\partial P_2^2}\right)_0 = 4\left[\frac{NRT_1(k-1)}{k}\right]^2 P_1^{\frac{2-8k}{3k}} P_4^{-\frac{2+4k}{3k}} \quad (4.11)$$

Similarly the cross derivative

$$\left(\frac{\partial^2 E}{\partial P_2 \partial P_3}\right)_0^2 = \left[\frac{NRT_1(k-1)}{k}\right]^2 P_1^{\frac{2-8k}{3k}} P_4^{-\frac{2+4k}{3k}} \quad (4.12)$$

Thus it has been demonstrated that

$$\left(\frac{\partial^2 E}{\partial P_3^2}\right)\left(\frac{\partial^2 E}{\partial P_2^2}\right) > \left(\frac{\partial^2 E}{\partial P_2 \partial P_3}\right)^2$$

which is condition (4.4) above. A true minimum value for total power, which in this case corresponds to maximum return on investment, will be obtained if P_2 and P_3 are specified by Eqs. (4.8). As might be guessed, the optimum interstage pressures correspond to an equal distribution of load among each of the three compressor stages.

Geometric Programming

The optimum finding methods based on the use of the differential calculus can become complex and unwieldy, involving a large amount of algebraic manipulation and arithmetic; for example, see Example 4.6. Under these conditions it is easy to become confused and to make errors. This undesirable situation can be avoided by a procedure due to Zener [5]; see also Sherwood [6] and Kermode [7]. This procedure has been termed "geometric programming"; see a text by Duffin, Peterson, and Zener [8].

Geometric programming is limited to cases where:

1. The cost (which is to be minimized) is expressible as a sum of terms where each term is a product of powers of the independent variable.

2. The number of such terms must be one greater than the number of independent variables.

These appear to be severe restrictions, but there are many practical cases where they are met. Sometimes some ingenuity is required in writing the terms, but it can often be done [Sherwood, Ref. 6, p. 95].

The geometric programming method employs several simple principles. (a) The cost U expressed as a function of the independent variables, may be represented by the relation

$$U(x_1, x_2, x_3, ..., x_n) = \sum_{i=1}^{i=n} T_i$$

and $T_i = a_i \prod_{j=1}^{j=m} x_j^{\theta_{ij}}$ with the restriction $n = m + 1$

Some thought will show that since each T_i is a product of powers of the x's, each T_i can be raised to a power δ_i such that $\prod_{i=1}^{i=n} T_i^{\delta_i} = K$, where K is a constant. The only way in which this can come about is for the exponents on each variable to be equal to zero. In that case, all the terms containing only variables will become unity and $\prod_{i=1}^{i=n} T_i^{\delta_i}$ will become a constant.

By combining the exponents on a variable their sum must equal zero, and an equation has been derived containing the δ's and the original exponents on the variables. Several such equations can be written, and it should be possible to solve for the values of the δ's.

Zener makes the condition that the sum of all the δ's shall be unity. Combining this equation with the others it is possible to solve for the δ's.

Zener also shows that the optimum point, where U is a minimum, is given by

$$U_{min} = \frac{T_1^{\delta_1} T_2^{\delta_2} T_3^{\delta_3} ... T_n^{\delta_n}}{\prod_{i=1}^{i=n} \delta_i^{\delta_i}}$$

Therefore, in order to find the optimum point it is only necessary to know $T_1, T_2, ..., T_n$ (which are known from the original cost equation) and the values of the various δ's.

Special Mathematical Techniques 159

Zener further points out that at the optimum point, the optimum value of each T_i can be obtained from the expression

$$T_i' = \delta_i U_{min}$$

where T_i' is the optimum value of that term in the cost equation. Thus, by a very simple calculation it is possible to find the contribution of that term to the total cost at the optimum. From this it is possible to derive interesting relationships among the variables at the optimum point.

An illustrative example will make this clearer.

Example 4.3

Consider again the problem of the three-stage compressor in Example 4.2. It is desired to minimize the amount of work required to compress a gas from inlet pressure P_1 to outlet pressure P_4 under the conditions set out in Example 4.2.

$$W = NRT_1 \frac{k}{k-1} \left[\left(\frac{P_2}{P_1}\right)^{\frac{k-1}{k}} + \left(\frac{P_3}{P_2}\right)^{\frac{k-1}{k}} + \left(\frac{P_4}{P_3}\right)^{\frac{k-1}{k}} - 3 \right]$$

where W is the total work required for the three-stage compression.

Obviously, some of these terms are constants: N, R, T_1, P_1, P_4, $k/(k-1)$, and 3. These constants can be combined in some simple way. Let

$k/(k-1) = \theta$

$W\theta/NRT_1 + 3 = W'$

$(P_2/P_1)^{(k-1)/k} = aP_2^\theta$, where $a = (1/P_1)^{(k-1)/k}$

$(P_4/P_3)^{(k-1)/k} = bP_3^{-\theta}$, where $b = P_4^{(k-1)/k}$

Then

$$W' = aP_2^\theta + (P_3/P_2)^\theta + bP_3^{-\theta}$$

By this arrangement a generalized polynomial has been derived wherein there are three terms and two variables.

Zener's method now requires that

$$T_1^{\delta_1} T_2^{\delta_2} T_3^{\delta_3} = \text{a constant}$$

Thus

$$(aP_2^\theta)^{\delta_1} (P_3/P_2)^{\theta \delta_2} (bP_3^{-\theta})^{\delta_3} = \text{a constant}$$

and

$$a^{\delta_1} b^{\delta_3} (P_2)^{\theta \delta_1 - \theta \delta_2} (P_3)^{\theta \delta_2 - \theta \delta_3} = \text{a constant}$$

It is apparent that for this to be true

$$\theta \delta_1 - \delta_2 \theta = 0$$

$$\theta \delta_2 - \theta \delta_3 = 0$$

also

$$\delta_1 + \delta_2 + \delta_3 = 1$$

Solving these three equations simultaneously

$$\delta_1 = \delta_2 = \delta_3 = 1/3$$

Now from Zener's definition of a minimum

$$W'_{min} = \frac{(a^{1/3} P_2^{\theta/3})(P_3/P_2)^{\theta/3} (b^{1/3} P_3^{-\theta/3})}{(1/3)^{1/3} (1/3)^{1/3} (1/3)^{1/3}}$$

$$W'_{min} = 3(ab)^{1/3}$$

Then, since

$$W' = (W \theta / NRT_1) + 3$$

$$3(ab)^{1/3} = (W \theta / NRT_1) + 3$$

$$W = 3NRT_1 (ab)^{1/3} / \theta - 3NRT_1 / \theta$$

Special Mathematical Techniques

$$W_{min} = 3NRT_1 \left(\frac{k}{k-1}\right)\left[\left(\frac{P_4}{P_1}\right)^{(k-1)/3k} - 1\right]$$

The value of each individual term at the optimum is

$$T_i' = \delta_i U_{min}$$

So the values of optimum work in each stage are

$$T_1' = (1/3)(W) \; ; \; T_2' = (1/3)(W) \; ; \; T_3' = (1/3)(W)$$

and the point of optimum operation is the point where the compressor load is equally divided between the stages.

If

$$W_1 = W_2 = W_3 = W/3$$

then

$$W_1 = W_2 = W_3 = NRT_1 \left(\frac{k}{k-1}\right)\left[\left(\frac{P_4}{P_1}\right)^{(k-1)/3k} - 1\right]$$

For the first stage the work would be

$$NRT_1 \left(\frac{k}{k-1}\right)\left[\left(\frac{P_2}{P_1}\right)^{(k-1)/k} - 1\right]$$

and the ratio of the work of the first stage to the total work would be

$$\frac{(P_2/P_1)^{(k-1)/k} - 1}{3(P_4/P_1)^{(k-1)/k} - 1} = \frac{1}{3}$$

Then

$$P_2 = (P_1^2 P_4)^{1/3}$$

In a similar fashion it can be shown that

$$P_3 = (P_1 P_4^2)^{1/3}$$

These are the same answers found in Example 4.2; Eqs. (4.8).

4.02 COMPLEX RELATIONSHIPS: THE USE OF SPECIAL METHODS

In Section 4.01 attention was centered on optimization problems that could be solved by application of the differential calculus. Frequently in process engineering the situation is too complex for the straightforward use of simple calculus. There are various situations that arise.

1. Relationships which, although analytical and continuous, are quite lengthy and complex so that differentiation would yield even more complicated expressions, and an analytical solution would be virtually impossible. This would be particularly true for more than one independent variable; see Ref. [3] for such complex equations.

2. Relationships which, though continuous, are not expressible analytically, for instance, as in Example 4.4 concerning distillation column economics. In cases of this type the relationship between the process variables may be expressed graphically and derivatives may be taken as the slopes of the lines.

3. Relationships which are limited by certain constraints on the variables involved so that the optimum can be located only within certain values of the variables and not at the true economic optimum. For instance, in the distillation of polymerizable monomers it often happens that the column bottom temperature must be below a certain value.

4. Relationships which are ill-defined and the data fragmentary. Here the optimum may exist outside of the realm of knowledge and an efficient method of extrapolation is needed.

5. Relationships which contain many variables so that a straightforward solution in a reasonable length of time is essentially impossible. A fast and adequately good approximation method is needed.

For these more difficult problems many "optimum finding" techniques have been devised. The general principle behind all of

Special Mathematical Techniques   163

these methods is that of "homing in," to use a phrase of Boas [1]. In the present age of guided missiles, all are familiar with the term "homing in." Essentially, these methods are trial-and-error procedures (or searches) starting from a point away from the optimum, and approaching closer and closer on successive trials until, after an acceptably small number of trials, the optimum point is reached.

These calculations, experiments, or trials require the determination (by calculations or by experiments) of the objective function, or its gradient, at successive points. The location of the optimum is determined by the comparison of a trial result with the one before. If the new value of the result gives a better answer the search is continued in that direction. If the new value of the result gives a worse result the procedure moves back to the point before and the search starts in a different direction.

A successful search technique, or homing-in procedure, is one which eliminates as much wasted effort as possible. A great deal of work has gone into devising and perfecting such methods, and there are now many available. Almost all of these methods have been programmed for high-speed digital computers because, in most cases, the calculations are lengthy, tedious, and expensive. For a comprehensive review of modern optimization theory and practice see Beveridge and Schechter [9].

Several of these "numerical" methods will be described here, here, illustrative examples given, and appropriate references cited. A comprehensive detailed treatment would be impossible within the confines of one chapter of one small book.

4.03 REGION ELIMINATION

For functions of one variable the method of "region elimination" is efficient.

A function is known as unimodal if it has one maximum or one minimum, but not both, in the region of interest. This is the type most commonly encountered in elementary process-engineering optimization problems. Consider Fig. 4.2. Here, C is a function of t with a suspected maximum between the two limits t_1 and t_2. The exact location of the maximum is unknown. If two more points, t_3 and t_4, are established, by calculation or experiment, it is clear that t_4 is higher (closer to the maximum) than

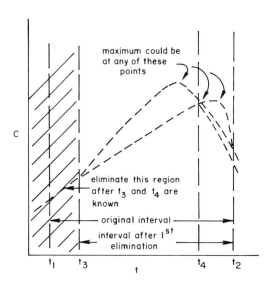

Fig. 4.2. Region elimination.

is t_3. From the concept of unimodality it is apparent that the shape of the curve dictates that the maximum must lie between t_3 and t_2. The curve can take any of three courses. The maximum could be between t_3 and t_4, close to t_4, or between t_4 and t_2. This forces the conclusion that the region to the left of t_3 cannot contain the maximum and may be eliminated from further analysis. This method of "region elimination" may be applied in a systematic fashion to narrow successively the region which might contain the maximum. Every time a point is calculated another region is eliminated. Using modern search methods the optimum can be located with a fair degree of precision with five or six calculations.

The definition of efficiency in a calculation of this type is "the fraction of the original interval within which the optimum lies after N calculations (or experiments) have been made." Obviously, the most efficient method of calculation (or experimentation) is the one which uses the least number of calculations (or experiments) to locate the optimum within a prescribed interval. Figure 4.3 shows a comparison between the various methods of optimum location by region elimination. It is apparent that the "golden mean" method and the Fibonacci number method are essentially equivalent and are much more efficient than other methods. For a more complete discussion see Boas [1]. A proof of the superior efficiency of the Fibonacci number technique is given by Denn [10].

Special Mathematical Techniques 165

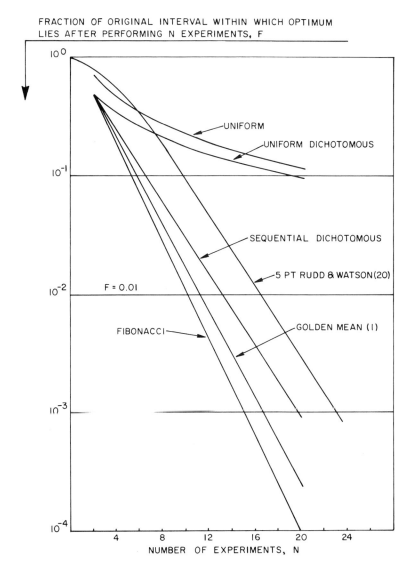

Fig. 4.3. Relative efficiency of various search methods.

Although the Fibonacci search technique is somewhat more efficient, its superiority over the golden mean method is slight, and an explanation is fairly complex. For present purposes, only the golden mean method will be considered; see Boas [1] and Denn [10] for discussions of the Fibonacci number method.

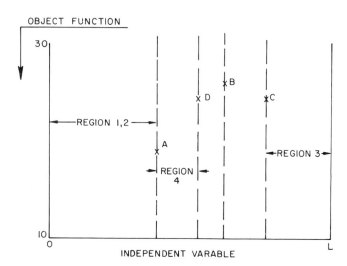

Fig. 4.4. Optimum location by region elimination golden mean method. Region 1,2: eliminated after first two calculations; region 3: eliminated after third calculation; and region 4: eliminated after fourth calculation.

The golden mean method involves the following steps:

1. Making two calculations (or experiments), one at a distance of 0.382 L from the left-hand side of an interval of length L, and the second at 0.618 L from the same left-hand side. Obviously, these two points are equidistant from the two edges of the interval. The two answers obtained may be represented on Fig. 4.4 as points A for 0.382 L and B for 0.618 L.

2. From the previous discussion of unimodality and region elimination it is apparent that the maximum cannot occur in the region to the left of A. Therefore, the region 0 to 0.382 L is eliminated and the maximum must be in the region from 0.382 L to L.

3. The calculation is repeated in the new (remaining) interval. Two calculations are made at 0.382(L - 0.382 L) and at 0.618(L - 0.382 L) in the remaining interval. It happens that one of these — the one nearest the eliminated interval — at point B, has already been made so only one new calculation is required. This result is shown as point C. In this method of calculation it will always be true that one of the new pair of calculations will have been done before. Consequently, only one calculation is required for each region elimination after the first.

Special Mathematical Techniques

4. By again invoking the principles of unimodality and region elimination it can be seen that the maximum cannot lie in the region to the right of C. Therefore, that region is eliminated and the region of interest now lies between A and C.

5. The calculation is repeated again at two points in the remaining interval. This interval, after elimination of the portion to the right of C, will be $L - 0.382 L - 0.382 (L - 0.382 L) = 0.382 L$. The two new points for calculation are $(0.382)(0.382 L)$ and $(0.618)(0.382 L)$ inside the interval AC. Again, the calculation at the point B, or $(0.618)(0.382 L)$, has been made, and only the calculation at $(0.382)(0.382 L)$ must be done. This result is shown as point D.

6. Again using modality and region elimination, it can be seen that the region AD is now excluded and that the maximum must lie in the region DC. The interval now remaining is $0.236 L$.

In four calculations the region wherein the maximum lies has been narrowed from L to $0.236 L$. This is precisely the number found from Fig. 4.3.

It is possible, but not illuminating, to continue with successive calculations and region eliminations. The more calculations the narrower the region containing the maximum. Figure 4.3 shows that many calculations using Fibonacci numbers or the golden mean will locate the maximum to within rather narrow limits. In engineering work it is seldom necessary to be so precise. However, even for the four or five calculations that engineers will usually make, the golden mean or Fibonacci number method of calculation is quite a bit more efficient than more conventional means. Denn [10] gives an example of a quite complex equation solved by the golden mean method. A computer was used, 16 calculations were made, and the minimum was located to with 0.1%.

4.04 ONE VARIABLE AT A TIME — WITH GRAPHICAL DIFFERENTIATION

A commonly encountered case is the one where there are two or more independent variables, and the relationship between them and the dependent variable is continuous, but not expressible analytically. A graphical construction can be made from well-known engineering methods so that the relationship is quite clear, and the partial derivatives must also be obtained graphically.

For such calculations it is usually necessary to do two things:

1. Express the profitability equation in terms of some dependent process variable other than the venture worth itself, as, for instance, the power consumption in Example 4.2.

2. Express the unit cost of each process variable as a simple function of the independent variables, as, for instance, the cost of a heat exchanger as a function of stream exit temperature.

The modified equation to be differentiated graphically may then be expressed as

$$W = aA + uU + vV + \ldots + zZ \tag{4.13}$$

where $A, U, V, \ldots, Z$ are the unit costs of each process variable expressed as simple functions of these variables. The term a is taken as a dependent variable and is the desired process result. The terms $u, v, \ldots, z$ are independent variables and are the operating conditions controlling the system and influencing a. The relationship between a and the independent variables must be known; it may be an equation, graph(s), or tabulation. Upon differentiation with respect to each of the independent variables, the following equations may be written:

$$\frac{\partial W}{\partial u} = a\frac{\partial A}{\partial u} + A\frac{\partial a}{\partial u} + u\frac{\partial U}{\partial u} + U + v\frac{\partial V}{\partial u} + \ldots + z\frac{\partial Z}{\partial u} = 0$$

$$\frac{\partial W}{\partial v} = a\frac{\partial A}{\partial v} + A\frac{\partial a}{\partial v} + u\frac{\partial U}{\partial v} + v\frac{\partial V}{\partial v} + V + \ldots + z\frac{\partial Z}{\partial v} = 0$$

$$\vdots \qquad \vdots \qquad \vdots \qquad \vdots \qquad \vdots \qquad \vdots \qquad \vdots \tag{4.14}$$

$$\frac{\partial W}{\partial z} = a\frac{\partial A}{\partial z} + A\frac{\partial a}{\partial z} + u\frac{\partial U}{\partial z} + v\frac{\partial V}{\partial z} + \ldots + z\frac{\partial Z}{\partial z} + Z = 0$$

A system for employing Eqs. (4.14), using the plan of one variable at a time, is as follows:

1. Assume a set of reasonable values for the independent variables and, from the relationship $a = f(u, v, \text{ and } \ldots z)$ (this relationship may be graphical, analytical, or tabular, but it must be known), a value for the dependent variable a may be computed.

Special Mathematical Techniques

2. Determine the unit costs A, U, V, and Z corresponding to the assumed values in Step 1. In doing this it is necessary to collect all the costs controlled by one variable into an expression which may be a function of that variable and of the others as well. There will be such a cost function for each variable. After these cost functions have been assembled it is necessary to find the incremental values of these costs. A, U, V, ..., and Z are the incremental costs associated with that variable. The incremental cost may be obtained by differentiating the cost-variable relationship.

3. From Eqs. (4.14) calculate appropriate values of $\partial a/\partial u$, $\partial a/\partial v$, $\partial a/\partial z$ which would exist if the initial assumptions of u, v, and z corresponded to the optimum.

4. From the relationship $a = f(u, v, z)$ between the variables find:

A new value for v, at the calculated $\partial a/\partial v$, if all the other variables remain at the first assumed values.

A new value for u, at the calculated $\partial a/\partial u$, if all the other variables remain at the first assumed values.

New values for the other variables are found in the same way. This procedure may require that the $\partial a/\partial u$, $\partial a/\partial v$, $\partial a/\partial z$, and others be obtained by graphical or numerical differentiation.

5. With the new values of the variables, Eqs. (4.14) can be solved again for new values of the partial derivatives, and the process repeated. This iteration process is continued until the assumed and estimated values check.

This procedure is subject to the same general limitations as the originally described procedure for differentiating Eq. (4.1).

A detailed example using two independent variables will make this clearer. Here, process correlations and engineering knowledge are available over the full range of interest of the variables involved.

Example 4.4 The Ethylbenzene-Styrene Separation by Fractional Distillation

Styrene is a valuable industrial chemical that is currently produced at rates in excess of 4 billion lb/yr. The plants that are used for this production are large, and the economics of the operation are stringent.

The process that is universally used is to produce ethylbenzene, purify this by distillation, dehydrogenate the ethyl side chain to the corresponding vinyl group (thus producing styrene or vinylbenzene), and then separating the ethylbenzene from the styrene by fractional distillation. The binary system ethylbenzene-styrene has a low relative volatility (1.37), the distillation must proceed under a substantial vacuum (column head pressure about 50 mm Hg), and a quite pure styrene must be produced. It is customary to specify that the styrene stream, which is the bottom stream from this distillation (ethylbenzene being slightly more volatile than styrene), is to contain 99.8 wt % styrene and 0.2 wt % ethylbenzene. In industrial practice this bottom stream will contain a small amount of tar and polymer, but this may be neglected for present purposes.

The concentration of the distillate stream coming from the top of the column will be controlled by the economic balance between the gain caused by the recovery of the styrene and the cost due to the size of the distillation column, the condenser, and the reboiler, and the cost of the steam and cooling water used to carry out the distillation.

It is apparent that the column may be made very tall, use a high reflux ratio, and recover a high percentage of the styrene, but at a high cost. It is also apparent that the cost of the column and the utilities may be kept low by using a short column and a low reflux ratio, but also losing a substantial percentage of the styrene and so increasing the cost of the distillation operation. Somewhere between the two extremes is the optimum point for the operation of the distillation.

It should be noted that styrene is temperature-sensitive and that the bottom temperature may not be allowed to go above 105°C. This limits the total pressure drop through the column, and since distillation column plates have some pressure drop through each plate, the total number of plates that can be used is limited. This point has caused much development work on plates of good efficiency and low pressure drop. For a discussion of this and other points concerning the ethylbenzene-styrene distillation the reader may consult the paper of Frank, Geyer, and Kehde [11]. For the present this temperature restriction will be disregarded, but it will be checked when the optimum point is found.

It is desired to design a distillation column to separate 5,000 lb/hr of an ethylbenzene-styrene solution. The following conditions will apply:

Special Mathematical Techniques   171

   Feed composition — 54 wt% ethylbenzene, 46 wt% styrene
   Bottoms composition — 0.2 wt% ethylbenzene, 99.8 wt% styrene
   Column top pressure — 50 mm Hg
   Tray efficiency (all trays) — 90%
   Tray spacing — 24 inches
   System relative volatility — 1.37 (constant, top to bottom)
   Feed thermal condition — liquid at its boiling point

Let the following symbols apply:

   F = feed rate (lb/hr); here a constant at 5,000
   D = distillate rate (lb/hr)
   x_D = wt fraction ethylbenzene in distillate
   B = bottoms flow rate (lb/hr)
   x_B = wt fraction ethylbenzene in bottoms; here a constant at 0.002
   R = reflux ratio
   n = number of theoretical plates in column
   P = % styrene recovered in bottoms stream
     = $(100)(0.998)(B)/(5,000)(0.46)$ = 0.0433 B

In this treatment the two independent variables will be the number of plates in the column, n, and the reflux ratio, R. The dependent variable, which will be solved for as a function of n and R, will be the percent recovery P.

The first step is to establish the relationship between P, n, and R. For the present case of a constant relative volatility this may be done in the following manner:

1. Assume P; calculate B, D, and x_D.

2. Knowing x_D and x_B and the relative volatility, use Fenske's equation to calculate the number of plates at total reflux.

3. Knowing x_D, x_F, and y_F, the minimum reflux ratio for that P may be calculated. This will be a constant for each P.

4. For each P assume certain numbers of theoretical plates. In this case assume n = 40, 50, 60, and 70. From Gilliland's corre-

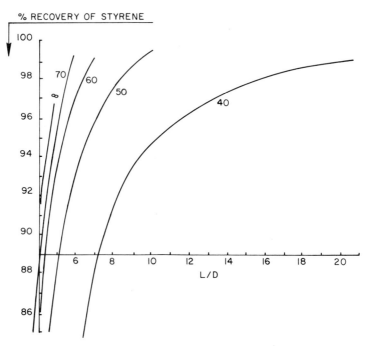

Fig. 4.5. Ethylbenzene-styrene distillation, 50 mm Hg; parameter is number of theoretical plates.

lation [12] the actual L/D for that n can be computed. This calculation relates the number of theoretical plates, the reflux ratio, and the percent recovery.

The results of this calculation are shown in two graphs: Figs. 4.5 and 4.6. One is the cross plot of the other. These graphs are the desired relationship between the percent recovery, the number of theoretical plates, and the reflux ratio.

In the present case the calculations were made simple because the system had a constant relative volatility. If this had not been so it would have been necessary to use graphical methods or to employ machine computations. Either procedure would have been expensive. In actual industrial practice machine computations are now frequently employed.

It is now necessary to establish the unit costs of each of the three variables:

Special Mathematical Techniques

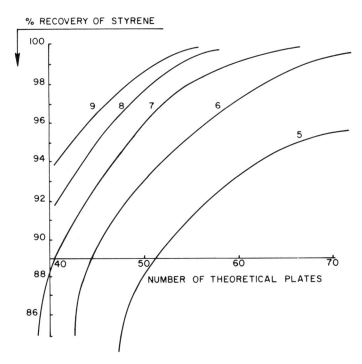

Fig. 4.6. Ethylbenzene-styrene distillation, 50 mm Hg; parameter is reflux ratio.

A = cost of styrene lost in distillate – $/(hr)(1% loss)

Styrene may be assumed to cost $0.08/lb. Then,

A = (5,000)(0.46)(0.01)(0.08) = $1.84/(hr)(1% loss).

The reflux ratio variable controls the following costs:

1. The cost of the plate which has a cross-sectional area sufficiently large to handle the required vapor flow. A higher reflux ratio will cause a higher vapor flow, which will, in turn, require a plate of greater cross section. This will be more expensive.

2. The cost of the reboiler heat necessary. The greater the reflux ratio the greater the amount of reboiler heat required.

3. The cost of the reboiler area necessary. A higher reflux ratio will require a larger reboiler.

4. The cost of the condenser area necessary.

5. The cost of the cooling water for the condenser.

The number of plates variable will determine the capital investment in the column due to its height. For each plate there will be a 24 in. tall section of column that will accompany it.

Before either of these costs can be determined it will be necessary to make a reasonable estimate as to the probable size and height of the column. The costs of process equipment often depend on their size, and it is often difficult to know the cost until the size is known; although, actually, it is the cost that is supposed to dictate the size. This awkward situation is most readily handled by making a rough calculation of the size of the equipment so that a fairly clear idea of the size-cost relationship may be obtained. Frequently, in a limited region, the cost-size relationship may be fairly simple. However, over a broad range of sizes the relationship may be too complex for an analytical expression. With respect to distillation columns using sieve plates and with 24-in. tray spacing this situation is discussed in Chapter 5.

A quick approximate calculation of the size of the distillation column would use the following assumptions:

1. Vapor velocity in the column – 8 ft/sec; see Appendix C

2. Reflux ratio – 7

3. Percent recovery – 98%

The results of this calculation give n = 55 and column diameter = 6 ft. This answer seems quite reasonable, and is helpful in simplifying the column cost-size relationship.

Using the cost data in Chapter 5 and in the correlations of Guthrie [13], it is possible to determine that the cost of a carbon steel column using sieve trays with 24-in. tray spacing is $550 d. Here d is the diameter of the column in feet. This is restricted to a range of diameters between 4 and 8 ft. In actuality, the cost-size relationship, even over this restricted range, is not precisely linear, but, for present purposes, it is precise enough. It must be remembered that this cost includes 24 in. of vertical column together with a sieve tray of diameter d ft.

It is easy to show that, at 8 ft/sec superficial vapor velocity in the column, the diameter may be related to the vapor flow rate by

$$d = 0.039 \, V^{\frac{1}{2}}$$

Special Mathematical Techniques   175

Therefore, the cost of a tray to handle V pounds per hour of vapor is

$$(550)(0.039)(V)^{0.5} \ \$$$

The incremental tray cost is

$$\frac{d}{dV}\left[(550)(0.039)(V)^{0.5}\right] = 10.7 (V)^{-0.5}$$

This is the cost of a tray to handle one additional pound per hour of vapor. Note that it is a fairly strong function of V. If there are n theoretical plates in the column, each of efficiency 0.9, the tray cost due to the vapor load will be

$$10.7(V)^{-0.5}(n/0.90) = 11.9\,nV^{-0.5} \ \$/(lb/hr \ vapor)$$

The reflux ratio variable also controls the areas of the reboiler and the condenser. These values, and their costs, may be estimated as follows:

Let the heat of vaporization of styrene be 153 Btu/lb. Then the reboiler heat load is 153 V Btu/hr. If the heat-transfer coefficient in the reboiler is 250 Btu/(hr)(sq ft)($°F$) and the temperature difference is 45°F, then the required reboiler heat-transfer area is

$$A = 153\,V/(250)(45) = 1.36 \times 10^{-2}\,V \ \text{sq ft}$$

From Chapter 5 it is learned that the installed cost of a reboiler of the simplest type is $346(A)^{0.62}$. In this case, it is $346(0.0136V)^{0.62}$. The unit cost of the reboiler is then

$$\frac{d}{dV}\left[346(0.0136\,V)^{0.62}\right] = (346)(0.62)(0.0136\,V)^{-0.38}(0.0136)$$

$$= 15\,V^{-0.38} \ \$/(lb/hr \ vapor)$$

In a similar fashion the cost of the condenser heat-transfer area can be estimated. If the column is adiabatic, and it should be, the heat load on the condenser is equal to the heat load on the reboiler. Cooling water may be assumed to enter the condenser at 90°F and to leave at 120°F. The condensing temperature of the process vapor is 136°F so that the mean temperature difference is 28.3°F. The heat-transfer coefficient in the condenser will be about 150 Btu/(hr)(sq ft)($°F$) so that the required condenser area will be $153\,V/(150)(28.3) = 3.6 \times 10^{-2}\,V$ sq ft. If it is assumed that condensers and reboilers are both simple heat exchangers of the shell and tube type, the costs for the two will be the same. In that case the unit cost of the condenser area will be found in the same way as for the reboiler. The unit cost of the condenser will be $27.2 \, V^{-0.38}$ \$/(lb/hr vapor).

The reflux ratio variable also controls the quantity of heat that must be supplied to the reboiler. If V is the quantity of vapor that is being produced every hour and the latent heat of vaporization of the process liquid is 153 Btu/lb, then the quantity of reboiler heat required is 153 V Btu/hr. The source of this heat will be steam at a fairly low pressure, and its cost may be taken as $\$0.75/10^6$ Btu. Therefore, the cost of the reboiler heat will be $(153)(0.75)/10^6 = \$1.15 \times 10^{-4}$/(hr)(lb/hr vapor).

The cost of the cooling water used in the condenser must also be taken into consideration. The cooling water rises 30° F and the heat given up by 1 lb of vapor when it condenses is 153 Btu. Therefore, the amount of cooling water used to condense 1 lb of vapor is 6.11×10^{-4} thousands of gallons. Cooling water costs between \$0.03 and \$0.10/1000 gallons for most industrial installations, and the cost has been rising steadily (and the availability declining steadily) in recent years. A cost of \$0.08/1000 gallons would seem to be a reasonable estimate. Using this cost for the cooling water, the charge for cooling water usage becomes $(6.11 \times 10^{-4})(0.08) = \$4.88 \times 10^{-5}$/(hr)(lb/hr vapor). It should be noted that this is close to one-half the steam charge. The cooling water charge is by no means negligible.

All the charges controlled by the reflux ratio variable are divided into two classes:

1. Charges due to capital investment. These will be the cost of the column area, the cost of the reboiler area, and the cost of the condenser area. Capital investment costs are converted into annual operating costs by multiplying the capital investment by the factor $m + d + i_m/(1 - t)$, where

m = annual charge for maintenance of equipment (frac/yr)
d = annual depreciation charge for tax purposes (frac/yr)
i_m = minimum annual return on incremental investment (frac/yr)
t = taxation rate, set by governmental authorities (frac/yr)

In this case the following values will be used:

$m = 0.05$ $i_m = 0.20$
$d = 0.10$ $t = 0.50$

Special Mathematical Techniques 177

Then $m + d + i_m/(1 - t) = 0.55$. This corresponds to a maximum acceptable payout time (as discussed in Chapters 2 and 3) of about two years.

This method of calculating the capital investment-related charges is the form given in Eq. (2.13). A preferred form is that of Eq. (3.47) where the relationship is $m + d + (i_m - 0.35i)/(1 - t)$. In principle, it is better to use Eq. (3.47), but, in practice, the difference is negligible.

The capital investment-related charges for this problem are

$$0.55[(11.9\,nV^{-0.5}) + (15\,V^{-0.38}) + (27.2\,V^{-0.38})]\frac{\$/yr}{lb/hr\ vapor}$$

2. Charges due to the consumption of steam and cooling water. These will be $1.05 \times 10^{-4}(8,000) + 4.88 \times 10^{-5}(8,000)$ $/(yr)(lb/hr vapor). The plant will operate 8,000 hr/yr.

The total unit cost due to the reflux ratio variable is obtained by adding the two charges derived above. The term V is easily shown to be related to the reflux ratio by the equation

$$V = (5,000 - 23\,P)(R + 1)$$

If the above sum is multiplied by $(5,000 - 23\,P)(R + 1)$ the total operating cost due to the reflux ratio is obtained; it is

$$0.55\{11.9\,n[(5,000 - 23\,P)(R + 1)]^{0.5}$$
$$+ 42.2[(5,000 - 23\,P)(R + 1)]^{0.62}\}$$
$$+ 1.312\,(5,000 - 23\,P)(R + 1)\ \$/yr$$

The cost due to the number of trays variable may be deduced as follows: At V lb/hr of vapor flow in the column, the diameter of the column is $d = 0.039\,V^{0.5}$, then the cost of that tray is $(550)(0.039\,V^{0.5})$. If there are n trays, each of 90% efficiency, then the cost of all the trays is $(550)(0.039\,V^{0.5})(n/0.90)$. Since this is a capital investment, the annual cost in dollars per year will be given by $0.55(550)(0.039\,V^{0.5})(n/0.90)$. Since $V = (5,000 - 23P)(R+1)$ the final result is

$$(0.55)(n)(23.8)(V)^{0.5}\ \$/yr$$

178   CHEMICAL PROCESS ECONOMICS

Now the total cost of conducting the distillation may be found by adding all the separate terms. There is obtained

$$W = 14{,}720(100 - P) + 6{,}560\,R - 30.2\,PR + 6{,}560 - 30.2\,P$$

$$+ 19.5\,n\,[5{,}000\,R - 23\,PR + 5{,}000 - 23\,P]^{0.5}$$

$$+ 23.2\,[5{,}000\,R - 23\,PR + 5{,}000 - 23\,P]^{0.62}$$

It is desired to minimize this function. Following Eqs. (4.14), W is differentiated with respect to R (with n constant) and with respect to n (with R constant). Each equation is set equal to zero, and there is obtained

$$\frac{\partial W}{\partial R} = 0 = \frac{\partial P}{\partial R}(-14{,}720 - 30.2\,R - 30.2) + (6{,}560 - 30.2\,P)$$

$$+ (5{,}000 - 23\,P - 23\,R\,\frac{\partial P}{\partial R} - 23\,\frac{\partial P}{\partial R})[9.8\,n(5{,}000\,R$$

$$- 23\,PR + 5{,}000 - 23\,P)^{-0.5} + 14.4(5{,}000\,R - 23\,PR$$

$$+ 5{,}000 - 23\,P)^{-0.38}]$$

$$\frac{\partial W}{\partial n} = 0 = \frac{\partial P}{\partial n}(-14{,}720 - 30.2\,R - 30.2) + (19.5)(5{,}000\,R - 23\,PR$$

$$+ 5{,}000 - 23\,P)^{0.5} + (-23\,R\,\frac{\partial P}{\partial n} - 23\,\frac{\partial P}{\partial n})[9.8\,n(5{,}000\,R$$

$$- 23\,PR + 5{,}000 - 23\,P)^{-0.5} + 14.4(5{,}000\,R - 23\,PR$$

$$+ 5{,}000 - 23\,P)^{-0.38}]$$

At this point the trial-and-error method of solution, obtaining $\partial P/\partial R$ and $\partial P/\partial n$ by graphical differentiation of the curves of Figs. 4.5 and 4.6, is started.

1. Assume a reasonable set of numbers for R, n, and P as a first trial solution to the above equations. The numbers used before to approximate the size of the column will be as good as any: $R = 7$, $n = 55$, and $P = 98$.

2. Substituting these into the two equations above and solving for $\partial P/\partial R$ and $\partial P/\partial n$ gives $\partial P/\partial R = 0.87$ and $\partial P/\partial n = 0.20$.

3. New values for R and n are found as follows:

Special Mathematical Techniques   179

   a. With n held constant at the original guess of 55 and with $\partial P/\partial R = 0.87$ it is found that R is now 8.

   b. With R held constant at the original guess of 7 and with $\partial P/\partial n = 0.20$ it is found that n is now 55 (no change).

   4. Using R = 8, n = 55, and P = 99, new values of $\partial P/\partial R$ and $\partial P/\partial n$ are calculated from the above equations. These new values are $\partial P/\partial R = 0.86$ and $\partial P/\partial n = 0.20$. These are essentially the same as the values found in the first calculation.

   5. New values for R and n are found by the same procedure as 3 above. These are R = 8, n = 55, and P = 99. Essentially perfect agreement with those results found from the calculations in step 3 above. With this excellent agreement it can be concluded that the problem is solved. The optimum point is located at R = 8, n = 55, and P = 99%.

   The trial-and-error aspect of this calculation is made much easier by a judicious choice of a starting point. If the original assumptions of R, n, and P are too far from the correct answer it may prove impossible to get the successive trials to converge to a solution. Instead, the successive answers will move back and forth across what appears to be the optimum point, but will not converge to a clean solution. When this occurs, it is much better to restart the calculation with a new original point that appears to be much closer to the optimum.

   The result achieved here is in good agreement with those of Frank et al. [11] and of Bodman [14]. Both of these authors have given extended discussions of this problem.

4.05 THE METHOD OF STEEPEST ASCENTS OR DESCENTS

   There are many situations in which neither an analytical nor a graphical representation is available for the function to be optimized for the entire range of the variables involved. These are called "poorly defined processes." The question then arises as to what course should be followed in order to find the optimum point. Any of several trial-and-error procedures may be used. However, if results are known only for a small area, some efficient method of extrapolation will help to minimize the amount of trial and error necessary. Such a method is the method of steepest ascents [1, p. 98].

For a situation involving the maximizing of a function of two independent variables, a contour map with the variables as horizontal coordinates and the profit function as the vertical coordinate will illustrate the relationship between the variables and the cost. Such a three-dimensional plot is called a "response surface." For systems with more than two independent variables a graphical plot is not possible and a system of equations is required.

The concept behind the idea of predicting a path of steepest ascent on the response surface is that a plane can be a good approximation to a curved surface over a limited area.

The best-fitting plane is obtained in a small area under investigation. Then, from the tilt of the fitted plane, the direction of further calculation or experimentation is established as the direction of steepest ascent (or descent). Calculations, or experiments, are made along this path until a decline in response is noted. Additional observations or calculations are made around this point to confirm whether a maximum (minimum) has indeed been reached or whether a new path of steepest ascent should be predicted.

In practice, the method of steepest ascents is generally applied to studies involving more than two variables. It is difficult to illustrate these cases graphically, but the mechanics is entirely similar to the two-dimensional approach. Specifically, the partial derivatives of the response W will indicate its rate of change with respect to a series of variables, x_1, x_2, x_3, ... From these derivatives the gradient ∇W is computed:

$$\nabla W = \underline{i}_1 \frac{\partial W}{\partial x_1} + \underline{i}_2 \frac{\partial W}{\partial x_2} + \underline{i}_3 \frac{\partial W}{\partial x_3} + \ldots \qquad (4.15)$$

where $\underline{i}_1, \underline{i}_2, \underline{i}_3, \ldots$ are unit vectors in the directions of coordinate axes representing the variables, each variable being associated with one dimension in a multidimensional space. The gradient ∇W is a vector which has the property of being directed along the path on which the response W increases most rapidly (i.e., the path of "steepest ascent"). The direction is along a path proportional to the rate of increase for a unit change of each variable. The condition $\nabla W = 0$ corresponds to the condition of Eq. (4.2).

The procedure outlined above is the basis for a number of search methods all of which are called "gradient methods." All gradient methods suffer from the defect that for a changing response surface (and it must change if a maximum or minimum exists), the direction of steepest ascent varies from point to point, and the true line of steepest ascent will be a curve rather than a series of

Special Mathematical Techniques   181

straight lines. Strictly speaking, the gradient established at the start of one search does not apply throughout all of that search. Despite this inaccuracy the direction of steepest ascent is widely used in various forms [9, p. 407].

It is also true that the nature of the response surface affects the problem. Surfaces with spherical contours give the fastest convergence and the most unequivocal answer. The closer the contours are to the spherical, the easier the problem is to solve. It is often possible to transform the variables so as to get coordinates which are more easily handled; see Boas [1].

The following simple example, similar to the compressor problem of Example 4.2, illustrates the way the procedure works.

Example 4.5

Consider the case of a three-stage compressor. It is desired to demonstrate the method of steepest ascents for $P_1 = 1$ and $P_4 = 10$ atm. Since the procedure implies knowledge of the work expended corresponding to only a limited area, rather than the complete range, the arbitrary assumption will be made that the intermediate pressures are taken at the start as $P_2 = 4$ and $P_3 = 7$ atm.

Solution. The optimum will be reached by the steepest descent to minimum total work, which corresponds to maximum profit for the operation employing such a compressor.

The answer will first be worked out from Eq. (4.5) from Example 4.2, to serve as a basis for comparison.

$$E = NRT \frac{k}{k-1} \left[\left(\frac{P_2}{P_1}\right)^{(k-1)/k} + \left(\frac{P_3}{P_2}\right)^{(k-1)/k} + \left(\frac{P_4}{P_3}\right)^{(k-1)/k} - 3 \right]$$

If NRT is set equal to 1, $k = 1.4$, $P_1 = 1$, and $P_4 = 19$, it is found by direct substitution that when $P_2 = 4$ and $P_3 = 7$

$E = 2.680$ ft-lb

Contours representing other values of E for values of P_2 and P_3 are shown in Fig. 4.7.

To obtain the path of steepest descent, increments of pressure must be taken proportional to these rates of change; thus

$$\frac{\Delta P_2}{\Delta P_3} = \frac{0.079}{0.009} = 8.8$$

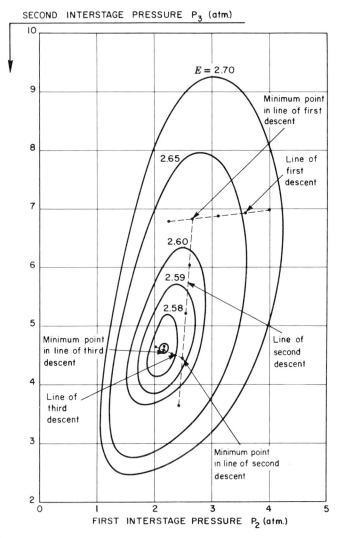

Fig. 4.7. Contours of constant work. Note: Path of steepest descent shown dotted. The direction of steepest ascent at $P_2 = 4$, $P_3 = 7$ is obtained from Eqs. (4.6) and (4.7):

$$\frac{\partial E}{\partial P_2} = [(4)^{-1/1.4} - (7)^{(1.4-1)/1.4}(4)^{(1-2.8)/1.4} = 0.079$$

and similarly

$$\frac{\partial E}{\partial P_3} = 0.009$$

Special Mathematical Techniques   183

If increments of P_3 are taken equal to 0.05, then the increments of $\Delta P_2 = 0.05 \times 8.8 = 0.44$. In this way the first descent is established at $P_2 = 4$ and $P_3 = 7$:

P_2	P_3	E
4.00	7.00	2.681
3.56	6.95	2.653
3.12	6.90	2.629
2.68	6.85	2.615 ← minimum point
2.24	6.80	2.622

Using the same procedure, a second descent is made starting at $P_2 = 2.68$ and $P_3 = 6.85$. This results in establishing the condition $P_2 = 2.50$ and $P_3 = 4.45$ with $E = 2.587$ as the minimum point. A third descent from this point establishes $P_2 = 2.14$ and $P_3 = 4.60$ with $E = 2.576$. These conditions are shown in Fig. 4.7. Three descents are sufficient to establish the optimum condition.

There are two limitations to the steepest ascent procedure:

1. Considerable precision in computation is required. Occasionally, in order to locate the optimum precisely, it may be necessary to use an analytical expression in the neighborhood of the optimum. This necessity comes about from the nature of the process response surface and from the nature of the optimum-seeking method.

In many chemical-process situations the response surface is not steep, and the optimum is located at the highest or lowest point of a long convex plateau or shallow valley. The optimum point may not be more than 5 to 10% larger or smaller than points far from the optimum. An examination of Fig. 4.7 shows that the optimum is only 4% smaller than the first point. In cases such as this the gradient will be small, and precise calculations will be necessary for significant differences to be seen; see also the paper of Reed and Stevens [3]. Modern computing machines can provide the necessary precision, but engineers must examine the process data and economic information carefully in order to make sure that elaborate calculations are worthwhile.

2. Many systems have restraints imposed on them by process requirements, materials specifications, or limits on invested capital. The steepest ascent method, as discussed above, does not take such restraints into consideration. The gradient may lead into an area in which operations are not possible.

4.06 OPTIMIZATION PROBLEMS INVOLVING RESTRICTIONS

In the methods just outlined it has been assumed that the controllable variables can be varied without any restraints. In practice this is often not possible, and variables may assume values only within ranges determined by such items as equipment limitations, safe operating practices, and limitations of availability of raw materials. These boundaries define an area within which the optimum must be established.

Thus, if the object function $W = f(x,y,z)$ is to be maximized, and the quantities x, y, and z are no longer three independent variables but are interrelated by another condition $\phi(x,y,z) = 0$, the situation, though not fundamentally different, sometimes results in awkward computations. In such situations Lagrange's method of undetermined multipliers may be employed; see Denn [10, pp. 18-32].

The condition for the maximum or minimum of a function W is

$$dW = \frac{\partial f}{\partial x} dx + \frac{\partial f}{\partial y} dy + \frac{\partial f}{\partial z} dz = 0 \tag{4.16}$$

The total differential of the restraining condition is

$$d\phi = \frac{\partial \phi}{\partial x} dx + \frac{\partial \phi}{\partial y} dy + \frac{\partial \phi}{\partial z} dz = 0 \tag{4.17}$$

Equation (4.17) is multiplied by an undetermined factor λ, a Lagrange multiplier. This product is then added to Eq. (4.16) to give Eq. (4.18).

$$\left(\frac{\partial f}{\partial x} + \lambda \frac{\partial \phi}{\partial x}\right) dx + \left(\frac{\partial f}{\partial y} + \lambda \frac{\partial \phi}{\partial y}\right) dy + \left(\frac{\partial f}{\partial z} + \lambda \frac{\partial \phi}{\partial z}\right) dz = 0 \tag{4.18}$$

If λ is assigned a value that makes

$$\frac{\partial f}{\partial x} + \lambda \frac{\partial \phi}{\partial x} = 0$$

$$\frac{\partial f}{\partial y} + \lambda \frac{\partial \phi}{\partial y} = 0$$

$$\frac{\partial f}{\partial z} + \lambda \frac{\partial \phi}{\partial z} = 0$$

$$\phi(x, y, z) = 0 \tag{4.19}$$

Special Mathematical Techniques   185

the condition for a maximum or a minimum is satisfied. Equations (4.19) constitute a set of four simultaneous equations with four unknowns: x, y, z, and λ. If the equations are linear they may be solved by algebraic methods to give numerical values of x, y, z, and λ. If the equations are nonlinear, special numerical methods must be used in their solution. The numerical values of x, y, and z that satisfy Eq. (4.19) are the values at which f(x, y, z) is a maximum or a minimum subject to the constraint that $\phi(x, y, z) = 0$.

This method may be extended for an arbitrary number of variables and an arbitrary number of subsidiary conditions. Although it does give an elegant formal procedure for determining the points where extreme values occur, it does not provide the sufficient conditions for establishing an actual maximum or minimum. Usually in practical cases it is known beforehand that an extreme point exists, so that the method provides a means for locating it exactly.

In addition, this presentation is confined to equality constraints: that is, $\phi(x, y, z) = 0$. It does not consider inequality constraints: that is, $\phi(x, y, z) >$ or $< C$. The use of inequality constraints makes the application of the Lagrange multiplier technique much more difficult.

Example 4.6

Natural gas is transported around many of the countries of the world by pumping the gas through steel pipelines of rather large diameter. Very large quantities of gas are handled, the initial capital investment is large, and, since the product has a low unit selling price, the economics of gas pipelines is of major importance. It is important to find the optimum values of the variables.

The cost of operating a gas pipeline may be considered a function of four independent variables, which are interrelated by the equations of fluid mechanics, together with suitable allowances for the costs of land acquisition, surveying, welding, and burying of the pipe. The four independent variables are:

1. The initial pressure to which the gas is compressed on entering the line and to which it must be recompressed at regular intervals along the length of the pipe. Obviously, this pressure may have almost any value. If it is low, the pipe diameter will be high, but its wall thickness (necessary to withstand the internal pressure) may be low and the pipe may not be heavy or expensive. On the other hand, a high pressure will result in a smaller diameter pipe, but one which has a greater wall thickness and may be more expensive to purchase and install.

2. The pressure to which the gas is allowed to drop during its passage through the pipeline. This may be low in comparison to the initial pressure so that the power required for recompression will be high, but the number of recompression stations will be small. Or, the lower pressure may be fairly close to the initial pressure, thus reducing costs for recompression, but increasing the number of recompression stations. The number of recompression stations is important since they are costly to build and consume manpower and supplies.

3. The distance between the recompression stations. If this is large the number of such stations is reduced and the capital investment and operating charges associated with them are reduced, but the recompression ratio is increased and the power consumption is made larger. If the distance is made small the reverse situation applies.

4. The diameter of the pipe. This may be made any size, but larger sizes are more expensive and must have a greater wall thickness in order to withstand the internal pressure. Smaller sizes can have a lesser wall thickness, are easier to handle in the field, but the pressure drop through them is greater, resulting in higher recompression costs.

The equations governing fluid flow through a pipe may be invoked to give a relationship between the quantity of gas to be moved and the system variables such as pipe length, pipe diameter, and pressure drop. This equation is

$$Q = 3.39 \left[\frac{P_1^2 - P_2^2}{fL} D^5 \right]^{\frac{1}{2}}$$

where Q = quantity of gas flowing (cu ft/hr); measured at 60° F and 1 atm pressure

P_1 = pressure to which gas is compressed before entering pipe line (psia)

P_2 = pressure to which gas is allowed to fall before leaving pipeline (psia); that is, the suction pressure on the gas compressor

D = diameter of gas pipeline (inches)

L = length of gas pipeline over which the pressure falls from P_1 to P_2 (miles)

f = friction factor of gas in ordinary steel pipeline. Here

Special Mathematical Techniques 187

Weymouth's equation may be used: $f = 0.008 \, D^{-0.33}$; see Sherwood [6, p. 89].

In the present case Q will be taken as 150 million cu ft/day; then

$$\frac{P_1^2 - P_2^2}{L} D^{5.33} = 2.69 \times 10^{10} \qquad (4.20)$$

By means of this equation the four variables are interrelated. This constitutes a restriction on the system; the four variables are not completely independent, but are restricted by Eq. (4.20).

The total yearly operating cost for a gas pipeline may be formulated in the following manner:

1. Determine the new capital investment, and from this, the annual fixed charges on this investment. The new capital investment will be made up of:

a. The cost of the compressors required to compress 1.5×10^8 cu ft/day of gas from P_2 to P_1

b. The cost of the pipe

c. The cost of acquiring land, surveying, ditching, welding, and other activities associated with placing the pipe in the ground.

The annual fixed charges on the new capital investment are determined by multiplying the new investment by the factor: $m + d + i_m/(1 - t)$. This factor has been discussed in Eq. (2.13). A preferred form is shown by Eq. (3.47).

In the present case, the maximum acceptable payout time, T_m (see Section 3.11) will be taken as five years. Since

$$T_m = 1 / \left(d + \frac{i_m}{1 - t} \right) = 5$$

then

$$d + \frac{i_m}{1 - t} = 0.2$$

In the case of the gas compressors the maintenance charge will be taken as 5%/yr so $m + d + i_m/(1 - t) = 0.25$. In the case of the pipeline the maintenance charge will be taken as 7%/yr so $m + d + i_m/(1 - t) = 0.27$.

The Installed Cost of Compressor and Motor

From the correlations of the cost of gas compressors in Chapter 5 it is found that the installed cost of an electric motor-driven centrifugal compressor will be $3.1(645)(Hp)^{0.8}$, where Hp is the compressor horsepower required. In the present case the compression ratio will be 1.5 or less (based on practical experience) so that the horsepower per compressor station will be about 3,800-4,000. Thus, a simple number for the incremental installed cost of the compressor, its electrical motor and drive, and its installation would be $320/Hp.

The power required for compression may be computed from the perfect-gas adiabatic compression formula for single-stage compression

$$Hp = Q\left\{\left(\frac{k}{k-1}\right)(RT_2)\left[\left(\frac{P_1}{P_2}\right)^{(k-1)/k} - 1\right]\right\}$$

$$Hp = \frac{1.16 \times 10^{-6} Q\rho_G RT_2}{(\text{mol wt})(550)}\left[\frac{k}{k-1}\right]\left[\left(\frac{P_1}{P_2}\right)^{(k-1)/k} - 1\right]$$

If it is assumed that the gas is methane and that $k = 1.28$, that the relative efficiency of actual compression to ideal compression is 75%, and that $Q = 1.5 \times 10^8$ cu ft/day, then

$$Hp = 4.07 \times 10^4 \left[\left(\frac{P_1}{P_2}\right)^{0.219} - 1\right]$$

From this, the installed cost of $320/hp, and the annual fixed charge factor of 0.25, the yearly fixed charge on the compressor-motor installation will be

$$0.25\left[\frac{320}{L}\left\{4.07 \times 10^4 \left[\left(\frac{P_1}{P_2}\right)^{0.219} - 1\right]\right\}\right] \text{ \$/(yr)(mile)}$$

Here L is the distance between compressor stations in miles.

The student should know that compressor-engine combinations are now, and have been, undergoing intensive development with the goal of reducing both capital and operating costs. Gas-fueled engines plus reciprocating compressors, gas turbines plus

Special Mathematical Techniques   189

centrifugal compressors, electric motor-driven reciprocating compressors, as well as electric motor-driven centrifugal compressors are all being built, in a variety of sizes, and subjected to long-term practical operation. For large installations (> 10,000 hp) a cost of about $100/hp is now contemplated.

The Installed Cost of the Pipeline

This cost will consist of the cost of the pipe itself, which depends on the pipe diameter and the pipe wall thickness plus allowances for installing the pipe. These charges may be computed in the following way:

For steel pipe the weight in tons per mile is given by

$$W = 28.2(D + t)t$$

where D = the pipe diameter (in.)

t = the pipe wall thickness (in.)

Also,

$$t = P_1 D/2S$$

where S = the maximum allowable stress in the pipe (psi)

= 20,000.

Therefore, the capital investment in one mile of pipe is

$$Y\left[28.2\left(D + \frac{P_1 D}{2S}\right)\left(\frac{P_1 D}{2S}\right)\right] \text{ \$/mile}$$

Y = purchased cost of pipe = \$240/ton

The cost of installing the pipe will be divided into three parts:

H dollars/mile for land acquisition, surveying, and clearing
(let H = \$5900/mile)

G dollars/ton for unloading, hauling, placing, and welding
(let G = \$30/ton)

N dollars/(mile)(in. diameter) for ditching, painting, back filling, and laying equipment
(let N = \$500/(mile)(in. diameter))

The total installed cost of the pipeline will be

YW + H + GW + ND $/mile

This is a capital investment of approximately $70,000/mile for a 28-in. diameter pipeline operating at 300 psia. This is probably too low by present standards, but it will suffice for purposes of illustration.

Using the values of Y, H, G, and N given above, and remembering that $m + d + i_m/(1 - t)$ for the pipeline is 0.27, the yearly fixed charges on the installed pipeline are

$$0.27(240 + 30)\left[28.2\left(D + \frac{P_1 D}{2S}\right)\left(\frac{P_1 D}{2S}\right)\right] + 0.27(500)D + 0.27(5,900)$$

The operating charges for the compressors and the pipeline will be composed of the following factors:

1. The fuel or power consumption for compression. In the present case the compressor will be driven by an electrical motor. Assuming a 90% efficiency in the motor the input electrical power will be

$$\frac{40,700}{0.90}\left[\left(\frac{P_1}{P_2}\right)^{0.219} - 1\right](0.746)(24)(360) = 2.91 \times 10^8\left[\left(\frac{P_1}{P_2}\right)^{0.219} - 1\right]$$

kw-hr/yr

If electrical power is available at $0.01/kw-hr the power charge for compression will be

$$\frac{2.91 \times 10^6}{L}\left[\left(\frac{P_1}{P_2}\right)^{0.219} - 1\right] \text{\$/(yr)(mile)}$$

2. The operating expenses for each compressor station may be taken as $4,000/month for labor, supervision, and salaries for a 4,000-hp station. This cost should be independent of size, so station operating costs are 48,000/L $/(yr)(mile).

3. The supplies consumed by the compressor station may be estimated as $3.00/(hp)(yr). The yearly cost of these items will be

Special Mathematical Techniques

$$\frac{3.00}{L}\left\{4.07\times 10^4\left[\left(\frac{P_1}{P_2}\right)^{0.219}-1\right]\right\} = \frac{1.22\times 10^5}{L}\left[\left(\frac{P_1}{P_2}\right)^{0.219}-1\right]$$

$/(yr)(mile)

4. An additional cost of $100/(yr)(mile) is added to account for the leakage of gas from the line.

It is now possible to write an expression for the total cost of operating a gas pipeline.

$$U = \underbrace{\frac{3.256\times 10^6}{L}\left[\left(\frac{P_1}{P_2}\right)^{0.219}-1\right]}_{\substack{\text{The fixed charges on the}\\\text{compressor-motor in-}\\\text{stallation}}} + \underbrace{\frac{2.91\times 10^6}{L}\left[\left(\frac{P_1}{P_2}\right)^{0.219}-1\right]}_{\substack{\text{the cost of electrical power}\\\text{to the compressor}}}$$

$$+ \underbrace{73\left\{28.2\left[\left(D+\frac{P_1 D}{2S}\right)\left(\frac{P_1 D}{2S}\right)\right]\right\} + 135D + 1{,}590}_{\text{the fixed charges on the installed pipeline}}$$

$$+ \underbrace{\frac{48{,}000}{L}}_{\substack{\text{the oper-}\\\text{ating cost}\\\text{of the com-}\\\text{pressor}\\\text{station}}} + \underbrace{\frac{1.22\times 10^5}{L}\left[\left(\frac{P_1}{P_2}\right)^{0.219}-1\right]}_{\substack{\text{the cost of supplies to}\\\text{the compressor}}} + \underbrace{100}_{\substack{\text{the cost}\\\text{of gas}\\\text{leakage}}} \qquad (4.21)$$

The units of U are $/(yr)(mile). There are four independent variables: L, P_1, P_2, and D. These variables are related by the flow equation relating pressure drop, pipe diameter, and flow rate

$$\frac{(P_1^2 - P_2^2) D^{5.33}}{L} = 2.69 \times 10^{10} \qquad (4.20)$$

Eq. (4.20) may be written in the general form $\psi(P_1, P_2, L, D) = 0$ by transposing the constant term to the left and setting the equation equal to zero; then

$$\frac{(P_1^2 - P_2^2)D^{5.33}}{L} - 2.69 \times 10^{10} = 0$$

Equation (4.21) above may be simplified to give

$$U = \frac{6.288 \times 10^6}{L}\left[\left(\frac{P_1}{P_2}\right)^{0.219} - 1\right] + 2{,}060\left[\frac{-P_1 D^2}{2S} + \frac{P_1^2 D^2}{4S^2}\right]$$

$$+ 135 D + \frac{48{,}000}{L} + 1690 \qquad (4.22)$$

The Lagrangian multiplier method may be used to solve for P_1, P_2, L, and D with the restriction that the flow equation applies. This would involve

$$\frac{\partial U}{\partial P_1} + \lambda \frac{\partial \phi}{\partial P_1} = 0$$

$$\frac{\partial U}{\partial P_2} + \lambda \frac{\partial \phi}{\partial P_2} = 0$$

$$\frac{\partial U}{\partial D} + \lambda \frac{\partial \phi}{\partial D} = 0$$

$$\frac{\partial U}{\partial L} + \lambda \frac{\partial \phi}{\partial L} = 0$$

These four equations plus the flow relationship, Eq. (4.20), give a system of five equations and five unknowns: P_1, P_2, L, and D with λ, the Lagrangian multiplier. It should be possible to solve these equations simultaneously for the five unknowns and so find the set of variables that will minimize the pipeline operating cost.

Performing the mathematical operations indicated above, there is obtained:

$$\frac{6.288 \times 10^6}{L}\left[0.219\, P_1^{-0.781} P_2^{-0.219}\right] + 2{,}060\left[\frac{-D^2}{2S} + \frac{2 P_1 D^2}{4S^2}\right]$$

$$+ \lambda \frac{2 P_1 D^{5.33}}{L} = 0 \qquad (4.23)$$

Special Mathematical Techniques

$$6.288 \times 10^6 \left[-0.219\, P_1^{0.219}\, P_2^{-1.219} \right] - \lambda \left[2D^{5.33}\, P_2 \right] = 0 \qquad (4.24)$$

$$2{,}060 \left[\frac{2P_1 D}{2S} + \frac{2P_1^2 D}{4S^2} \right] + 135 + \lambda \left[\frac{5.33\, D^{4.33}(P_1^2 - P_2^2)}{L} \right] = 0 \qquad (4.25)$$

$$-48{,}000 - 6.288 \times 10^6 \left[\frac{P_1^{0.219}}{P_2} - 1 \right] - \lambda \left[(P_1^2 - P_2^2) D^{5.33} \right] = 0 \qquad (4.26)$$

$$L = \frac{(P_1^2 - P_2^2) D^{5.33}}{2.69 \times 10^{10}} \qquad (4.27)$$

By means of simple algebraic manipulation it is possible to reduce the five equations above to three.

$$\frac{3.7 \times 10^{16}}{(P_1^2 - P_2^2) D^{5.33}} \left[P_1^{-0.781} P_2^{-0.219} - P_1^{1.219} P_2^{-2.219} \right]$$

$$+ 1{,}030\, D^2 \left[\frac{1}{S} + \frac{P_1}{S^2} \right] = 0 \qquad (4.28)$$

$$2{,}060 \left[\frac{P_1 D}{S} + \frac{P_1^2 D}{2S^2} \right] + 135 - \frac{9.9 \times 10^{16}\, P_1^{0.219}\, P_2^{-2.219}}{D^{6.33}} = 0 \qquad (4.29)$$

$$-48{,}000 - 6.288 \times 10^6 \left[\left(\frac{P_1}{P_2} \right)^{0.219} - 1 \right]$$

$$+ 6.9 \times 10^5\, P_1^{0.219}\, P_2^{-2.219} (P_1^2 - P_2^2) = 0 \qquad (4.30)$$

It is apparent that any further reduction would result in serious complication, so it is better to revert to a trial-and-error solution.

Step 1. Assume various values of P_1 and solve Eq. (4.30) for a corresponding value of P_2.

Step 2. Using these values of P_1 and P_2 solve Eq. (4.29) by trial and error for D.

Step 3. Using these values of P_1, P_2, and D solve Eq. (4.27) for L.

Step 4. Using these values of P_1, P_2, D, and L solve Eq. (4.22) for U.

This calculation gives a value for the minimum cost of operating the pipeline if the initial pressure is P_1. For instance, if P_1 = 400 psia, then P_2 = 343 psia, D = 25 in., L = 45 miles, and U = $23,930/(yr)(mile)$.

This procedure may be repeated for various values of P_1, both lower and higher than 400, until it is observed that the cost, U, goes through a minimum. The following table gives the results of all these calculations.

P_1 (psia)	P_2 (psia)	D (inches)	L (miles)	U [$/(year)(mile)]
200	174	30	28	26,810
300	254	28	48	23,520
400	343	25	45	23,927
600	505	21	49	24,816

If U is plotted against P_1 on an expanded scale the curve shows a sharp minimum at 290 psia. Therefore, the solution to the problem is

$P_{1,opt}$ (psia)	$P_{2,opt}$ (psia)	D_{opt} (in.)	L_{opt} (miles)	U_{min} [$/(yr)(mile)]
290	248	28	43	23,550

Actually, as can be seen from the tabulation above, the differences are rather slight and the minimum at 290 psia is not much smaller than operations at 300, 400, or 500 psia. The increase in cost is rather sharp as the pressure is reduced below 290 psia, but the increase in cost is rather small as the pressure is increased toward 600 psia.

Special Mathematical Techniques 195

As has been mentioned before, long flat minima such as this are frequently encountered in industrial optimization problems.

At the present time the operators of long gas pipelines are greatly concerned with reducing their costs. Some of the items under test, with the view of reducing costs, are the following:

1. Very high strength thin-walled pipe. Pipe with a yield strength of 65,000 psi is now in use and even higher strengths should be available in the future.

2. Gas turbine-driven centrifugal compressor units. These have lower first costs and are cheaper to operate.

3. Larger diameter, thinner-walled pipe. Diameters as large as 36 in. are under consideration.

4. Greater automation of the installation work; particularly welding.

5. Increased use of automatic controls on the compression equipment.

4.07 LINEAR PROGRAMMING

The general problem of finding a maximum, as treated thus far, assumes that some partial derivative of the function $W = f(x,y,z)$ becomes equal to zero within the range of interest of the variables. However, it is apparent that many functions, as for instance, an exponential, the S-shaped growth curve, or a quadratic, have derivatives which do not become zero. The best example of a function which does not go through a maximum or a minimum is a straight line; see Fig. 4.1. In all of these cases the optimum solution will occur at the boundary, or limiting, value of the variables — what are called the constraints on the variables. In cases of this type an explicit analytical solution is not possible. In simple cases an easy graphical or numerical method may be used. In more complex cases a systematic trial-and-error procedure utilizing high-speed computers is necessary.

Linear programming deals with the determination of an optimum solution of a problem where the objective function, the process model, and the constraints are all expressed in linear relationships.

The objective function, W, takes the form

$$W = c_1 X_1 + c_2 X_2 + \ldots + c_j X_j = \sum_j c_j X_j$$

where c_j is the profit (or cost) per unit of X_j used.

The relationship between the variables (the process model) must also be expressed as a set of linear equations or inequalities. The variables must be otherwise independent and must exceed the number of equations. In addition, no quantity can be negative; that is, $X_i \geq 0$ and $b_i \geq 0$. This requirement ensures that only useful (positive or zero) answers will be obtained. For instance,

$$a_{11} X_1 + a_{12} X_2 + a_{13} X_3 = b_1$$
$$a_{21} X_1 + a_{22} X_2 + a_{33} X_3 = b_2$$

Here the number of variables is X_1, X_2, and X_3, which is one more than the number of equations. In general,

$$\sum_j a_{ij} X_j = b_i; \quad i = 1, 2, \ldots, m; \quad j = 1, 2, \ldots, n; \quad n > m$$

where the sum of the coefficient a_{ij} multiplied by the value of each variable X_j equals a requirement b_i in the i-th equation. Since the number of variables exceeds the number of equations, an infinite number of solutions is possible. The best one is selected.

Often the relationships are expressed as inequalities. This comes about because practical problems can seldom be expressed as true mathematical equalities because of the need for freedom of action. A specification on the purity of a product would be impractical if it said "exactly 0.01% impurity." What serves the same purpose, and provides needed flexibility, is to say "a maximum of 0.01% impurity"; smaller amounts are permissible. Such inequalities may be converted into equalities (equations) by the introduction of "dummy" or "slack" variables. When this has been done well-known mathematical methods may be used to solve the system of equations. Unless all inequalities are converted to equations by the use of "slack" variables, linear programming will not work.

The boundary values, or constraints, on the variables must also be noted. These will take the form: $10 \geq x_1 \geq 0$, or $x_1 + 2x_3 \leq 16$, and the problem must be solved within, or on, the bounds of these restraints.

The coefficients a_{ij} are, in effect, the partial derivatives of W or b with respect to the X_j's, respectively. Since these are never zero the optimum is not smooth and continuous, but lies on

Special Mathematical Techniques 197

the topmost vertex of the polygon formed by the mesh of intersecting straight lines defined by the equations. One of the advantages of linear programming derives from this fact. Since the optimum must lie on a constraint, it is not necessary to examine the region between restraints in order to find the optimum. This reduces, by a very large amount, the effort involved in finding the optimum.

Linear programming has several other advantages:

1. The equations are linear algebraic equations and can be solved by simple methods, or by well-known and easily available computer programs. Such programs have been carefully compiled and can accommodate hundreds of equations.

2. The linear-programming solution always finds the economic optimum. Thus, it makes feasible the comparison of alternatives on a consistent basis.

3. For only minor deviations from linearity the system may be regarded as linear, and the problem solved by the linear-programming technique. In economic analysis it frequently happens that the cost data are not precise, and a reasonable approximation by a straight line is adequate. In such cases linear programming is an acceptable substitute for a more rigorous approach.

Linear programming has several weaknesses:

1. It can handle only linear problems. Many industrial problems are nonlinear and cannot be solved by linear methods.

2. A practical linear-programming problem may consume large amounts of computer time. Also, computing time increases approximately as the square of the number of equations.

3. A typical linear program, calculated on a computer, will continue to search for the optimum even though the improvement from one iteration to the next is very slight and not worth finding. In this way expensive computer time is used for no good purpose.

4. Linear programming will solve only problems of a static or steady-state nature. If the time element is important, as in a batch process, the linear-programming technique cannot yield an answer.

For problems involving many variables the solution is carried out by an iterative process, which at each stage of calculation assigns either zero or positive values to all the X_j variables. As the calculation proceeds, the values selected, while meeting the

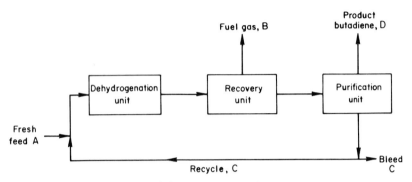

Fig. 4.8. Butadiene plant.

constraints imposed, will tend to increase the summation to be maximized at each stage. If a solution exists, a maximum profit case will finally be realized.

The most widely applicable method of linear programming is the simplex method described by several authors [9, pp. 290-322; 15]. This method is suitable for blending and scheduling problems involving different products, operations, and feed supplies, where the profitability requirements can be reduced to a mathematical formulation. Linear programming has also been applied to game-theory problems, to the evaluation of new business ventures and competitive bids, and to planning optimum inventories against a variable demand. Problems of this type are treated in any of several books on linear programming [16,17].

The nonlinear case has been considered. The problem of obtaining numerical answers is quite complex.

The following very simple example serves to illustrate how process problems can be handled by the linear-programming technique. The reader interested in further details of procedure should consult Refs. [9, p. 323; and 17].

Example 4.7

It is desired to determine the optimum operation of a plant for the production of butadiene by the catalytic dehydrogenation of butylene. The steps involved are shown schematically in Fig. 4.8. Basis: one 24-hr day.

It is assumed that such items as labor, steam, and fuel consumption will remain relatively constant so that the profitability of

Special Mathematical Techniques 199

Table 4-1   Plant Tests for Butadiene Production

A Feed (bbl/day)	B Fuel (bbl/day)	C Bleed (bbl/day)	D Butadiene (bbl/day)	E % conversion of C_4	F % n-butane in total feed
7,880	1,300	3,500	2,960	30	15
6,740	900	3,120	2,630	25	15
12,900	1,380	8,460	3,034	20	11

one system of operation as compared with another will be a function of the feed required (at \$8.40/bbl), the fuel gas produced (equivalent to \$0.90/bbl of butene degraded), the bleed from the purification unit (at \$6.32/bbl), and the butadiene product (at \$32.80/bbl). The profit function to be maximized is therefore

$$32.8\,D + 6.32\,C + 0.90\,B - 8.40\,A$$

Note: The authors are well aware that these prices may not be comparable to those prevailing today. Butadiene, especially, has been steadily reduced in price over the years. Its price is now nearer \$25/bbl. However, relatively minor changes in the prices do not affect the utility of the example.

The operation of the plant is assumed to involve only two variables subject to independent selection: the conversion of butylene per pass through the catalytic dehydrogenation system, and the percent of n-butane (which is assumed to be inert) allowed to build up in the total feed to the dehydrogenation unit (that is, fresh feed plus recycle). Additional restrictions due to process limitations are that the fresh feed A cannot exceed 9,000 bbl/day, that the build-up F of n-butane in the total feed stream must not exceed 17%, and that the conversion E of n-butylenes per pass must be at least 25%. Calculations based on plant tests gave the results shown in Table 4.1.

These values of A, B, C, and D may be plotted against values of E and F. The resulting graphs, using E as abscissa, are difficult to interpret when the entire range of E from 20 to 30% is considered. There is either (1) a sharp minimum in the lines at E = 25%, or (2) the data scatter badly and, with only three points, no conclusion can be drawn about the functional relationship. However, when attention is focused on the region $E \geq 25\%$ it is

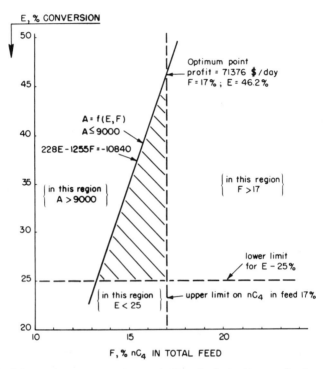

Fig. 4.9. Linear programming solution to butadiene plant problem.

possible to say that there is a linear relationship with a positive slope.

Similar plots for A, B, C, and D against F show that a linear relationship is probably correct, although in some cases the numbers are not precise. In the region $F \leq 17$ the slope is negative for A, B, C, and D.

Equations for A, B, C, and D may be written

$$A = 7{,}880 - \left[\frac{7{,}880 - 6{,}740}{30 - 25}\right](30 - E) + \left[\frac{12{,}900 - 7{,}880}{15 - 11}\right](15 - F)$$

$$B = 1{,}300 - \left[\frac{1{,}300 - 900}{30 - 25}\right](30 - E) + \left[\frac{1{,}380 - 1{,}300}{15 - 11}\right](15 - F)$$

$$C = 3{,}500 - \left[\frac{3{,}500 - 3{,}120}{30 - 25}\right](30 - E) + \left[\frac{8{,}460 - 3{,}500}{15 - 11}\right](15 - F)$$

Special Mathematical Techniques   201

$$D = 2{,}960 - \left[\frac{2{,}960 - 2{,}630}{30 - 25}\right](30 - E) + \left[\frac{3{,}034 - 2{,}960}{15 - 11}\right](15 - F)$$

In addition, there are the restrictions that $A \leq 9{,}000$, $F \leq 17$, and $E \geq 25$.

The problem is readily solved by simple methods:

1. The profit function, which is to be maximized, is

$$W = 32.8\,D + 6.32\,C + 0.90\,B - 8.40\,A$$

The above equations for A, B, C, and D may be substituted into the equation for W giving

$$W = -1024 + 800\,E + 2080\,F$$

2. A graphical figure, Fig. 4.9, is prepared. Here, a horizontal line at $E = 25\%$ shows that the area below this line cannot be used in the solution because of the constraint that $E \geq 25\%$. Similarly, a vertical line at $F = 17\%$ may be drawn, and this line shows that the area to the right of $F = 17\%$ cannot be used because of the restraint that $F \leq 17\%$. Also, since $A \leq 9{,}000$, and $A = 7{,}880 - 228(30 - E) + 1{,}255(15 - F)$, there results an additional restriction on the relationship between E and F

$$228\,E - 1{,}255\,F \leq -10{,}840$$

If this is written as an equation a straight line results, as shown on Fig. 4.9. From an inspection of the inequality it can be seen that points to the left of the line $A = f(E, F)$ can exist only if $A > 9{,}000$. Since this violates the restriction on A it can be seen that the area to the left and above the line $A = f(E, F)$ cannot be used in the solution.

As a result of these relationships and restrictions the plot shows that the solution can be only in the triangular area, shown cross-hatched, or on one of the lines, or at one of the intersections of the lines. Since

$$W = -1024 + 800\,E + 2080\,F$$

it is apparent that the optimum (maximum) point will occur at the maximum allowable values of E and F. From the plot on Fig. 4.9, these are $E = 46.2\%$ and $F = 17\%$, and, substituting these in the profit equation,

$$W = -1{,}024 + (800)(46.2) + 2{,}080(17)$$

$$W = \$71{,}376/\text{day}$$

It should be noted that, as with all linear optimization problems, the optimum appears on the constraint line.

If more than two independent variables were involved, it would not be possible to employ a geometrical construction, and more elaborate mathematical techniques would be required which in effect accomplish the same thing analytically.

To solve the subject problem by algebraic methods, the constraints must be changed to equalities by the introduction of slack, or "dummy," variables S_1, S_2, and S_3 to obtain the following form:

$$228\,E - 1{,}255\,F + S_1 = -10{,}865$$

$$-E + S_2 = -25$$

$$F + S_3 = 17$$

This array of simultaneous equations may be solved by the methods of matrix algebra, instead of by a graphical method, as in Fig. 4.9. However, if more than a few equations are involved, and in a practical problem there are nearly always many variables and many restraints, it is necessary to follow a systematic procedure or "drill" to obtain a solution.

The procedure, or algorithm, most commonly used is the simplex method (previously mentioned). When this algorithm is used with high-speed computing machines, answers can be obtained rather easily. Good general purpose linear-programming computer programs are readily available, and can be used by persons with no training or experience in computer work.

A description and an example of the simplex algorithm would be lengthy and complex, and would only add to the bulk of this book without contributing anything to the value of it. There are many descriptions in readily available references [9, pp. 290-322; 15, pp. 361-368; 18, pp. 726-731].

Special Mathematical Techniques

4.08 DYNAMIC PROGRAMMING

For any but the simplest systems the search for the optimum condition can be time-consuming and expensive. Many variables must be investigated at several levels, and the resulting response surface can contain a very large number of points. In order to reduce the effort necessary to find the optimum point of the system many clever approaches have been devised. One of the most important among these is the concept of utilizing some special characteristic of the system to systematize the search for the optimum. The concept of linear programming is one of these. Here the linear relationships cause the optimum to be found at one of the constraints and, in addition, the techniques of matrix algebra may be used to process thousands of linear algebraic equations.

The concept of dynamic programming is another such device. There are systems in the process industries that have individual elements or sections which are connected in series and where there is no recycle of matter, energy, or information from the product end of the system to the feed end. In such systems, known as acyclic systems, there is no flow of information from tail to head, only from head to tail. As a consequence, any change in the conditions imposed on one section can affect only that section and the ones downstream from it. The sections upstream from the subject section are not affected at all.

This property of acyclic systems leads to the thought that such a system might be optimized in sections starting from the tail end. Since such a downstream section is influenced only by the feed that it receives from upstream, it is apparent that the downstream section can be suboptimized with respect only to the feed that it

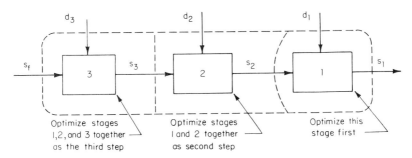

Fig. 4.10. Optimization of a three stage process by dynamic programming.

receives, and without respect to the operations in the upstream sections. Successively larger subsections of the system, starting from the tail end, could be optimized until all of the system had been optimized.

The basic principle behind the dynamic programming method is the "principle of optimality" [19]. "An acyclic system is optimized when its downstream components are suboptimized with respect to the feed they receive from upstream" [20].

Consider the schematic diagram in Fig. 4.10. Here, the three stages – the stages may be many or few and need not be of the same kind – are connected by a flow of material which is characterized by the numerical value of a state variable, s. This may be temperature, concentration, pressure, or some similar variable which is widely used in process engineering. There may be more than one state variable. The variables d_3, d_2, and d_1 are design or decision variables, and they may be specified to influence the operation of the stage; for instance, percent recovery of the more volatile component in a distillation column.

The first stage in applying dynamic programming to the system shown is to consider the tail end stage – usually called stage 1 – and to calculate the optimum point for this stage based on the feed stream which it receives – s_2, as a function of the design variable d_1. This problem may be written as

$$\max_{\{d_1\}} [\, U_1 \,(s_2, d_1)] = \max_1 (S_2)$$

where U_1 is the objective function to be maximized. A distillation column might be optimized with respect to the feed composition using percent recovery as a design variable.

Because the exact numerical value of s_2 is not known, the upstream stages not yet having been optimized, it is necessary to calculate an optimum point for stage 1 for several – perhaps many – values of s_2. It should be apparent that this can mean a good deal of calculation. There may be as many as five calculations to find the optimum for each s_2 and there may be five or ten more values of s_2. Thus, for only a simple search involving one state variable and one design variable it may be necessary to do 25 or more design-type calculations for one stage. If there are more design variables and more state variables the amount of computation increases exponentially. In such cases the memory and speed of even an electronic digital computer can be taxed.

Special Mathematical Techniques

This heavy computational load is the principal disadvantage to dynamic programming. Even for a simple system many engineers will be discouraged by the labor involved.

After the first (tail end) stage has been optimized for a number of values of s_2, attention is shifted to the combination of stages 1 and 2. These are now to be optimized together with respect to s_3, the feed to stage 2. This problem may be written as

$$\max{}_2 = \max_{\{d_1, d_2\}} [U_2(s_3, d_2) + U_1(s_2, d_1)]$$

However, the last term has just been evaluated for s_2 and d_1, so the new problem reduces to

$$\max{}_2 = \max_{\{d_2\}} [U_2(s_3, d_2) + \max{}_1]$$

This simplification helps enormously because the number of variables has been reduced from two, $d_1 + d_2$, to one, d_2. Thus, stage 2 may be optimized with respect to s_3 in a normal fashion considering only stage 2.

When the optimization of stage 2 has been completed for several, or many, values of s_3 then the term $\max_{\{d_2\}} [U_2(s_3, d_2)]$ will be known for each s_3 and may be added to corresponding values of $\max_{\{d_1\}} [U_1(s_2, d_1)]$ to obtain $\max{}_2$. It must be noted that these numerical values may be added only when s_2 is the same for both stages, as it would be in actual operation.

A similar optimization is now done for stages 1, 2, and 3 together by considering only stage 3, with respect to the feed composition. When $\max{}_1$, $\max{}_2$, and $\max_{\{d_3\}} [U_3(s_f, d_3)]$ are added, the optimum point for the entire system has been determined.

The advantage of dynamic programming is that the optimization of a complex multivariable problem has been reduced to a series of much less complex single-variable problems. A very great reduction in complexity is achieved. However, the computation load is still heavy, and for most practical problems a computer is used.

Fig. 4.11. A three stage gas compressor: A dynamic programming flow plan.

Example 4.8

Consider the three-stage gas compressor problem that was discussed in Chapter 3 and in Example 4.5. In Chapter 3 the problem was solved analytically by the differential calculus, and in Example 4.5 the problem was solved by the steepest ascents method. Here, it will be shown that the dynamic-programming method may also be used to solve the same problem.

In the previous discussion of the dynamic-programming method it was pointed out that two considerations were most important:

1. The system must not have any recycle of material, energy, or information from the tail end(s) of the process to the front end(s).

2. The system is optimized by optimizing successively larger subsections, with respect to the feed stream entering each section, starting with the tail end section.

It is apparent that the three-stage compressor problem, where each stage receives the stream from the stage before (and does not receive anything else), is an acyclic system and may be optimized by dynamic programming.

The problem posed by the three-stage compressor is to adjust the intermediate pressures, P_2 and P_3, so that the cost of compressing a given quantity of gas from P_1 (known) to P_4 (known) is a minimum.

Consider the schematic flow diagram in Fig. 4.11.

The work requirement for a perfect gas undergoing adiabatic compression, with interstage cooling back to the entering temperature, after each compression is

$$W = NRT \frac{k}{k-1} \left[\left(\frac{P_2}{P_1}\right)^{(k-1)/k} + \left(\frac{P_3}{P_2}\right)^{(k-1)/k} + \left(\frac{P_4}{P_3}\right)^{(k-1)/k} - 3 \right]$$

Since N, R, T, and k are all constants this reduces to

Special Mathematical Techniques 207

$$W = A\left[\left(\frac{P_2}{P_1}\right)^b + \left(\frac{P_3}{P_2}\right)^b + \left(\frac{P_4}{P_3}\right)^b - 3\right]$$

The optimization problem becomes

$$\min W = \min_{\{P_2 P_3\}} \left[\left(\frac{P_2}{P_1}\right)^b + \left(\frac{P_3}{P_2}\right)^b + \left(\frac{P_4}{P_3}\right)^b - 3\right]$$

To minimize W it is necessary to find the values of P_2 and P_3 that will make the total work input of the system a minimum. The fixed charges may be regarded as constant so that the energy requirements will determine the cost of compression.

This system minimization may be done by utilizing the principles of dynamic programming.

1. Consider the last stage, stage 3. Here the gas is compressed from P_3 to 10 atm. If the gas is considered to be air and k = 1.4, b = 0.286. Since A is constant, the work W_3 = $A[(10/P_3)^{0.286} - 1]$, and A may be given the value unity. Since P_3 cannot be known in advance of the final solution it is necessary to calculate W_3 for several values of P_3; say, 8, 6, and 4.

P_3	P_3/P_4	$(P_4/P_3)^{0.286}$	$W_3 = (P_4/P_3)^{0.286} - 1$
8	1.25	1.066	0.066
6	1.67	1.158	0.158
4	2.50	1.300	0.300

2. Consider the second stage, stage 2. Here the gas is compressed from P_2 to P_3. Since W_3 has been calculated for P_3's of 8, 6, and 4 it is necessary to calculate W_2's for several values of P_2 for each value of P_3. Thus, the results in the following tabulation are obtained.

P_2	$P_3 = 8$ W_2	$P_3 = 6$ W_2	$P_3 = 4$ W_2
6	0.085	0	—
4	0.220	0.123	0
2	0.486	0.370	0.063

3. Consider the first stage, stage 1. Here the gas is compressed from 1 atm to P_2. Since P_2 is not known, but P_1 is, it is necessary to calculate W_1 for each P_2 given above. It should be noted that P_1 is fixed at 1 atm so the number of calculations for W_1 is limited.

P_2	W_1 (at P_1 = 1 atm)
6	0.67
4	0.486
2	0.220

It is now necessary to find the combination of P_1, P_2, P_3, and P_4 that will make the term $W = W_1 + W_2 + W_3$ a minimum. This is most easily accomplished, and visualized, by a form of decision tree diagram. See Fig. 4.12.

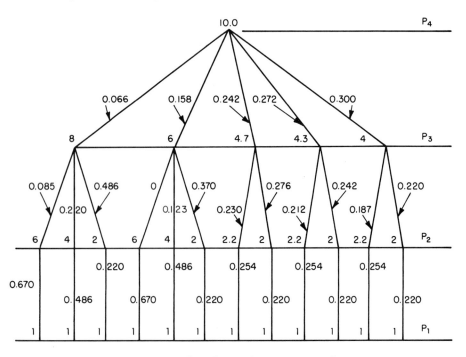

Fig. 4.12 A decision tree for dynamic programming.

Special Mathematical Techniques

From the summary of several possible paths it is apparent that those with one large pressure increase and two small pressure increases should not be used. The various paths and the energy consumptions can be tabulated as follows:

Path	Total work
1 - 6 - 8 - 10	0.821
1 - 4 - 8 - 10	0.772
1 - 2 - 8 - 10	0.772
1 - 6 - 6 - 10	0.828
1 - 4 - 6 - 10	0.767
1 - 2 - 6 - 10	0.748

Apparently what is wanted is a path where the pressure changes (ratios) are more equal. Therefore, a search will be carried out using P_3 between 4 and 5 and P_2 between 2 and 2.5. This calculation gives

Path	Total work	
1 - 2 - 4 - 10	0.740	
1 - 2.2 - 4 - 10	0.741	
1 - 2 - 4.3 - 10	0.734	
1 - 2.2 - 4.3 - 10	0.738	
1 - 2 - 4.7 - 10	0.738	
1 - 2.2 - 4.7 - 10	0.726	Minimum

Here the pressure ratios are $P_2/P_1 = 2.2/1 = 2.2$; $P_3/P_2 = 4.7/2.2 = 2.14$; $P_4/P_3 = 2.12$. Apparently the answer is to have the pressure ratios the same for each stage of compression. Thus, $P_2/P_1 = P_3/P_2 = P_4/P_3$. This was the answer found by the other methods.

This has been an extremely simple example without much calculation and without finding a true optimum for each successive stage of the process. However, the basic principle has been shown.

For more detailed explanation and examples see Beveridge and Schecter [9, pp. 672-701] and Rudd and Watson [20, pp. 212-244].

It should be noted that dynamic programming is quite a versatile technique. It can handle simple processes, such as that shown in the example. With equal ease more complex systems, such as multistage processes where each stage is different from the stages before and after, can also be computed. It is not necessary to have a mathematical equation for each stage; tables and graphs are good enough. Furthermore, since each stage is optimized separately from the other stages, any desired method of optimization may be used on any stage. Despite these advantages the computation load can become quite heavy for all but the simplest systems, and it is usually true that a high-speed computer is required.

REFERENCES

1. A.H. Boas, Chemical Engineering, February 4, 1963, p. 107.
2. L.J. Hvizdos, Chemical Engineering Progress, 60, No. 11, 64 (1964).
3. L.A. Reed and W.F. Stevens, Canadian Journal of Chemical Engineering, 41, 182 (1963).
4. R. Courant, Differential and Integral Calculus, Vol. 1, Interscience, New York, 1937.
5. C. Zener, Proceedings of the National Academy of Science, 47, 537 (1961).
6. T.K. Sherwood, A Course in Process Design, The MIT Press, Cambridge, Mass., 1963, p. 245.
7. R.I. Kermode, Chemical Engineering, December 18, 1967, p. 97.
8. R.J. Duffin, E.L. Peterson, and C. Zener, Geometric Programming, John Wiley and Sons, Inc., New York, 1967.
9. G.S.G. Beveridge and R.S. Schechter, Optimization: Theory and Practice, McGraw-Hill Book Co., New York, 1970.
10. M.M. Denn, Optimization by Variational Methods, McGraw-Hill Book Co., New York, 1969.
11. J.C. Frank, G.R. Geyer, and H. Kehde, Chemical Engineering Progress, 65, No. 2, 79 (1969).

Special Mathematical Techniques 211

12. E.R. Gilliland, in Chemical Engineers' Handbook, 4th ed., McGraw-Hill Book Co., New York, 1963, pp. 13-43.

13. K.M. Guthrie, Chemical Engineering, March 24, 1969, p. 114.

14. S.W. Bodman, The Industrial Practice of Chemical Process Engineering, The MIT Press, Cambridge, Mass., 1968, pp. 125-150.

15. F. Scheid, Numerical Analysis, Schaum's Outline Series, McGraw-Hill Book Co., New York, 1968, p. 361.

16. G. Hadley, Linear Programming, Addison-Wesley Publishing Co., Inc., Reading, Mass., 1962.

17. S.I. Gass, Linear Programming, McGraw-Hill Book Co., New York, 1958.

18. M.S. Peters and K.D. Timmerhaus, Plant Design and Economics for Chemical Engineers, McGraw-Hill Book Co., New York, 1968, Appendix A, pp. 719-732.

19. R. Bellman, Dynamic Programming, Princeton University Press, Princeton, N.J., 1957.

20. D.F. Rudd and C.C. Watson, Strategy of Process Engineering, John Wiley and Sons, Inc., New York, 1968, p. 213.

Chapter 5

NOTES ON COST ESTIMATION

5.00 INTRODUCTION

Thus far in this book certain equations and techniques have been developed for the economic analysis of financial investment in industry. In order to apply these procedures it is necessary that there be a systematic method for the estimation of the new capital investment and the factory cost of the product. Elaborate analyses of future money payments are of little value if the estimation of the capital investment and the manufacturing cost of the product are not made realistically.

In making a preliminary economic analysis of a new venture it must be recognized that the problem is quite complex. It involves many items of a nontechnological nature which cannot, at that time and perhaps ever, be resolved satisfactorily. These items are either neglected, or simplistic solutions are assumed in the preliminary analysis. In this case, the problem of economic evaluation reduces to the determination of appropriate values to substitute in the expression for venture worth, W, Eq. (3.9).

The estimation of investment I, working capital I_w, and gross return R_k are discussed in the present chapter. The material includes some well known and reasonably reliable methods for estimating these items. However, it must be recognized that the results may vary considerably depending on the skill and experience of the investigator and the information available to him. Items such as allowance for depreciation for taxes d_k, and the tax rate t, are prescribed by law and may be computed accurately, as briefly described in Chapter 3. The minimum acceptable return rate i_m, the return on invested capital i, and the venture life n will be discussed at some length in the following chapter. It should be understood that all these items could be treated much more extensively than space allows in this book.

The heart of the venture worth analysis is a realistic estimate of the new capital investment and the manufacturing cost of the product.

When contemplating a new enterprise the financial management of any industrial corporation concerns itself with two questions:

1. What total amount of new capital investment will be required before a new project will produce profits? In other words—how much money must be spent (in a rather short period of time) before a new project will function profitably, perhaps several years later? The desire of financial officers to know this number is perfectly understandable. On a smaller scale such a question is asked by everyone every day. A familiar example is the purchase of an automobile. If the machine is to cost $500 the purchase might be thought reasonable, but if the price were $1500 the purchase might be postponed or abandoned. The magnitude of the new capital investment is a matter of the greatest importance.

The purchase of a new chemical plant for $1,000,000 might well be regarded as a good investment, but the expenditure of $5,000,000 for the same plant might be looked upon as an unreasonable risk.

2. What will be the cost of producing the new product in the new plant? As discussed in Chapter 1, a product selling for less than 30¢/lb might well command a market of many millions of pounds a year and so justify the expenditure of millions of dollars for a production facility. However, if the selling price were 40¢/lb it might have no market at all, and there would be no point whatsoever in proceeding with the project.

Financial officers must have a good estimate of the total new investment and the manufactured cost of the product. With this knowledge they can quickly evaluate the probable future marketability of the material and the attractiveness of the investment.

It is impossible to overemphasize the importance of the preliminary cost estimate. It is on the results of this work that company managers will decide the future course of a project: abandonment, more laboratory work, pilot plant development, or more detailed and reliable cost estimation.

5.01 THREE STEPS IN COST ESTIMATION

The first step in the preparation of a cost estimate (both for new capital and for the manufacturing cost) is to lay out a process flow sheet showing all the major items of equipment, including flow lines and instrumentation. The heat and material balances should be computed so that temperatures, pressures, and stream compositions will be known. Despite the fact that there may be a lack of data or experience, this preliminary process-engineering work should be done as carefully as possible. A useful check list of items to be considered is given on p. 26-16 of the Chemical Engineers' Handbook [1].

The second step is to calculate the size and shape of the pieces of equipment and to specify the materials of construction. The last activity must be carefully thought out because the material of construction strongly influences the mechanical design and the cost of the equipment.

The calculation of the size and shape of process equipment will utilize the design procedures learned in university unit operation courses. Some of these techniques are complex and can be cumbersome and time-consuming. For a preliminary design both time and effort will be saved by using fast approximation design procedures, such as those outlined in Appendix B. More sophisticated and comprehensive design methods usually result in only minor changes in the design made from the approximate calculations. In a preliminary evaluation the extent of the knowledge and experience available is not adequate enough to justify elaborate calculations.

After the first two steps have been completed a good deal of information concerning the technical aspects of the process has been gathered. By a careful study of the flow sheet and the equipment for the process an experienced engineer can tell—essentially at a glance—those steps in the process that are likely to be troublesome, expensive, or unusual. Every preliminary design should be carefully examined for the technical problems likely to arise in the design, fabrication, transportation, and operation of the equipment.

A technical evaluation should always accompany the economic evaluation.

The third step in the making of a preliminary cost estimate consists of two parts: (1) Estimate the purchased costs of all the equipment shown on the flow sheet, and then estimate the installed "ready to run" cost of the complete plant. (2) Estimate the manufacturing cost of the product from the process heat and material balances and the costs of raw materials, utilities, labor, and depreciation.

5.02 EQUIPMENT COST ESTIMATION

There is an extensive literature on the costs of equipment and plant construction. This literature is quite valuable to the process engineer, but it must be used wisely. There are several points to be considered.

Equipment Cost-Time Indices

The costs of chemical equipment are not constant; they change with time. As a general rule costs become greater as time passes. Many attempts have been made to correlate equipment costs with some economic index which would reflect this change with time. Figure 5.1 shows a plot of several of these indices against time. Note the logarithmic vertical scale.

The most widely known index is the Engineering News-Record Index. This number is compiled and published by the magazine Engineering News-Record. This was the first of the indices and is most applicable to the construction industry. It is a weighted average of the costs of steel, lumber, cement, and labor. The swift increase in construction costs, of great importance to the chemical industry, is apparent from the plot. Many correlations of chemical equipment costs [2] are based on an ENR index applicable to some year, 1957—say. This indexing method was widely used for chemical equipment in the years through 1957, but since that time the costs of construction have risen more rapidly than the costs of process equipment. At present, it is not recommended that the ENR index be used for predicting equipment costs.

A pertinent point concerning modern chemical plant costs can be made here. The ENR index for construction costs is now rising much faster than the indices concerned with equipment costs. The

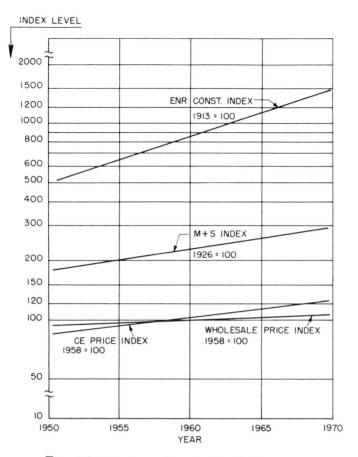

Fig. 5.1. Equipment cost-time indices.

conclusion reached from an analysis of these indices, and reinforced from practical experience, is that the costs of chemical plants are rising rapidly. This is due primarily to the rapid rise in the cost of construction work (steel, concrete, site preparation, and equipment installation) rather than to a rise in the costs of the process equipment [3].

The other indices are

a. The Marshall and Stevens Index, now renamed the Marshall and Swift Index, which is a weighted average of equipment costs in eight different process industries. This index is published regularly in Chemical Engineering.

b. The Plant Cost Index which is published regularly in Chemical Engineering magazine. A description is given by Norden [4].

c. The Wholesale Price Index of the U.S. Department of Commerce, which is published regularly in the pamphlet Business Statistics. This number, together with the Consumer Price Index, is generally regarded as a measure of the purchasing power of the U.S. dollar. It should be the most reliable measure of the cost of process equipment that exists.

From the plot it can be seen that the indices (a), (b), (c) are rising at almost the same rate, and that any one of them may be used to correct equipment costs to the present day, or further, by the ratio

$$\text{Present day cost} = \text{past cost}\left(\frac{\text{present value of index}}{\text{past value of index}}\right)$$

It must be emphasized that these costs are purchased equipment costs. They are usually quoted as FOB the manufacturer's warehouse. After purchase the chemical manufacturer must pay all costs for transportation to the plant site and for erection and connecting into the plant system. The charges for transportation, erection, and connection can be several times the equipment purchase cost. It must be apparent that a piece of chemical equipment must be connected into the plant system before it is of any value. Consequently the costs of transportation, installation, insulation, piping, wiring, and instrumentation must be added to the cost of the device before any meaningful analysis can be made.

For use in economic-balance calculations, as shown in Chapter 8, the installed equipment cost must always be used.

Equipment Cost-Size Relationships

Another point which must be considered is that the costs of process equipment, at one point in time, change with the size, power, or weight of the equipment. In general, a larger or more powerful machine, tank, or vessel costs more than a smaller one, but not in direct proportion. Usually a size-cost relationship of the following form applies:

$$\text{Cost} = K(A)^n$$

or

$$(\text{Cost})_L / (\text{Cost})_S = (A_L / A_S)^n$$

where A = some characteristic size measurement such as volume, area, or horsepower

K = a constant, the value of the cost when A is unity

L and S = subscripts designating larger and smaller sizes

A relationship of this form, when plotted on double logarithmic coordinates, is a straight line. On such coordinates it is possible to relate the costs and sizes of a particular type of equipment over a wide range of sizes—say, 100 or more times. Thus, such plots or equations are quite useful for the presentation of a large amount of information in a compact form. The cost literature of chemical engineering has many plots of this type. Two are shown in Figs. 5.2 and 5.3. Figure 5.2 is a copy of Guthrie's [5, Fig. 3] figure showing

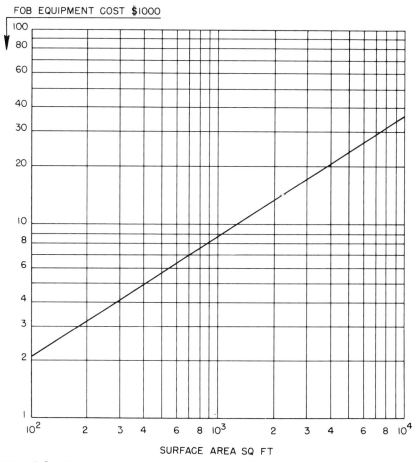

Fig. 5.2. Purchased cost shell and tube heat exchangers; carbon steel, 150 psi, mid-1968.

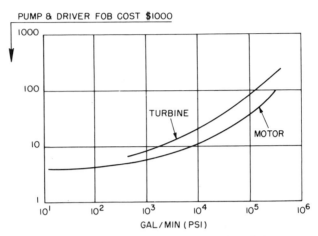

Fig. 5.3. Purchased cost centrifugal pumps plus drivers; cast iron, 150 psig, 250°F, mid-1968.

the purchased cost of shell and tube heat exchangers of the simplest type as of the middle of 1968. The line is straight, over a range of sizes of 100 times, and the exponent n is 0.65. In this case, because the line is straight, the value of n is constant, and the simple equation shown above applies over the whole range of sizes.

However, the line is not always straight and the value of n may be considered constant only over a rather small range of sizes. Figure 5.3 shows Guthrie's Fig. 6, in which the purchased costs of centrifugal pumps plus their electrical motor drives are plotted against the product of the pump capacity and the discharge pressure (a measure of the required horsepower). The line is not straight, so that the exponent n is not a constant. For all practical purposes the curved line may be subdivided into four essentially straight lines each of which would have its own value of n. This change in n, as the size changes, introduces certain complications in economic-balance calculations.

Some engineers consider n to be a constant for all sizes and for all types of equipment. Statements are sometimes made concerning the "six-tenths" or the "seven-tenths" rule, meaning that n may be regarded as constant at 0.6 or 0.7. This is just true enough to be an attractive approximation, but not good enough for correct work. Every type of equipment has its own value of n, and this value may or may not be a constant. Modern writers [5-8] recognize this situation and point out that the proper value of n must be known.

Guthrie [5] has an extensive collection of size exponents for chemical equipment.

Equipment Cost-Business Relationships

A third point about the compilations of costs which can be found in the literature is that they take no consideration of special business relationships which may exist. Suppliers of chemical equipment may change their prices up or down from the average values shown in accordance with the following:

1. The nature of the competition for sales which exists in their business. Some types of equipment are sold by many suppliers and the competition for sales is intense. The prices quoted by the various sellers will be very close to one another, quite competitive. However, a seller may be willing to reduce his prices by as much as 30% if he believes, or is led to believe, that by so doing he may receive a large order. On the other hand, some kinds of equipment are made by only one or two firms. They enjoy a monopoly and sell only for high prices with a high profit.

Business competition, or the lack of it, strongly influences the purchased cost of chemical equipment.

2. The overall state of the economy and the amount of business being done by the sellers. Manufacturers of equipment have heavy commitments in plant, machinery, and skilled workers. If business is poor, such manufacturers are often willing to quote low prices simply to obtain an order that will keep their plant and work force busy and intact. Also, occasionally, a company may seek to enter a new field or reestablish itself in an old one by lowering prices to obtain business. In such cases the prices quoted may be below their own costs and therefore lower than would be expected. On the other hand, if the manufacturers have a great deal of profitable work, that is, their business is good, they may well quote high prices, not really caring whether they get an order or not. In recent years, because of the heavy pressure of chemical plant construction, this has often been the case.

3. The nature of the business relationship between the equipment manufacturer and the equipment purchaser. It sometimes happens, and in the present age of mergers and conglomerates it happens fairly frequently, that both supplier and purchaser will be

parts of the same large corporation. In such cases the purchaser will have no alternative but to buy the supplier's equipment. Competition from other suppliers will be excluded.

5.03 PRECISION OF THE PRELIMINARY COST ESTIMATE

Cost estimates of the preliminary kind, and especially those made by process engineers working without the assistance of professional cost estimators, are not precise. A probable range of precision would be -35% to +25%, with the error more likely to be on the low side.

A "final" reliable cost estimate should be made only in cooperation with professional cost estimators, and only after price quotations have been received from equipment manufacturers. The preparation of a "final" cost estimate is rather a long and complex business and is not really the province of the process engineer. Jordan [9, pp. 28-32] has a more complete discussion of cost estimation and the degree of reliability offered by cost estimates done rapidly by process engineers or more slowly by professional cost estimators.

Despite the imprecision which accompanies a preliminary cost estimate, such an estimate must be done. It must be done by the process engineers using the process data and information available at the time. Nothing more will be done on a project until the preliminary cost estimate has been completed.

5.04 A COMPILATION OF EQUIPMENT COSTS

Table 5-1 contains both the purchased and installed costs of 16 separate kinds of chemical-process equipment. The costs are corrected to a time basis of July 1970. This list of costs was compiled by searching the work of Guthrie [5], Peters and Timmerhaus [7], and Bauman [6]. In addition, the authors' friends supplied much valuable information. The costs are tabulated or expressed in such a way that a cost for a particular size may be easily calculated.

Notes on Cost Estimation

Table 5-1 Purchased and Installed Costs of Process Equipment (A-Q)
Time Basis: July 1970

A. Pumps

Carbon steel centrifugal pumps with electrical motors

Motor horsepower	Purchased cost ($)	Installed cost ($)
1	600	2,270
10	1,400	5,300
100	6,000	22,700

The average size exponent, n, is 0.52, but actually changes from 0.37 to 0.63 as size increases. See Guthrie [5] for factors that account for changes in pump material.

B. Heat Exchangers

Carbon steel, shell and tube, fixed tube sheet, 100 psi

Heat-transfer area, A (sq ft)	Purchased cost ($)	Installed cost ($)
100	1,880	6,180
1,000	7,870	25,850
10,000	31,400	103,300

Purchased cost = $105(A)^{0.62}$ \$

Installed cost = $346(A)^{0.62}$ \$

Incremental installed cost = $214(A)^{-0.38}$ \$/sq ft

Note that the costs shown above for the exchangers are quoted in total dollars—not dollars per square foot. The incremental installed cost is given in dollars per square foot.

C. Columns

Some data on columns are tabulated on the facing page.

This table is only for columns 6 ft in diameter and varying in height from 28 ft to 108 ft. Similar tables may be prepared for other diameters and other heights. The limited space available here prevents a more extensive tabulation. The numbers given here are for columns of a size commonly encountered in the chemical industry. The costs of empty columns of other diameters may be calculated using a size exponent of 0.65. The size exponent for the trays is 1. For a more comprehensive study see Guthrie [5].

An inspection of this and a more extensive compilation for towers of other diameters reveals several important facts:

1. The installed cost of the column plus the trays is a complex function of the number of trays, the tray spacing, and the column diameter.

2. The installed cost of the empty column is much greater than the cost of the trays. If the cost of the trays were totally neglected the final column cost would be reduced by only a small amount.

3. The installed cost of the column depends on the spacing between the trays. If the tray spacing is reduced the column height will be lowered and the column cost will become smaller. The tray spacing should always be made as small as possible. However, conventional practice is to use 18 to 24 in.

4. The incremental cost of one tray plus 2 ft of column is also dependent on both the number of trays and the column diameter. This value, expressed in dollars per square foot, can range from 300 to 45. A value of 90, as used in Section 8.04, is a fair average.

5. It is difficult to express these costs by a single algebraic equation. When preliminary economic analyses are being done it is probably best to estimate the size and the number of trays from an approximate calculation and then use a single number to represent the column cost.

Columns

Distillation, absorption, or extraction

Carbon steel columns; carbon steel sieve trays; 1 atm press; 70-300°F; tray spacing 2 ft; column height extended 4 ft above top tray and 4 ft below bottom tray

No. of trays	Cost of empty column		Column diameter - 6 ft			Average installed cost[a] 1 tray plus 2-ft column		Incremental cost[b] 1 tray plus 2-ft column	
			Cost of all trays	Total cost column plus trays					
	Pchsd ($)	Instd ($)	($)	Pchsd ($)	Instd ($)	($)	($/sq ft)	($)	($/sq ft)
10	11,000	47,800	1,400	12,400	49,200	4,910	174	4,250	150
20	19,400	84,400	2,800	23,200	87,200	4,360	154	3,060	108
50	35,200	153,000	7,000	42,200	160,000	3,200	125	1,960	70

[a] Total cost/no. of trays.
[b] Slope of total cost-no. of trays curve, dC/dN.

D. Air-Cooled Heat Exchangers

Tubes are 1-in. OD, 14 BWG, with 9-1/2-in. diameter aluminum fins per inch of tube length. Tubes are mounted in bundles on 2-1/8-in. triangular pitch. The heat-transfer area, A, is the inside tube area calculated from the equation: $q = (U)(A)(\Delta T)(F)$; $U = 90$ Btu/(hr)(sq ft)(F); ΔT = the log mean temperature difference; F = temperature difference correction factor, see Kern [6, p. 549].

	Purchased cost ($)	Installed cost ($)	Incremental installed cost ($/sq ft)
Carbon steel tubes	$78A^{0.75}$	$180A^{0.75}$	$135A^{-0.25}$
316 Stainless steel tubes	$144A^{0.75}$	$324A^{0.75}$	$244A^{-0.25}$

The installed cost includes tube bundle, fan and motor, and field erection.

E. Column Packings

1 inch

	Raschig rings ($/cu ft)		Pall rings ($/cu ft)
Porcelain	7	Polypropylene	26
Carbon steel	17		25
Stainless steel	70		100

For a 6 ft diameter column of carbon steel packed with 1-in. carbon steel Pall rings

Ht of packing (ft)	Installed cost of empty column ($)	Cost of packing @ $25/cu ft ($)	Total cost of installed column and packing ($)	Incremental installed cost of column and packing ($/ft)
20	47,800	14,150	61,950	2,870
40	84,400	28,300	112,700	2,500
100	153,000	70,750	223,750	1,850

F. Crystallizers

Continuous; forced circulation; carbon steel

$$\text{Purchased cost} = 11{,}600(T/D)^{0.55} \text{ \$}$$

$$\text{Installed cost} = 2.4(11{,}600)(T/D)^{0.55} \text{ \$}$$

T/D = capacity in tons/day of dry solid product

G. Filters

a. Plate and frame type; cast iron frames

$$\text{Purchased cost} = 389(A)^{0.58} \text{ \$}$$

$$\text{Installed cost} = 2.68(389)(A)^{0.58} \text{ \$}$$

A = filtration area (sq ft)

b. Continuous rotary vacuum filter; carbon steel; effective filtration area, $A_e = 0.25 A_T$; A_T = total peripheral area of drum (sq ft)

$$\text{Purchased cost} = 1{,}620(0.25 A_T)^{0.63} \text{ \$}$$

$$\text{Installed cost} = 2.4(1{,}620)(0.25 A_T)^{0.63} \text{ \$}$$

H. Dryers

Continuous rotary (kiln type); hot air heat; 300 °F; carbon steel; cost includes motor, drive, fan and motor, dust collector, and solids feed. Cost does not include air heater.

$$\text{Purchased cost} = 195(A)^{0.8} \text{ \$}$$

$$\text{Installed cost} = 3.0(195)(A)^{0.8} \text{ \$}$$

A = peripheral area of dryer (sq ft)

I. Blowers and Fans

Centrifugal turbo blower; carbon steel; electric motor drive; 1,000 to 10,000 cu ft/min at 70 °F and 1 atm total pressure; fan discharge pressure up to 10 psig. Cost includes fan and motor.

$$\text{Purchased cost} = 6.7(\text{cfm})^{0.68}$$

$$\text{Installed cost} = 2.78(6.8)(\text{cfm})^{0.68}$$

J. Centrifuges

a. Batch type; top suspended; electric motor drive; carbon steel

$$\text{Purchased cost} = 354D \ \$$$

$$\text{Installed cost} = 2.78(354)D \ \$$$

$$D = \text{diameter of basket (in.)}$$

The cost includes electric motor and drive. Note that the size exponent is 1. Cost is directly proportional to basket diameter.

b. Solid bowl; electric motor drive; stainless steel construction; helical conveyor discharge

$$\text{Purchased cost} = 2{,}250(Hp)^{0.73}$$

From p. 19-92 of Ref. [1] the following relationship between bowl diameter and motor horsepower is taken:

Bowl diameter (in.)	Motor horsepower (hp)
14	15
24	30
32	60
40	100
54	150

$$\text{Installed cost} = 2.4(2{,}250)(Hp)^{0.73}$$

K. Waste Heat Boilers

The cost of these units was developed from consideration of packaged steam boilers minus fuel pumps and burners; see Guthrie [5, p. 138].

$$\text{Purchased cost} = 4.2[(\text{lb steam/hr}) \times 10^{-3}]^{0.7}$$

$$\text{Installed cost} = 7.6[(\text{lb steam/hr}) \times 10^{-3}]^{0.7}$$

$$\text{Incremental installed cost} = 5.3[(\text{lb steam/hr}) \times 10^{-3}]^{-0.3}$$

All above costs are in thousands of dollars.

L. Ball Mills and Pulverizers

a. Ball mills, carbon steel, complete with ball charge, motor, and gears

$$\text{Purchased cost} = 4{,}000(\text{tons feed/hr})^{0.7} \text{ \$}$$

$$\text{Installed cost} = 7{,}350(\text{tons feed/hr})^{0.7} \text{ \$}$$

1 ton/hr to 10 tons/hr

b. Pulverizers: high-speed hammer mills; carbon steel with motor horsepower from 10 to 100; capacities up to 25 tons/hr; cost includes motor and drive

$$\text{Purchased cost} = 3{,}220(\text{tons/hr})^{0.39} \text{ \$}$$

$$\text{Installed cost} = 2.5(3{,}220)(\text{tons/hr})^{0.39} \text{ \$}$$

M. Storage Tanks

Carbon steel; 1 atm pressure; 300°F; length/diameter ratio = 1
1,000 gal - 10,000 gal

$$\text{Purchased cost} = 1{,}250(\text{gal} \times 10^{-3})^{0.6}$$

$$\text{Installed cost} = 5{,}000(\text{gal} \times 10^{-3})^{0.6}$$

N. Process Vessels

Process vessels may be used either with or without agitation. Furthermore, they may be made of various materials and of various strengths. The costs of several alternatives are given below.

Vessels without Agitation

Material	1 atm		500 psi	
	Pchsd cost ($)	Instd cost ($)	Pchsd cost ($)	Instd cost ($)
Carbon steel	$15(\text{gal})^{0.82}$	$65(\text{gal})^{0.82}$	$22(\text{gal})^{0.82}$	$93(\text{gal})^{0.8}$
Stainless steel	$57(\text{gal})^{0.82}$	$100(\text{gal})^{0.82}$	$82(\text{gal})^{0.82}$	$123(\text{gal})^{0.8}$
Glass-lined steel	$660(\text{gal})^{0.5}$	$1400(\text{gal})^{0.5}$	—	—

[Process Vessels, cont'd]

If agitation is considered to be by a flat-bladed turbine driven by an electrical motor, the cost of the agitator assembly (turbine, shaft, drive, and motor) attached to the vessel will be

a. turbine and shaft of carbon steel $1,070(Hp)^{0.56}$ \$

b. turbine and shaft of stainless steel $2,000(Hp)^{0.56}$ \$

where Hp = horsepower input to motor

A figure of 5 hp/1,000 gal of vessel capacity is a reasonable number for preliminary work. For a discussion of power consumption in agitated vessels see Jordan [9, Chapter 3]. To obtain either the purchased or the installed cost of a process vessel with agitation add either (a) or (b) above to the cost of the unagitated vessel.

The authors are well aware that glass-lined steel vessels do not use turbine agitation. However, they do use agitation, and a reasonable approximation to the cost may be obtained by assuming turbine agitation.

O. Water Cooling Towers - 15°F Cooling Range

$$\text{Purchased cost} = 476(\text{gal water/min})^{0.6} \ \$$$

$$\text{Installed cost} = 834(\text{gal water/min})^{0.6} \ \$$$

The installed cost includes field installation plus pumps and motors.

P. Mechanical Refrigeration Units

For temperatures near +40°F

$$\text{Purchased cost} = 1,750(\text{tons refrigeration})^{0.55}$$

$$\text{Installed cost} = 1.42(1,750)(\text{tons refrigeration})^{0.55}$$

For temperatures near -20°F

$$\text{Purchased cost} = 4,550(\text{tons refrigeration})^{0.55}$$

$$\text{Installed cost} = 1.42(4,550)(\text{tons refrigeration})^{0.55}$$

Notes on Cost Estimation

Q. Process Gas Compressors

Centrifugal compressor with electric motor drive; motor horsepower 50 to 500. For a quick approximate calculation of compressor horsepower assume isothermal compression; use

$$Hp = 0.0044 \, P_1 Q_1 \, \ln[P_2/P_1]$$

where P_1 = inlet pressure (psi)

Q_1 = inlet gas flow rate (cfm)

P_2 = outlet gas pressure (psi)

Purchased cost = $645(Hp)^{0.8}$ \$

Installed cost = $3.1(645)(Hp)^{0.8}$ \$

The installed cost includes compressor, motor, and drive.

This list is probably no more useful or accurate than many others, but it is fairly compact and may be used for student problems and for rapid preliminary estimates.

In Table 5-1 (A-Q) each item of equipment has been assigned two costs:

1. The purchased cost, which is the cost quoted by the equipment manufacturer. It does not include any of the charges which will be incurred before the equipment is installed in the plant, ready for operation.

2. The installed cost, which is the purchased cost of the equipment plus all the charges which will be incurred while the equipment is being connected into the plant system and made ready for operation.

The costs have been tabulated in this way for two reasons:

1. For economic-balance calculations, as shown in Chapter 8, it is necessary to use the incremental installed cost of that particular piece of equipment. The purchased cost by itself does not reflect the true cost of placing the equipment into a functioning chemical plant.

2. There are several rapid approximation procedures available for estimating the final plant cost from the total purchased

equipment cost. These procedures are quick and good enough for preliminary estimation purposes. They are widely used. In order to use them the worker must know the purchased equipment cost. Furthermore, it is convenient to have an estimate of the purchased equipment cost so that the costs quoted by the suppliers may be checked.

5.05 ESTIMATING THE FINAL PLANT COST

In the first edition of this book Happel presented a method for estimating the final erected plant cost from the total of the purchased equipment costs. Although this method is now fairly old, it still works quite well, giving acceptable answers. A detailed description, together with an illustrative example, are given below.

The method is based on the list of process equipment to be used in that process. It must be apparent that omission of any necessary item of equipment will cause the estimate to be too low. All items of process equipment must be on the list. The drawing of a detailed engineering flow sheet and the consulting of an equipment check list such as Table 26-6 of the Chemical Engineers' Handbook [1] are strongly recommended as helpful aids in the cost estimation procedure.

Happel's method follows the outline given below and the form shown in Table 5-2.

1. A list is drawn up of all the pieces of process equipment. Table 5-2 lists only those pieces of equipment that are commonly encountered in petrochemical plants. For other types of equipment the engineer may use the installed costs, as shown in the illustrative example, or he may supply his own numbers for Labor and items H through O. The installed costs for special equipment, taken from Table 5-1, may be labeled Q and added to P in Table 5-2.

2. Opposite each piece of equipment in the list its purchased cost is written under Material, and under Labor some fraction of the purchased cost, as shown. As has been mentioned before, in recent years there has been a faster increase in labor costs than in material costs. If there is a range of possible labor costs it would be wise to use the higher cost.

3. The sum of all the equipment purchase costs is designated "Sum of Key Accounts" and is lettered G (or whatever).

Notes on Cost Estimation

4. The cost of the material required for constructing the plant and installing the equipment is then estimated as being some fraction of G, as shown. This is entered under the Material column.

5. The cost of the labor associated with each installation is then calculated as some fraction of each one, and listed under Labor, as shown.

6. The Material and Labor columns are then added to give the sum of Materials and Labor. This is labeled P (or whatever), as shown.

7. If special mechanical equipment, for instance a ball mill, is to be used, its installed cost is estimated and labeled Q. Then P and Q are added to give R, the installed cost of all the process equipment.

8. The value of R is then multiplied by a factor of about 1.6, as shown, to give the final completed plant cost.

The final plant cost will often range between 3 and 5 times the sum of the purchased equipment costs. This fact is sometimes not realized by those inexperienced in cost estimation.

It must be understood that this method is a quick approximate method and should not be used in situations where accuracy is needed. However, this method does give surprisingly good results, and it has the advantage that all of the items contributing to the final cost are separately entered into a compilation. This "separate entry" feature is very important because future work on the estimate will undoubtedly involve criticism of and changes in one or all of the items. To be able to study each item separately is a great advantage.

There are several points worthy of careful consideration:

1. The list of key accounts should contain every item of process equipment necessary. Only a few of the many possible are shown in Table 5-2. Even though the exact size, shape, and material of construction of the device are not known, it should be in the list. When it is listed it will carry a cost, perhaps not the correct cost, but certainly more correct than zero. Cost estimates nearly always tend to be too low. Careful attention to the equipment list will help to prevent low estimates.

The omission of pieces of equipment from the list of key accounts is an error frequently made by inexperienced estimators.

2. The percentages for labor shown in Table 5-2 are based on carbon steel as a material for the construction of the equipment.

If the equipment is made of stainless steel or titanium or ceramic the labor cost will be a smaller fraction of the purchase price. For simple methods of correcting for the cost of material, see Guthrie [5].

3. The piping cost estimate in Table 5-2 is 40 to 50% of the sum of the key accounts. In some chemical plants, especially those involving the handling of solids, this percentage will be too high. On the other hand, if a substantial part of the piping is made of glass, ceramic, plastic, or stainless steel the percentage may be too low. Guthrie [5] discusses piping costs in some detail. A useful approximation is the following one:

Type of plant	% of key accounts as piping
Solids processing	10
Mixed solids and fluids processing	20
Fluids processing	50

The well-known and widely used factors of Lang [10] provide a quick way of finding the final plant cost from the sum of the purchased equipment costs, item G in Table 5-2. There are three of these factors, one for each of the types of plants mentioned above.

To find the final plant cost multiply item G in the equipment list by one of the factors listed below:

For a solids processing plant	3.1
For a mixed solids and fluids processing plant	3.6
For a fluids processing plant	4.7

The procedure of Lang is somewhat faster and somewhat easier to use than that due to Happel. However, it is not recommended that the Lang factors be used because the multiplying factors can obscure details in the cost estimate which should be clearly set apart so that they may be studied individually.

4. The fraction of the key account allotted to foundations, buildings, and structures is based on an average value. For construction on poor soil, or in a severe climate, or for heavy equipment these values could be doubled.

Notes on Cost Estimation 235

Table 5-2 Equipment Cost Schedule—Process Unit

Item	Material	Labor
Vessels	A	10% of A
Towers, field fabricated	B	30 to 35% of B
Towers, prefabricated	C	10 to 15% of C
Exchangers	D	10% of D
Pumps, compressors and other machinery	E	10% of E
Instruments	F	10 to 15% of F
Key accounts (Sum of A to F)	G	
Insulation	H = 5 to 10% of G	150% of H
Piping	I = 40 to 50% of G	100% of I
Foundations	J = 3 to 5% of G	150% of J
Buildings	K = 4% of G	70% of K
Structures	L = 4% of G	20% of L
Fireproofing	M = 1/2 to 1% of G	500 to 800% of M
Electrical	N = 3 to 6% of G	150% of N
Painting and cleanup	O = 1/2 to 1% of G	500 to 800% of O
Sum of material and labor	P	
Installed costs of special equipment	Q	
Sum of P and Q	R	
Overheads		30% of R
Total erected cost		130% of R
Engineering fee (10% of erected cost)		13% of R
Contingency fee (10% of erected cost)		13% of R
Total investment		156% of R

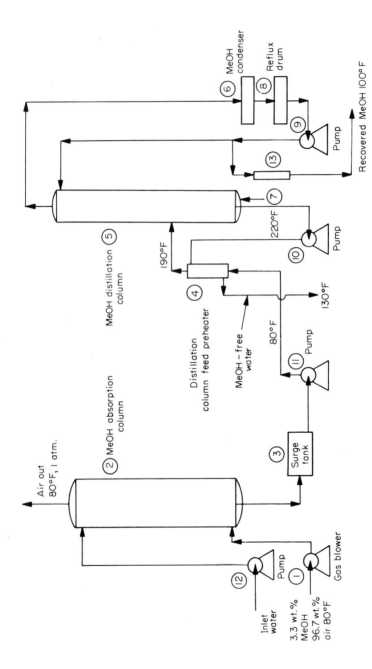

Fig. 5.4. Flow sheet for methanol recovery plant.

Notes on Cost Estimation 237

5. The items labeled "engineering fees" and "contingency" may vary depending on the competitiveness of business and on the accuracy of the estimate. The 10% figures given should not be regarded as fixed.

"Contingency" is an allowance to cover the costs of items not estimated directly but known to exist in a process plant.

The "engineering fee" may vary from 0% in a desperately competitive business situation to 25% in a situation where the engineering contractor can command premium prices for his services.

Special items of this type must always be carefully considered, but the successful solution of them may elude even the expert. The student should note that the problem exists and discuss the matter with the instructor.

A close examination of the work of Guthrie [5] and that of Happel, discussed above, shows that they are basically similar. Guthrie starts with a purchased equipment cost, applies various factors to that cost to estimate the material and labor required to insert that particular piece of equipment into a functioning plant, and then adds all the factors together to get a material and labor factor that will convert the purchased equipment cost into the installed equipment cost. These costs may then be added to get a total installed equipment cost similar to item R of Table 5-2. Therefore, it should be possible to add the installed costs from Table 5-1 to obtain the same number as item R of Table 5-2. From item R, or the sum of all the installed equipment costs, a simple multiplication by about 1.6 should give the final plant cost.

Thus there appear to be at least three methods for estimating the final plant cost from the purchased equipment costs: (1) Happel's method, based on Table 5-2, (2) the Lang factors, and (3) the sum of all the installed equipment costs, based on Guthrie's work and values such as those in Table 5-1. The illustrative example below shows these various methods.

Example 5.1

Estimate the installed cost of a solvent recovery plant.

Figure 5.4 is a flow sheet for a simple solvent recovery plant. Methanol is recovered from an air-methanol mixture by absorption in water in a packed column. The effluent water-methanol solution is pumped from a surge tank through a heat exchanger into a methanol-water distillation column. Pure methanol is distilled overhead and essentially methanol-free water is removed from the bottom of the distillation column. For the sake of ease and simplicity the instrumentation is not shown, although a practical engineering flow sheet would show it.

The material balance and equipment design calculations are not shown. The material balances are elementary and the design calculations follow the approximation methods given in Appendix B.

Equipment List for Methanol Recovery Plant

	Purchased cost ($)	Installed cost ($)
1. Gas inlet blower; carbon steel, 3,000 cfm @ 80°F & 1 atm, 10 hp	18,400	51,200
2. Gas absorption tower; carbon steel, 28 ft tall, 4.6 ft diam, 20 ft packed ht, 1 atm, 80°F	8,050	35,000
Tower packing; 1-in. carbon steel Pall rings	8,350	8,350
3. Absorption tower bottoms surge tank; carbon steel, 2,000 gal	1,900	7,570
4. Distillation column feed preheater; carbon steel, shell and tube heat exchanger, 458 sq ft	4,650	15,500
5. Methanol-water distillation tower; 3.4-ft diam, 17 actual sieve plates, 24-in. tray spacing, 42 ft tall, carbon steel	9,700	39,100
	(tower plus trays)	
6. Methanol condenser; carbon steel shell and tube heat exchanger, water-cooled, 563 sq ft	5,350	17,650
7. Distillation column reboiler; no reboiler—use open steam	0	0
8. Reflux accumulation drum; 500-gal carbon steel vessel	2,420	10,460

Notes on Cost Estimation

9. Pumps; carbon
10. steel; centrifugal
11. 1 hp; all pumps   2,400   9,080
12. the same   4 @ 600   4 @ 2,270
13. Distillation column product
    cooler (small)   120   500

    Total purchased cost $61,340

    Total installed cost (includes instrumentation) $194,410

The costs above were taken from Table 5-1.

Happel's Method

Item			Material ($)		Labor ($)
Vessels: 3 + 8	A		4,320	10% of A	432
Towers, prefabricated: 2 + 5	B		26,100	15% of B	3,910
Exchangers: 4 + 6 + 13	C		10,120	10% of C	1,012
Pumps, compressors and other machinery: 1 + 9 + 10 + 11 + 12	D		20,800	10% of D	2,080
Instruments: 15% of A + B + C + D	E		9,180	15% of E	1,372
Key accounts: Sum of A through E	G		70,520		
Insulation	H	10% of G	7,052	150% of H	10,540
Piping	I	50% of G	35,260	100% of I	35,260
Foundations	J	5% of G	3,526	150% of J	5,290
Buildings	K	4% of G	2,820	70% of K	1,980
Structures	L	4% of G	2,820	20% of L	564
Fireproofing	M	1% of G	705	800% of M	5,640
Electrical	N	6% of G	4,230	150% of N	6,340
Painting and cleaning	O	1% of G	705	800% of O	5,640

[Happel's Method, cont'd]

Sum of material and labor	P	$198,892
Special equipment	Q	0
Total installed cost of equipment	R	198,892
Overheads 30% of R		59,700
Total erected cost		258,592
Engineering fee 10% of erected cost		25,859
Contingency 10% of erected cost		25,859
Total investment		$310,310

From the equipment list: the sum of all installed costs = $194,410

1.6(194,410) = $310,500

From these two calculations it would appear that it is acceptably precise to use either Happel's method or the sum of all the installed costs, multiplied by 1.6, to obtain the final erected plant cost.

The Lang factor method gives

4.7 × 61,340 (the total purchased equipment cost) = $288,000

This is close enough to the other two answers to be useful for preliminary purposes.

From all this it may be concluded that the sum of the process equipment costs, either purchased or installed, may be used as a basis for a rapid preliminary estimate of the final plant erected cost. Happel's method has the advantage that separate calculations are made for each item, the calculations are separately listed, and the numbers may be changed as the estimators see fit. Unfortunately, the number of possible items of equipment is so large and the material and labor factors so varied, that a list similar to Table 5-2 would be quite complex and beyond the possibilities of this book. For this reason a list of installed costs, such as Table 5-1, is more useful, although this list, too, is short, undoubtedly in some error, and subject to change. Guthrie's paper [5] should be consulted for details.

The method of Lang, while apparently oversimplified, does give answers which are acceptably close to the answers given by the other two methods.

Notes on Cost Estimation

In process engineering work two terms are often used:

1. Battery limits. This term describes the entire amount of process equipment which is actively involved in producing the product, but it does not include units such as storage tanks, utility producers (such as steam boilers, electrical switchgear, and water coolers), or personnel service units such as parking lots, cafeterias, and medical centers.

The term "battery limits" is derived from the early days of petroleum refineries when the distillation columns (heated by direct firing and all distillation done batchwise) were constructed in parallel connected "batteries" of individual units. Naturally enough the term "battery limits" came to include only those units directly involved with the main processing stream.

2. Grass roots. This term means essentially what it implies-- a chemical plant built from the ground up (from a field containing nothing but grass), and needing all facilities: process, utility, roads, site development, rail lines, and fences.

Table 5-2, and the cost estimation methods discussed previously, are concerned only with "battery limits" plants. For almost all plants, provision for storage of raw materials, intermediate products, and final products must be made separately from the provisions for process equipment.

The total investment in a complete new "grass roots" petroleum refinery or petrochemical plant will include additional items amounting to 150-200% of the investment in process equipment. For chemical plants these additional items may be as low as 70% and for new units added to existing plants only 25-50% additional investment will usually be required. The distribution of investment in refinery-type plants is given approximately in Table 5-3.

Table 5-3 Investment Distribution—Complete Plants

	% of total investment, range	% of total investment, average	% of process equipment, average
Process units	30-45	35	100
Utilities	10-25	20	57
Offsite	30-45	35	100
Buildings	5-20	10	29
		100	286

Offsite facilities in Table 5-3 include tankage, roads, fencing the site, and product-shipping facilities.

From the above two simple rules may be deduced:

1. For a "battery limits" chemical plant the final installed cost of the plant will be

$$1.6R$$

or

$1.6 \times$ (the sum of the installed costs)

2. For a "grass roots" refinery-type plant—and many "chemical" plants are being built in this manner today—the final installed cost will be

$$R + 1.8R = 2.8R$$

where R = sum of material and labor cost in Table 5-2. It is also equal to the sum of the installed costs.

5.06 SOURCES OF COST DATA

Whenever a cost estimate is made the results will be questioned. Engineering supervisors and financial officers will question the engineer about the source and validity of his cost data. This is a perfectly correct action on their part. They cannot accept the conclusions of the engineer without satisfying themselves that the data, both technical and economic, on which the conclusions are based are as good as can be expected. Consequently, the engineer must have a method by which he can obtain current, and correct, cost data. He must be able to justify these data.

If the engineer is a member of a professional design or cost-estimating group, the group will (or certainly should) have a file of cost data which was originally compiled, and now kept up to date, as a part of the routine duties of the members of the group. These data should be useful. If the engineer is not a member of such a group he must rely on cost data published in the technical press. These data must be regarded with suspicion and used with discretion.

Although it is frequently stated that it is desirable for the engineer to keep his own file of cost data, this activity is not recom-

Notes on Cost Estimation

mended. It is too difficult to do adequately well, too time-consuming, and too liable to serious error.

Serious cost estimation is serious business and should only be done by professional cost estimators. Rapid preliminary cost estimation should be regarded as not accurate, having a precision of -35% to $+25\%$.

There are, in general, two sources of cost data for rapid preliminary cost estimates. For final definitive cost estimates the quotations from the equipment vendors will be used.

1. The company records of monies actually paid for various pieces of equipment and for various kinds of work. Practically all engineering contracting firms have cost-estimating groups that continally search current invoices for up-to-date prices. These groups keep elaborate records and should be in a position to make an accurate cost estimate fairly quickly.

Chemical manufacturing companies sometimes have cost-estimating groups and sometimes they do not. For the engineer working in development and making a preliminary cost estimate, it may be difficult or impossible to obtain accurate cost data from his company's records. He must use other compilations of cost data.

2. The literature of chemical engineering contains a great deal of published information on the costs of chemical equipment. Probably the most comprehensive and most useful publication is that of Guthrie [5]. There are several books that may be consulted. The most recent and most comprehensive of these are listed here:

Table 5-1 of this book, although this compilation is by no means complete

Peters and Timmerhaus, Plant Design and Economics for Chemical Engineers [7]

Bauman, Fundamentals of Cost Engineering in the Chemical Industry [6]

Several of the chemical engineering journals publish cost data fairly frequently. These are Chemical Engineering, Hydrocarbon Processing and Petroleum Refiner, Oil and Gas Journal, and Industrial and Engineering Chemistry (an annual review).

In addition to these, Zimmerman and Lavine [11] publish a journal called Cost Engineering that, over a period of years, has published cost information on many different kinds of process equipment.

In 1960 and 1970 Chemical Engineering published two books made up of cost-engineering articles from Chemical Engineering magazine. These books [2] and [12] are valuable compilations of data and methods in the cost estimation field.

All of these references are useful, but, as discussed earlier, they must be used with discretion. The student cannot be expected to be successfully critical of the published data because of his lack of experience. The professional engineer should work back and forth among the publications, and utilize all of his knowledge and experience before selecting a cost.

5.07 THE MANUFACTURING COST OF THE PRODUCT

The manufactured cost of a product is found by adding together a rather large number of small charges which account for the cost of all the materials, labor, utilities, and other matters that enter into the manufacturing of the final product. Some of the more important items that must be accounted for when calculating the factory cost of the product are raw materials, labor and supervision, utilities, maintenance and repairs, depreciation, packaging.

In order that all these charges be properly taken into account it has been found best to prepare a detailed check list with all the items named and with spaces provided in the list for costs to be written next to the item's name. Figure 5.5 shows such a list, taken from Jordan [9].

By utilizing such a diagram several important goals are achieved.

1. All of the items contributing to the final cost are evaluated, and there exists some reason for that cost. Essentially nothing of importance is omitted. The simple act of filling in all the spaces almost ensures that a reasonably correct cost will be obtained.

2. The individual cost items are set apart by themselves where they can be easily identified and questioned or changed without changing any other item. Changes will occur, and it is convenient to be able to do so with a minimum of trouble and recalculation.

The diagram is arranged in such a way that the cost for each of three separate rates of production may be calculated. Also, the cost of each item is given in cents per pound of finished product and

Manufacturing Cost Estimate

Estimator _____ Date _____
Product _____   Production rate (lb/yr) _____
                        Plant capital invst ($) _____ _____ _____
Plant location _____ Producing days/yr _____
                        Calculation of raw material costs
                        Basis: _____ lb/yr

Raw material	Unit of material	Price per unit	Usage (lb/lb of prod)	Total units (lb/yr)	Total cost ($/yr)	Cost per lb product (¢/lb)
	tons	$				
	pounds	¢				
	M cu ft	¢				
			Total raw material cost			

Production rate (lb/yr)			
Labor requirements			
Operators/shift			
Shifts/week			
Hours per week/man			
Supervisor hours/week			

	$/yr	¢/lb	$/yr	¢/lb	$/yr	¢/lb
Total raw material cost 1						
Other factory costs						
Direct labor hr @ $ /hr						
Supervision @ % of DL						
Total labor costs 2						
Utilities costs						
Steam @ ¢/Mlb						
Electricity @ ¢/kw-hr						
Process water @ ¢/M gal						
Cooling water @ ¢/M gal						
Air @ ¢/M cu ft						
Gas @ ¢/M cu ft						
Refrigeration @ $ /ST						
Total utilities cost 3						
Indirect costs						
Payroll overhead @ 35% DL						
Stores and supplies @ 20% DL						
Laboratory control @ 25% DL						
Security @ 20% DL						
Yard @ 10% DL						
Process improvement @ 15% DL						
Total indirect costs 4						
Maintenance @ % of PI 5						
Total processing cost						
2 + 3 + 4 + 5						
Total raw material and						
processing cost						
1 + 2 + 3 + 4 + 5						
Plant general expense						
Property taxes, Insurance						
@ 1% of PI						
Depreciation						
Other Plant expense						
Total plant general expense 6						
Total factory cost						
1 + 2 + 3 + 4 + 5 + 6						
Less credit for by-products						
Net factory cost in bulk						
Packaging cost						
Net factory cost						

Fig. 5.5. Manufacturing cost estimate.

in dollars per year. It is advisable to calculate the cost in these ways because financial officers will always ask questions about the effect of changes in the production level on the product cost. It is very helpful to know just how many dollars per year are involved. In some of the modern "chemical refinery" type plants producing styrene or ethylene the value of the product may amount to $75 to $100 million a year.

In Chapter 1 the economic benefit gained from the construction and operation of large plants, as opposed to small plants, was discussed in some detail. Very large plants, with productive capacities greater than 100,000,000 lb/yr, enjoy a considerable economic advantage over smaller plants. In recent years many chemical plants of very high productive capacity have been constructed.

Raw Materials and Products

Raw materials and products alike may be divided into two categories: (1) Those purchased or sold in the open market, and (2) those available from sources within the company or sold to markets controlled by the producing company. These markets are called "captive" markets.

For materials purchased or sold on the open market, listings of current prices may be found in technical magazines such as Chemical and Engineering News and The Oil, Paint and Drug Reporter; these are published at frequent intervals. For projects that are expected to materialize in a relatively short time, actual quotations and bids may be obtained from suppliers and purchasers. If the product is new, or is expected to displace a substantial proportion of present production, a detailed market survey and analysis will be desirable. In cases where competition exists, a rule of thumb is to discount published prices by 20%.

Raw materials and products are frequently purchased and sold on the basis of long-term contracts which have been negotiated by the companies' business officials. These contracts, especially for raw materials such as natural gas and sulfur, provide for the low-cost purchase of very large quantities of the raw material over a period of 5 to 10 years. Similarly, a producer may contract to supply a large amount of product at a certain price (below the open-market price) to special customers. It is quite advantageous for companies to have such contracts, and many do. The estimator should consult company officials about the existence of such contracts before attempting a detailed cost estimate.

When considering the use of captive raw materials, or of constructing facilities for the production of raw materials (such as a phenol plant to supply a phenol-formaldehyde resin plant), chemical companies face difficult and confusing problems in setting a price for the raw material. There are several different methods that can be used:

1. The open-market sales value of the material may be assumed if the material could be sold as a finished product (as for instance, phenol or sulfuric acid), and if the market could absorb the volume indicated at the market price. This procedure has the difficulty that there would then be no economic benefit from the captive supply since the material could be purchased in the open market as cheaply as from the captive unit. The advantage of this method is that the economic performance of the captive material producer is easily evaluated. From the manufacturing viewpoint it is desirable to have a captive source of raw material. In some instances this may be the only source.

2. The manufacturing cost of the raw material could be used. This could be done whether the material is to be produced in currently idle facilities or in a plant especially constructed for the purpose. This procedure has the difficulty that there would be no profitable sales, and therefore no return on investment.

3. An incremental cost; if the raw material could be produced in a unit currently operating below capacity.

For plants of even moderate complexity this problem quickly becomes involved, and an estimate of the proper price can become somewhat arbitrary. Depending on the method of allocation selected, there will be different profitabilities for a given project. The estimator is usually not in a position to decide upon the proper price for a captive raw material. Company accountants should be consulted.

For many chemicals the cost of the raw materials is by far the largest part of the manufacturing cost. The estimator should pay careful attention to this item.

Labor and Supervision

As a general rule chemical plants do not employ much direct operating labor. It is seldom that this cost exceeds 15% of the cost of the product. The nature of the work in the chemical industry plus a heavy use of instrumentation and materials-handling equipment has made it unnecessary for the industry to use large quantities of labor. The modern trend to automation, extensive instrumentation, and centralized computer control has reduced still further the need for operating labor. In many instances the necessity for safety coverage, that is two operators working together so that each provides care and protection for the other in case of accident or illness, determines the number of operators required. The nature of the task may not require so many men.

It is difficult to predict the amount of operating labor required unless the estimator has had extensive plant operating experience or is assisted by company employment records. Some companies have such records and have arranged them into "manning" tables for practically every task imaginable. For the estimator working without such assistance the best procedure is to study the engineering flow sheet, and employing his own and his colleagues' experience, estimate the number of operators needed.

Although the cost of the operating labor by itself is not a major factor in the total manufacturing cost, there are six other items in the cost list (see below) which are calculated from the operating labor cost. Thus, it becomes a fairly important matter to make a good estimate of the number of operators required and of the wages paid. Experienced engineers and company accountants are the best source of such information. Data on current labor rates may be obtained from the U.S. Bureau of Labor Statistics. As a first approximation the hourly wage rate for chemical plant operators may be taken as $4.00/hr. A more detailed analysis must consider premiums paid for shift work and holiday overtime pay.

The cost for supervision may be taken as 15% of the operating labor cost.

For approximation purposes, a correlation devised by Wessel [13] is useful in predicting the direct operating labor requirement of chemical process plants. Table 5-4 is based on this method.

Table 5-4 Direct Operating Labor (100 tons/day capacity)

Type	Daily labor (man hours/step)	Size extrapolation	
		Power	Range (tons/day)
Large equipment, high instrumentation or fluid processing	33	0.25	2 - 1500
Average	48	0.25	2 - 1500
Duplication of small units; batch equipment	77	0.25	2 - 1500

Notes on Cost Estimation

In order to apply Table 5-4 it is necessary to make a count of the primary operational steps, such as filtration, electrolysis, distillation, and drying. Then, depending on the type of plant, the daily labor requirement per step is determined. Multiplication by the number of steps gives the total direct labor requirement per day for the base 100 tons/day size. Extrapolation to larger or smaller sizes follows the 0.25 power.

Labor-Related Costs

There are several items in the list of costs, primarily under Indirect Costs, which are calculated as percentages of the direct labor cost. These are (1) Payroll Overhead, (2) Stores and Supplies, (3) Laboratory Control, (4) Protection or Security, (5) Yard, and (6) Process Improvement.

1. Payroll Overhead. This item includes such costs as state unemployment and disability payments, vacations, sick leaves, pensions, group insurance, accident and liability insurance, and profit-sharing programs. These are the so-called "fringe benefits" attached to the wages of employment. The trend over a period of years has been for these benefits to increase. A useful figure to employ is 35% of the direct labor cost.

2. Stores and Supplies. This is a term describing the cost of maintaining storeroom(s) of tools and supplies of all kinds: large wrenches, work gloves, laboratory glassware, and stationery materials plus storerooms and warehouses of spare parts such as pumps, motors, valves, and piping. A useful figure is 20% of direct labor.

3. Laboratory Control. Many chemical plants must have chemical laboratories for analysis, testing, and process improvement work. The capital investment in such laboratories may be added to the capital investment in the plant, but there will be a charge for manpower and material to operate such laboratories. A useful figure would be 25% of the direct labor charge.

4. Protection or Security. Every process plant must have guards for protection against trespass, theft, and vandalism and for traffic control. These guards work every hour of every day and usually at least two are on duty always. A useful figure is 20% of the direct labor charge.

5. Yard Costs. The plant will usually have a labor force working outside the process area. Their tasks will include tank and freight car unloading, yard, fence, and parking lot maintenance, and assisting

in large-scale plant maintenance work. This cost may be taken as 10% of the direct labor cost.

6. Process Improvement. Many large plants have process improvement groups whose task is to work on the improvement of the plant procedures. For brand new plants, which may have start-up troubles, such process improvement charges may be heavy. However, they are usually considered as part of the start-up expense, which may be a separate appropriation, and are not added to the unit cost of the product. For a plant which is running smoothly, enough process improvement charges will be added to the product cost. They can be assumed to be equal to 15% of the direct labor charge.

The total of the labor-related charges is 125% of the direct labor charge.

Utilities

Under the heading of utilities are included expenses for the consumption of fuel, steam, electricity, water, air, refrigeration, and other facilities supplied to the process unit from outside

Table 5-5   Utilities Price Ranges[a] January 1975

Item	Price
Fuel (oil)	$1.30 - 1.80 / MM Btu
Steam (500 psia)	$1.00 - 1.50 / M lb
Cooling water (tower)	$0.06 - 0.10 / M gal
Process water	$0.25 - 0.45 / M gal
Electrical power	$0.02 - 0.04 / kw-hr
Compressed air	$0.20 - 0.50 / M cu ft[b]
Refrigeration	
34°F	$0.75 - 1.40 / ton day[c]
-20°F	$2.00 - 3.00 / ton day

[a] The rapid increase in present day energy costs makes it almost impossible to put a precise value on the cost of any utility.

[b] Cu ft measured at 1 atm and 60°F.

[c] 1 ton day = 288,000 Btu removed.

sources. The equipment necessary to supply utilities to the battery limits of a unit which is part of a larger plant is usually amortized over a longer period of time than the processing equipment. In addition, there is an allowance for profit on the capital involved. Thus, for preliminary estimation purposes, it may be assumed that utilities are priced as if they were purchased from an outside organization. A range of typical prices is given in Table 5-5. The lower range of these items is applicable in regions where fuel (natural gas, coal) is cheap, or, in the case of electrical power, where hydroelectric power is available. At the present time, and into the foreseeable future, water is a sensitive item. The estimator should make sure that ample supplies of water are available to the project. They may not be.

Property Taxes and Plant Insurance

Every chemical plant must pay taxes to the local municipality, and it must have one or more insurance policies as protection against the loss of property and income from fire and other destructive events. These charges may be assumed as 1% of the total plant investment per year.

Maintenance

Maintenance is an item of great importance to every chemical plant. A well-organized program of preventive maintenance and repair must be conducted. The details of repairs or maintenance labor and materials are difficult to predict accurately. An average value for the items of labor (4%) plus materials (2%) equals 6% of the plant capital investment per year for repairs and maintenance. For plants involving low temperatures, few moving parts, and low down time, figures as low as 2 to 3% per year are common. Overall costs as high as 10% per year are encountered in cases of high temperatures and pressures, corrosive materials, and processes involving frequent replacement of parts or changing of catalyst. Glauz [14] has summarized some of the methods employed for obtaining more accurate estimates.

The maintenance required by the plant will change with time. Usually maintenance will be higher during initial operation, while minor changes are being made in equipment, and operators are unfamiliar with the process. Again, toward the end of the useful life of the equipment, maintenance costs will rise, owing to replacements caused by wear and corrosion. For the purpose of making economic studies on existing units (as compared with new-plant studies) it is probably more accurate to consider maintenance proportional to plant throughput or power consumption [15]. Also, maintenance costs have been shown to be directly related to plant age [2, pp. 329-334].

The simple fixed percentage rule is useful enough for preliminary estimates, but it should be remembered that the percentage will vary with plant operating conditions, plant design, and plant age.

Depreciation

The depreciation charge has been discussed and manipulated in various ways in earlier chapters. In the venture worth analysis the depreciation charge is set out by itself, separated from other cash flow items. For the purpose of arriving at a factory cost for preliminary economic evaluation it is customary to add the depreciation charge to all the other cost items; see Fig. 2.2, Chapter 2, for a discussion of this method of accounting for the depreciation charge. For present purposes the simple straight-line, 10-year depreciation charge is acceptable.

By-Product Credit

All chemical plants produce by-products. Frequently these can be sold or processed and a money credit made. Also frequently the by-products are poisonous or objectionable and money must be spent to render them harmless. The conservative approach to by-product credit is to assume no credit or debit, but to calculate the quantity involved, and to call attention to the problem.

Packaging Cost

Many chemical plant products are shipped to the consumer in boxes, drums, bottles, or barrels. All of these packages cost money, and this cost must be included in the cost of the product. It is not possible to give a single useful figure for the packaging cost. Two widely used kinds of packages are 55-gal steel drums, which cost about 2¢ per pound of product, and railroad tank cars, which cost about 0.5¢ per pound of product.

Other Plant Expense

"Other Plant Expense" is an item which should cover fairly unusual costs. Royalty payments are an example.

Manufacturing Cost

All of the above items are added together to obtain a value for the manufacturing cost. It is important to recognize that the sum will include the depreciation charge. In the normal cost estimation practice in industry the depreciation charge is included, but in the venture worth analysis it is not. This distinction should always be kept in mind.

The form for calculating the manufacturing cost contains a figure for the number of producing days a year. There are 8,760 hours in a 365-day year. Some large chemical plants and many petroleum refineries will run more than 95% of the time, or at least 8,320 hours a year. However, many chemical plants will be shut down for at least 1 month each year for maintenance, cleaning, catalyst change, and vacations. In the chemical industry (say phenol manufacture) it is common practice to base cost calculations on an operating period of 8,000 hours. Many other smaller operations such as insecticide or plastic manufacture will operate for much shorter periods. The estimator should be quite sure that his time basis is a realistic one.

Working Capital, I_w

"Working capital" is a name used to designate the amount of money required by the company to finance the daily operations of the project. As the plant operates, raw materials and supplies must be purchased, inventories of supplies and products must be stored, purchasers accounts (accounts receivable) must be carried for several weeks, taxes must be paid when due, and some cash must be kept on hand (in a checking account) to pay small operating expenses. Working capital is considered to be liquid, not fixed, it may vary with the sales rate, and is recovered at the end of the project. Working capital is not depreciated over the life of the project and it is not an expense which is tax deductible.

A useful formula for computing the necessary amount of working capital is the following:

I_w = 30 days raw material cost + 30 days out-of-pocket operating expense + 30 days finished product inventory value + 30 days accounts receivables

Other, simpler formulas are the following two:

I_w = 25% of total annual expense

I_w = 10% of new capital investment

Gross Return, R_k

The gross return is equal to the difference between the sales price, s, and the net factory cost, c (not including the depreciation charge) multiplied by the total amount of material sold.

$$R_k = Q_k (s - c)$$

SUMMARY

Techniques have been outlined for making estimates of items entering into an economic evaluation up to the determination of gross income or profit. Though there is some variation in the methods used, depending on the data available and the accuracy of the estimate required, in general the figures obtained from one study to another should be comparable. Certainly they offer a sound basis for discussion and further study. Up to this point in the economic consideration, few assumptions that are matters for management decision enter the picture. The payout time (capital investment/gross income) is therefore a simple index which can be employed as a basis for further more elaborate calculations taking into account such items as depreciation and taxes. Maximum acceptable payout time is thus a convenient index to employ in making economic-balance calculations in the design of process equipment, and it is employed for this purpose in later chapters.

REFERENCES

1. Chemical Engineers' Handbook, 4th ed., McGraw-Hill Book Co., New York, 1963.
2. C.H. Chilton, ed., Cost Engineering in the Process Industries, McGraw-Hill Book Co., New York, 1960.
3. R.C. McCurdy, Chemical Engineering Progress, 65, No. 5, 19 (1969).
4. R.B. Norden, Chemical Engineering, p. 134, May 5, 1969.
5. K.M. Guthrie, Chemical Engineering, p. 114, March 24, 1969.
6. H.C. Bauman, Fundamentals of Cost Engineering in the Process Industries, Reinhold Publishing Corp., New York, 1964.
7. M.S. Peters and K.D. Timmerhaus, Plant Design and Economics for Chemical Engineers, McGraw-Hill Book Co., New York, 1968.
8. D.F. Rudd and C.C. Watson, Strategy of Process Engineering, John Wiley and Sons, Inc., New York, 1968.
9. D.G. Jordan, Chemical Process Development, Interscience Publishers, John Wiley and Sons, Inc., New York, 1968.
10. H.J. Lang, Chemical Engineering, June 1948, also p. 12 of Ref. [2].
11. O.T. Zimmerman and I. Lavine, Cost Engineering, Industrial Research Services, Dover, New Hampshire.
12. H. Popper, ed., Modern Cost Engineering Techniques, McGraw-Hill Book Co., Inc., New York, 1970.
13. H.E. Wessel, Chemical Engineering, July, 1952.
14. R.L. Glauz, Chemical Engineering Progress, 51, 122 (1955).
15. D.E. Pierce, Chemical Engineering Progress, 44, 249 (1948).

PROBLEMS

5.1. Company purchase records indicate that in 1960 a carbon steel distillation column 6 ft in diameter, containing 25 actual trays on 24-in. spacing, cost $20,300 before installation.

(a) What is the estimated cost of the same column in 1970?

(b) What is the estimated cost of a column for the same service in 1970 that would be 9 ft in diameter?

5.2. A chemical reaction is to be conducted continuously in a single-stage stirred tank reactor. The reaction is exothermic and the heat is to be removed by water cooling in an external shell and tube heat exchanger.

Heat-exchange calculations indicate that 2,000 sq ft of heat-exchange area will be required.

There are two possibilities: (1) One 2,000-sq ft exchanger may be used, or (2) two exchangers, connected in parallel and each of 1,000-sq ft area, may be employed. In the latter case the reaction may be conducted at a reduced rate even though one exchanger has been withdrawn for cleaning and repair. What is the installed cost of (a) the one 2,000-sq ft exchanger, and (b) the two 1,000-sq ft exchangers.

5.3. A topic of much interest in present day process engineering is the use of air-cooled heat exchangers instead of water-cooled exchangers. Consider the use of air-cooled heat exchangers to condense 10,000 lb/hr of benzene from a distillation column. What would be the installed cost of (a) a water-cooled condenser of conventional design, and (b) an air-cooled condenser?

5.4. The erected cost of a unit to manufacture ethyl chloride from ethylene at a capacity of 120,000,000 lb/yr is known to have been $2,500,000 in 1958.

(a) What would be the erected cost of the same plant in 1970?

(b) In 1958 the operating labor for this plan was 5 men per shift. What would be the annual labor, supervision, and indirect costs for this unit in 1970?

5.5. In December 1966 the Imperial Chemical Industries, Ltd., of England started a plant for the production of methanol using a new process. This process employed the standard $CO + 2H_2 \rightarrow CH_3OH$ reaction but used a new catalyst and operated at a much lower reaction pressure. The combination of the lower pressure and the different catalyst was quite successful. The new "low pressure" methanol process is now widely accepted.

P.L. Rogerson and R.J. Kenard, with N.W. Nimo, describe the process and its economics in Chemical Engineering Progress Symposium Series 98, vol 66, 1970.

For a 500 ton/day plant producing pure methanol on a 350 day/yr basis, the following data are given:

New capital investment	$ 9,700,000
Working capital	$ 850,000
Total investment	$10,550,000
Raw materials	lb/lb methanol
Natural gas; 1,050 Btu/cu ft	0.74
Catalyst and other chemicals ($122,000/yr)	
Utilities	
Electrical power	0.014 kw-hr
Process water	1.15 gal
Cooling water	37 gal

The depreciation may be taken as 10%/yr; straight line.

From these data calculate (a) the manufacturing cost of the product; (b) the percentage of the manufacturing cost that is contributed by the raw materials, the utilities, the maintenance, the depreciation charge, and the labor cost.

5.6. Using the data of the preceding problem calculate the effect on the manufacturing cost of a change in design capacity to (a) 200 tons/day, and 1,000 tons/day. Also calculate the contribution of each of the principal cost items to the total cost, as in Problem 5.5.

5.7. Plot a graph showing the effect of changes in the cost of the raw material on the cost of the final product for each of the three production levels mentioned above.

5.8. In Section 5.02 the equipment-size cost relationship is given by the expression: Cost = $K(A)^n$, where A = some measure of the size or capacity of the equipment, and n is 0.5 to 1.0, with 0.7 being quite common.

(a) If the cost-size relationship is given by the above equation, plot a generalized graph showing the cost of a multiple-unit installation as the ratio to the cost of a single unit of the same total capacity versus the number of smaller units, for exponents of 0.5, 0.7, and 1.0.

(b) A filtration operation can be done by one rotary vacuum filter with a total peripheral area of 500 sq ft. What would be the installed cost of the 500-sq ft filter and of two 250-sq ft filters of the same type?

Chapter 6

RISK, RETURN RATE, AND
INVESTMENT RECOVERY

6.00 INTRODUCTION

The basic equations developed previously are believed to be correct in principle. Therefore, the judgment needed to arrive at a sound conclusion is largely confined to the choice of the numerical values of the pertinent variables to be substituted in the equations. Here company policy and the judgment of experienced financial, engineering, and sales personnel must be used in order to select proper values of the variables.

6.01 RISKS

New capital investments are made with the expectation of a substantial annual return from the investment plus the recovery of the entire investment, or perhaps more, after a rather small number of years. Together with the hope of a future return, there is the knowledge that the project may not be completely successful, or may fail altogether, and there will be little or no return and a large or total loss of all of the investment. The possibility of a loss is always present. This possibility is called the "risk" accompanying an investment.

Table 6-1 Projects, Riskiness, and Minimum Acceptable Return Rate

Type of project	Riskiness	Minimum acceptable return rate %
Installation of more efficient equipment; neutralization of acidic waste stream; solvent recovery by absorption and distillation	Low	10-15
Increase of present capacity; change of catalyst; change in automatic instrumentation; installation of new facilities for an old product	Moderate	15-25
New facilities for a new product	High	20-50, or more

Some projects are "riskier" than others, and, in general, the greater the risk the larger must be the expected annual return and the shorter the expected length of time for the return of the investment in order that the project attract the necessary capital. Table 6-1 lists several typical projects with their "riskiness" and the associated minimum acceptable rates of return.

Risks are of two types: (1) The technological risk, that is, the risk that the project may fail because of technical reasons such as excessive corrosion or catalyst poisoning. (2) The business risk, that is, the project may not be economically feasible because of costs, competition, or lack of sales.

Present-day technology is rather advanced, and there is a great deal of experience and ability available for work on technical problems. Difficult technical problems can usually be solved, or delineated so that their effect can be recognized and countered in some way, such as improving the catalyst or changing the material of construction of the heat exchangers. As a consequence, the technological risk is not great, but it must be studied carefully because the solution may involve heavy expenditures of time and money.

On the other hand, the business risk can be greater, less easily defined and controlled, and more serious in its effect than the technological risk. Factors such as raw material costs, sales prices, competition, the general level of business activity, and the policies of government are all business risks, and are not entirely under the control of the investing corporation.

In Chapters 2 and 3 the venture worth concept was introduced and it was explained that this was the best method for analyzing the economic attractiveness of a proposed venture. Also in Chapter 3, a detailed discussion of each of the six terms that comprise the venture worth was given.

The venture worth calculation, and any other economic analysis, involves the worker in the prediction of the future. Such predictions are uncertain and carry a business risk with them. Among the predictions that must be made are gross sales return (for each year as far as 10 years into the future), capital investment, interest rates, probable plant life, and the plant salvage value. In making an analysis such as the venture worth, the analyst must remember that the business risks mentioned above are important to the value of W that is calculated. Whatever numbers are selected for quantity sold, sales price, manufacturing cost, new capital investment, and plant life will influence the value of the venture worth, and so influence decisions regarding the future of the project. An analysis of the sensitivity of the venture worth to changes in variables such as capital investment, sales return, and interest rates was given in Section 3.03.

6.02 GROSS SALES RETURN, $R_k = Q(s - c)$

As shown in Chapter 3, gross sales return is one of the two most important items in the venture worth equation. The whole purpose of a commercial operation is to buy low and sell high; that is, make s - c as large as possible. The selection of realistic values for Q and s is the province of those skilled in sales and market research, and their advice should be sought. In fact, many companies have a standard procedure for the selection of sales volumes and prices that includes the thoughts of sales people. Approximate values for the sales price can be obtained by one, or both, of two methods:

1. Consult Fig. 1.3. This plot shows, in an approximate way, the price-sales volume relationship for industrial chemicals. By a simple inspection of the graph a sales price for the new substance can be found. This procedure has the merit of preventing the selection of a sales price for the new material that is not consistent with the experience of the industry.

2. Use the device explained in Example 3.2. Here the venture profit, V, is set equal to zero, and a value for the selling price is calculated using reasonable values for the other variables. The selling price so obtained is that which will make $W = 0$ at the end of the n-th year.

There is a very real business risk attached to assigning a value to s, the sales price. There are two problems involved: (1) the selling prices of chemicals always trend downward with time (see Chapter 1), and (2) today, every product must compete with older, better established products that have long records of successful application. As a consequence, the predicted sales price of a new chemical may be quickly depressed by competitive forces. A venture worth calculation extending over a period of 10 years into the future and using a constant sales price may well be overly optimistic. In Section 3.07 quantitative methods for accounting for a declining sales return are discussed.

Another business risk associated with R_k is the annual volume of sales which is predicted. If this is smaller than expected the project will be unprofitable. If it is larger than predicted, the plant capacity may be insufficient and operations may be forced into an uneconomic range, or sales contracts will not be fulfilled and penalties will be incurred.

The term $\sum_1^n R_k (1 - t)/(1 + i)^n$ in the venture worth expression is composed of several parts. Two of these—the sales rate and the sales price—are interrelated, difficult to predict, almost certain to change over a 10-year period, and crucially important to the value of the venture worth.

The greatest business risk of all is the prediction of the sales demand and the sales price.

6.03 TOTAL NEW CAPITAL INVESTMENT, I

The term I appears as a negative value in the venture worth equation. If I is large the possibility of a strongly negative venture

Risk, Return Rate, and Investment Recovery   263

worth is always present unless the negative I term is counterbalanced by a large positive value for R_k. Projects with large capital investment must have correspondingly large sales revenues. It is very easy to invest large sums of money in a project, and very hard to generate the required sales revenues necessary to make the project profitable.

The numerical value of I enters into several of the economic profitability indices other than the venture worth. In every case, the higher the value of I the harder it is for the project to appear attractive.

In addition to the danger that a large capital investment presents to the profitability of a given project, there is the fact that a large capital investment may imperil the existence of the company. Companies, like individuals, are limited in the amount of funds that they can invest either from surplus or from borrowing. Many corporations can consider the investment, and the possible loss, of $1,000,000, perhaps one-tenth as many can consider the investment of $5,000,000, and almost none can consider an investment of $50,000,000. Financial officers must always balance the magnitude of the possible loss against the benefits of the possible gain to their particular company. A large gain is splendid, but a large financial loss must be avoided--because that may mean disaster.

The total new capital investment is an important business risk.

6.04 INTEREST RATES

Chemical companies usually find the capital for new projects from their own excess funds (corporate surplus) or from borrowing from banks and insurance companies. The business risks attached to borrowing are (1) the inability to meet the interest payments, and (2) the inability to retire, extend, or refund the loan on the date of maturity.

In the first case, loans made at a high rate of interest are more difficult to repay than those made at a low rate. Interest payments are an expense and therefore damaging to the profitability of a project. If a project must be financed by borrowing at high interest rates, that project must have a high rate of return before it can be profitable.

However, because interest is a business expense, payments may be deducted from the gross income, in exactly the same way as raw material costs are, before the taxable income is obtained. The taxable income, after the deduction of the annual interest charge, is $R_k - d_k I - iI$ (see Fig. 2.1), and the taxes are $(R_k - d_k I - iI)t$. Therefore, because of the interest payments the tax paid is lower by iIt and the effective interest charge then becomes $iI(1 - t)$. Today, corporation taxes are essentially 50% of the revenues after expenses. Consequently, it is a good enough approximation to say that the interest rate that a company pays on borrowed money actually costs the company one-half the stated rate. In recent years interest rates have been near 8%; historically, this is very high, but actually, due to the corporation income tax structure the real interest rate is nearer 4%. This situation makes financing of new ventures by borrowing rather attractive, and chemical companies do borrow, some more than others.

But, as always, there is another side to the picture. In the first place, despite the reduction in interest rates which results from the taxation allowances, interest is an expense and must be paid and does reduce the net income. If the same amount of financing came from the company's surplus no expense would be involved. Furthermore, debt financing can be a hazard. The interest and the capital must be paid, there is no alternative, and such payments take precedence over both preferred and common-stock dividends. The servicing of the debt has first call on the company's earnings. Consequently, when business is poor and earnings low, there may be only enough money to pay the interest charges, and none will remain for stock dividends. This will cause a drop in the common stock price and dissatisfaction among the common-stock holders. Their dissatisfaction can lead to serious arguments with management. Financing a project with high interest rate loans constitutes a business risk that cannot be ignored.

The second business risk, that of the inability to repay the loan on the maturity date, can have even greater difficulties for the corporation. Many loan agreements involve obligations such as sinking fund payments, common-stock dividend restrictions, the immobilizing of fixed assets, and the maintenance of working capital above a stated minimum. Failure to meet any of these obligations may precipitate the full maturity of the loan with embarrassing consequences.

Almost all chemical companies avoid having a large portion of their financial structure as debt. Nonetheless, many companies do have 20 to 30% of their financial structure in the form of bonds or loans. In recent years, the decline in the prices of the common stocks of practically all chemical companies has substantially raised the percentage of debt to debt plus common-stock value.

The balance sheet of the Dow Chemical Company, shown in Example 6.1, indicates that long-term debt comprised 40.4% of the total capitalization in 1968. This is a sharp increase from 1962.

These words were not sooner written than the price of Dow common stock increased in value by a factor of more than 2 in just 2 years. This substantial increase in the value of the common stock reduced the long-term debt fraction of the capitalization to a more normal level of 20%. As this book goes to press, Dow is borrowing a large sum of money through the sale of debentures.

6.05 THE ROLE OF GOVERNMENT POLICY

It must be remembered that the U.S. government manipulates the basic interest rate in the country in order to stabilize the economy. In times of inflation, heavy capital investment, and large cash flows—as during the period 1965-1969—the Federal Reserve Board raises its discount rate. This is done with the deliberate intention of restricting credit, reducing new capital investment, and slowing the pace of business. By July 1970 the Federal Reserve Board discount rate had risen to about 5.5% and the prime interest rate on loans by banks to their best customers to about 8%. These were very high interest rates, so high that new capital investment in the chemical industry was substantially reduced. This, of course, was exactly the effect the government wanted. Conversely, when business activity slows, capital investment declines, and the rate of inflation decreases the Federal Reserve Board lowers its discount rate, the commercial banks and insurance companies reduce their interest charges, and new capital investment becomes more attractive. In the winter-spring of 1970-1971 the U.S. government did exactly this; interest rates were lowered, despite a continuing inflation, in the hopes of stimulating business activity.

Substantial changes in interest rates can occur over rather short periods of time.

Interest rates on borrowed money are a substantial business risk. Financial managers must constantly adjust their thinking to actual or imagined changes in governmental policy which will change interest rates.

In addition to manipulating the country-wide interest rate the government can raise or lower taxes, increase or decrease the allowable depreciation rates, and impose stricter requirements on air and stream pollutants. All of these, and many more, governmental activities affect industry strongly. So strongly, in fact, that many chemical companies maintain offices in Washington, D.C. and in many state capitals in order that they may be closer to the making of laws that are so important to them.

6.06   ESTIMATED PROJECT LIFE

As has been mentioned before, the evaluation of the economic attractiveness of a project involves the evaluator in the prediction of the future. Not the least of these predictions is that of the probable life of the project, the variable "n."

The predicted life of the project represents the best guess that the managers can make as to the length of time that can be permitted to pass before the new capital investment is recovered from the proceeds of the operation. The number n corresponds to the number of years over which the initial capital investment will be recovered.

The selection of the value of n constitutes a considerable business risk. It is conservative (low risk) to have n small so that the investment will be recovered quickly. However, this will create heavy capital recovery charges against the project, and this, in turn, will create a higher selling price, which may make the project noncompetitive, or it may reduce the amount of money available for dividend payments. On the other hand, it is fairly risky to pick a value for n that is large, because of the very real danger that presently unknown forces will rise in future years to change the competitive position of the new project. However, a large value for n will result in lower capital recovery charges, perhaps a lower price for the product, and more cash for dividend payments and the creation of a surplus.

There are several restrictions on the value that n can have:

1. The Internal Revenue Service has set certain rules about the minimum length of time which may be used for the depreciation of equipment and buildings. Because of the tax advantages to be gained from the depreciation allowances, no manager will choose a value of n that is larger than the depreciation time allowed by the Internal Revenue Service, although a value that is smaller might be chosen. For chemical industry processing equipment the time

Risk, Return Rate, and Investment Recovery

allowed by the government is 11 years. For buildings that might house the equipment, or serve as warehouses, the time is about 30 years. Other items of equipment, such as steam generators, have depreciable lives of 28 years. However, since so much of the capital investment in a chemical plant is for process equipment, it is common practice to use 10 years as the depreciable life in chemical-process economic studies. This is a length of time that is convenient for computation purposes, but which is close enough to the allowable depreciation time.

2. Technical and economic changes of major proportions occur in the chemical and petroleum industries with considerable speed. It is perfectly possible, and several instances have been given in Chapter 1, for a process to become technically and/or economically obsolete within 5 to 8 years. Newer processes, lower cost raw materials, new air and stream pollution abatement regulations, and other forces largely outside the control of the project managers may force the shutdown of the subject process before the equipment has deteriorated to such an extent as to need replacement and before the government-allowed depreciation charge is completed.

Because of this technical and economic depreciation factor, it is considered prudent for managers to assign projects lives that are shorter than the official depreciable life.

The degree of risk associated with a project will be reflected in the value of n assigned (high risk, low n). The values of n actually used in industry vary between 5 and 10 years. Because of the convenience to preliminary analysis of having the depreciable life and the project life the same, it is common practice to let both equal 10 years. This is not wrong enough to cause difficulties. Other difficulties are more important.

6.07 DEGREES OF RISK

In Section 6.01 it was stated that in the chemical industry there are projects that are more risky than other projects. What makes one project more risky than another? Primarily, the degree of difficulty of achieving the predicted sales quantity at the price desired. This simple statement obscures a host of questions which have answers that are only partially satisfactory. Roughly speaking, there are three degrees of risk. These may be called low, moderate, and high; see Table 6-1.

A low-risk project is one which is essentially certain to operate properly and which has a known sales outlet at an accurately known price. Two typical examples of such low-risk projects are the following:

1. The installation of a new steam generator which will supply steam to the chemical plant at a certain pressure, temperature, and rate. Present-day steam technology is so advanced, and there is so much experience in its use, that there can be very little risk in an investment in a new steam-generating plant.

2. The recovery of a valuable solvent, such as methanol, by water absorption and subsequent distillation. An example of this is given in Chapter 5. In such a case the technology is well known, and the experience and knowledge in the field are great. There is little chance that the solvent recovery system will fail.

In both of the cases discussed above the technology, the markets, and the sales prices are well known, and there is only a small chance of error or failure. The risk is low.

A moderate-risk project is one where new capacity is to be constructed to serve an existing market, or where a new product, only slightly different from existing ones (but with superior properties) is to be introduced. Hopefully, this new capacity, or the new product, will increase production efficiency and/or sales revenue. However, when building new capacity for an existing market there is always the risk that the costs of the raw materials and, especially, the sales price will change in ways that will adversely affect the project. The risk is increased if new technology is to be used in manufacturing, for instance, the use of the fluid bed instead of the fixed bed in the production of aniline.

A high-risk project is one where a new product is to be made by a new process. Here all of the technological and business risks discussed previously exist. Despite extensive research, development, and marketing work a large element of uncertainty is present. High-risk projects are, in fact, so risky that the number of successful operations of this sort is much smaller than is commonly believed.

A current example of a high-risk project is the cyclic hydrocarbon production plant of the Columbian Carbon Division of the Cities Service Company. Here, butadiene is rearranged into cyclododecatriene and cyclooctadiene. These are new chemicals, never before produced commercially, and they are produced in a 25,000,000-lb/yr plant which is also new. These chemicals have great promise in many different ways, but the risk taken in moving into this entirely new field is very great.

The years 1965-1970 have been so bad for the chemical industry that almost all United States chemical companies are avoiding high-risk projects, and, instead, are concentrating their technical efforts on improving the processes already in operation. See Chapter 1 for a further discussion of such trends in the chemical business.

6.08 RATE OF RETURN

In the venture worth equation the term i is defined as the average rate of return on the company's capital investment. A larger discussion of the meaning of "i," and of ways of computing it is required.

In the venture worth equation the term i appears in two roles.

1. As the rate of interest used in discounting future cash flows to the present.

2. As the rate of interest used in computing the capital recovery factor in Eq. (3.5), where $e = i/[(1 + i)^n - 1]$, the sinking fund deposit factor.

In case 1 the rate of interest is used to express the time value of money. In its simplest form, this means the ability of money to earn compound interest when deposited in a bank savings account. However, in modern corporate practice money is seldom put into a savings account, but is invested in more lucrative ways. The rate of interest used in the time value of money should reflect this; see below.

In case 2 the rate of interest means the return on the total amount of corporate funds, regardless of how this amount may be invested. Referring to Fig. 2.1, it can be seen that the cash flow rate eI is returned to the corporation, is mixed with all other corporate funds, and is available for investment in whatever way the corporation's officers deem best. It may even be kept as cash. This means that investments may be made in stocks and bonds (of governments or corporations), in inventories of raw materials, intermediate, or final products, in real estate, in bank deposits, in new plant and equipment, or in any other promising field (including cash). Present-day corporations regard the whole world of capital investment as their province. As a consequence, the i for use in the capital recovery factor should be the estimated value of the rate of return that the corporation may earn on all of its invested capital during the next 5 or 10 years. The invested capital would include securities, inventories, plant and equipment, loans, real estate, and cash. Cash earns no return, but corporations cannot operate without some cash available for the payment of small currently due bills.

Essentially, the total invested capital is the total assets of the company minus the total liabilities plus any long-term debt. Subtracting total liabilities from the total assets yields the dollar value of the company owned by the stockholders (both preferred and common). By adding back the amount of long-term debt to the difference between total assets and total liabilities the quantity of invested capital, or total capital worth, of the company is obtained.

It is important to note the position of the company's own bonds or other forms of long-term debt. Bonds are sold, or loans are incurred, for the purpose of increasing the company's quantity (sometimes called the "pool") of capital. This borrowed money is at once invested in any of the several ways described above (including being kept as cash). As a consequence, the long-term debt becomes a part of the total assets of the company, and the interest payments on the debt become part of the company's earnings.

With this background it is possible to examine several of the definitions of i and to select the one most suitable for use in the venture worth equation. The financial report of the Dow Chemical Company for the year ending December 31, 1968 is employed to provide actual numbers that may be used in illustrating this discussion. A condensed version of this report is shown in Example 6.1.

Example 6.1

The Dow Chemical Company published the following summary of its operations for the year ending December 31, 1968 [1].

	$
Assets	2,312,225,000
Cash and receivables	476,095,000
Marketable and other securities	287,929,000
Inventories	276,329,000
Plant and equipment	2,060,427,000
Less depreciation fund	933,404,000
Net plant and equipment	1,127,023,000
Other assets	144,849,000
Total assets	2,312,225,000
Liabilities	
Total current liabilities	571,392,000
(includes: notes payable to banks, provision for income taxes, long-term debt maturing currently, accounts payable, and dividends payable)	
Long-term debt	684,584,000
Minority interest in subsidiaries	41,626,000
Foreign investment less reserve	5,000,000
Total liabilities	1,302,602,000
Net worth	1,009,623,000
Common stock ($5 par)	156,153,000
Earned surplus	358,830,000
Capital surplus	562,171,000
Total stockholders equity	1,077,154,000
Less treasury stock at cost	67,531,000
Net stockholders equity	1,009,623,000
Total liabilities plus net worth	2,312,225,000

Income account or profit and loss statement

Sales income	1,652,492,908
Cost of goods sold	1,207,233,707
Selling and administration expenses	222,834,271

	$
Operating profit	222,424,930
Other income	70,845,666
Total income	292,910,596
Interest charges	54,187,580
Other income charges	6,781,154
Net income before income tax	231,941,862
U.S. and foreign income taxes	92,000,000
Minority interest	3,982,023
Net earnings	135,959,839
Common dividends (cash)	72,450,494
Other surplus items	3,510,035
Earned surplus for year	139,469,874

The financial statement shown in Example 6.1 is somewhat condensed from the original, but some idea of the complexity of the operation of a large diversified chemical company operating internationally can be obtained from it. The almost incredible size of the operations should also be noted. These numbers are so large as to be almost impossible to comprehend—and, it must not be forgotten that there are several chemical companies even larger than this.

The analysis of financial statements is a well-developed technique. There are many texts explaining the methods and the meaning of the various numbers developed. This text will not venture into this field, but the interested reader may consult the books of Graham, Dodd, and Cottle [2] and Jaedicke and Sprouse [3]. Statistical services such as Moody's Industrials [1] will provide the required data although such data are nearly always between 1 and 2 years late. Current information can be obtained from the quarterly financial statements of the various companies, although they are fairly succinct and are not audited. Also, Moody's Handbook of Common Stocks [4], published quarterly, provides much background material. There is no lack of published information about industrial companies and their business results. Almost all modern municipal libraries subscribe to such publications.

Using the data in Example 6.1, the definitions of i can now be illustrated:

1. Probably the simplest and most easily understood definition is given by the ratio net earnings/total plant and equipment. In the case of the Dow Chemical Company in 1968 this ratio equals;

$$\frac{135{,}959{,}839}{2{,}060{,}427{,}000} = 0.066$$

This definition suffers from two defects: (a) It makes no allowance for the fact that a substantial fraction of the total investment in plant and equipment has already been recovered. In many chemical companies today between 40 and 60% of the total investment in plant and equipment has been recovered through the accumulations of the stream(s), eI. Since the recovery has already been achieved, the recovered portion should not be counted as capital investment.
(b) It makes no allowance for the fact that the company may have capital invested in things other than plant and equipment. As stated before, other investments also make money and can be important to the company's financial position. Since all, or part, of the capital recovery cash flow could be put into any of these various investments it would seem reasonable to include these in the total invested capital.

2. Another possible definition is

Net earnings/undepreciated plant and equipment

In the case of Dow in 1968 this would be

$$\frac{135{,}959{,}839}{1{,}127{,}023{,}000} = 0.121$$

This definition has the advantage of accounting for the fact that a large part of the capital invested in the plant and equipment has been recovered. However, it still does not account for the other kinds of capital the company has such as bonds, stocks, loans, inventories, and others.

3. Still another possible definition is

Net earnings/total undepreciated invested capital

The total undepreciated invested capital would be defined as total assets minus cash and receivables. This definition could be defended on the grounds that cash and receivables are not invested capital, although they are counted as assets. In the subject case of the Dow Chemical Company this would be

$$\frac{135{,}959{,}839}{1{,}836{,}130{,}000} = 0.074$$

Total invested capital	$
Marketable and other securities	287,929,000
Inventories	276,329,000
Net plant and equipment (after depreciation allowance)	1,127,023,000
Other assets	144,849,000
Total	1,836,130,000

This definition has the advantages that it includes all the invested capital of the company and also makes an allowance for that part of the plant and equipment which has been recovered. The definition has the disadvantage that it does not include the interest payments on long-term debt as part of the company's earnings, although it does include the capital resulting from the investment of such borrowings. Furthermore, it may be objected that cash and receivables are a legitimate form of invested capital because a corporation cannot function without them.

4. A fourth possible definition is

Net earnings plus interest payments/total invested capital

In the subject case of Dow in 1968 this would be

$$\frac{135{,}959{,}839 + 54{,}187{,}580}{1{,}836{,}130{,}000} = 0.104$$

This ratio represents the return earned by the company's activities on its total quantity of invested capital. This definition has the advantage that it includes the interest payments as part of the company's earnings. However, a disadvantage might be that total invested capital should not be defined as total assets minus cash and receivables.

5. Probably the best way to define i is

$$\frac{\text{Net earnings plus interest payments}}{\text{Total assets - total liabilities + long term debt}}$$

In the reference case of Dow in 1968 this would be

$$\frac{135{,}959{,}839 + 54{,}187{,}580}{1{,}694{,}207{,}000} = 0.112$$

This definition is commonly referred to by financial analysts as the return on total capital. It does consider depreciated equipment, payments on bonds and loans as part of earnings, and long-term debt as part of capital. By defining invested capital in this way, cash and receivables are considered as part of capital. However, most of this amount is canceled because short-term liabilities are subtracted. This definition has another advantage in that it can be easily calculated from some of the principal items in the company balance sheet. Furthermore, it is a number that is widely used by financial analysts and is reported and tabulated in many financial statistical reports. For instance, Forbes Magazine, January 1, 1971, p. 88, lists this ratio, in the form of a 5-year average, for 21 well-known chemical companies. Five of these are tabulated below:

Company	% return on total capital (5-year average: 1966-1970)
Dow Chemical Co.	10.2
American Cyanamid Co.	11.3
Rohm and Haas Co.	10.8
Diamond-Shamrock Co.	8.2
Allied Chemical Co.	6.5
Industry Median	8.7

It should be noted that this method of defining i is entirely a balance sheet method and makes no reference to the market value of the company's bonds, preferred stock, or common stock.

Definition 5 is used in this book and is recommended as the most reasonable definition of i. It must be emphasized that the value of i, calculated from the company's annual report for any one year, may not be typical of the company's operations over a longer time—such as 5 years. Before selecting a value for i it would be wise to determine this ratio for at least 5 years into the past. Even then, a simple numerical average may not be the correct number to use. Judgment must be employed in selecting a proper value for i. For instance, in the chemical industry in 1970 the general trend of i was down. Forbes Magazine reports that the value of i decreased, as an average for the whole chemical industry, at an annual rate of 4.4% for the 5 years 1966-1970. When selecting a value for i for use in the venture worth equation, it is wise to know about such trends.

The chemical industry is now going through a confusing and disconcerting period. No one can foretell accurately what may happen. Yet this is exactly what must be done in venture worth analysis. The value of i used should be based on an analysis of past performance coupled with a professionally cold-blooded estimate of future prospects.

6.09 ESTIMATION OF THE MINIMUM ACCEPTABLE RETURN RATE, i_m

The rate of return required for management to consider that a new project is attractive must be somewhat higher than the current average earnings rate, although the latter is a useful guide to establishing a lower bound for i_m. The definition and calculation of the average earnings rate were considered in some detail in Section 6.08, and this material may be used as a starting point for the determination of i_m.

The average earnings rate i reflects both the company's successful and unsuccessful ventures. The failures lower the average; hence new ventures must be better than average (because some new ventures will fail) if the average rate of return is to remain constant, or be improved.

Risk, Return Rate, and Investment Recovery

In Section 6.07, degrees of risk were discussed and in Table 6-1 the minimum acceptable return rate usually desired for each of these degrees of risk was stated. This is no more than an approximation, and individual workers may wish to alter these numbers to suit their own purposes. Almost always the final result of such investments is a return rate that is closer to the lower limit of the range given in Table 5-1. The reason for this is simple; things seldom work out the way people wish them to. Some projects are failures, some are only marginally profitable, and in some the rate of return falls off badly after a few years. The successful projects must make up for the unsuccessful ones, and this requires that a high rate of return be required in the original analysis.

In Chapter 3 a discussion was given of the role played by i_m in the venture worth. It was pointed out that the critical point is the difference between i_m and i. When this is zero, that is $i_m = i$, the venture worth will be increased and the project made to appear more appealing than it really is. If i_m is made quite high the venture worth will be reduced and the project made to appear unattractive. The assignment of unrealistically high or low values to i_m can result in misleading values of W. However, possible values of i_m are confined to the relatively narrow range of 15 to 30%. The realities of the chemical business really do not permit the use of values much different from this.

6.10 DEVELOPMENT PROJECTS

In the chemical-process industries new projects may be divided into two classes: (1) those requiring development work—these may be called "development projects"—and (2) those not requiring such work. An example of the second class is the purchase of the rights to an established process from another company. The seller guarantees that the new plant will operate successfully. Consequently, development and design problems are not the concern of the purchasing company, and the money and time required for development need not be considered. Under such circumstances an economic evaluation of the new project would take the form of Example 3.2, where no consideration was given to the time and the cost of development work.

The first class of projects is widely encountered in the chemical industry. Development work on these processes can be expensive and time-consuming, and excessive amounts of it may seriously penalize the future profitability of a project. On the other hand, if the development work has not been done properly, and not enough has been learned about the process and the product, the production plant may be seriously delayed in starting, sales may not be as high or as profitable as desired, and difficult problems may be encountered in the operation of the plant. Production problems of this kind are to be avoided, and the proper operation of a development program may help to avoid them. The whole area of how much development work to do, whether or not to build a pilot plant, where to concentrate the development activity, what risks are involved, and how the development activity will affect the economics of the project is a most difficult one. Extended discussions of development problems have been given by many authors (see Jordan [5,6]).

Project managers must not only make a first decision about the nature of the development work to be done, but, as the development work proceeds and more information is gathered, they must make decisions, at periodic intervals, about the nature of future work. Such decisions are difficult because of the large number of alternative courses of action. Jordan [5] has described a decision-tree type diagram which assists in making such decisions.

The venture worth concept can be helpful in choosing between the various alternatives. Consider the decision which is usually the most difficult: to build and operate a pilot plant in order to obtain more information and experience with the process and then, several years later, build a production plant. Or, to build the production plant at once, dispensing with the pilot plant. It can be helpful to calculate the venture worth for both choices, and then compare the values.

Let W_1 = the venture worth for the process including the cost of the development work, W_2 = the venture worth for the process not including the cost of the development work, and D = the present worth of the development work cost. Then, if

$$(W_1 - W_2)/D > 1$$

the development work should be done. However, if

$$(W_1 - W_2)/D \leq 1$$

the development work should not be done.

The expressions above are only attempts to quantify the common-sense principle that any development work done must not exceed in cost the expected long-term profits created by such development work.

Example 3.2 described the calculation of W for a project that did not include any development work. If a similar calculation for this project were to include the cost and the time for such work the costs must be discounted back to the starting time of the project. This will mean that the costs and profits of the production plant will be discounted even more than in Example 3.2. This more extensive discounting of the cash flows of the project, which has had a period of development, lowers the value of W_1. This is in accord with experience and common sense, which would dictate moving into production at the earliest possible date. However, there are other considerations: without development work there might be no plant at all [5,6].

6.11 INVESTMENT RECOVERY

As has been stated before, the investor (corporation, bank, insurance company, or single individual) has two objects in mind when making an investment: (1) a substantial annual return in the form of dividends, interest, or profit shares every year for many years, and (2) the repayment or recovery of the initial investment after some years into the future. No one will invest so much as $1.00 in a venture unless he is reasonably sure that this $1.00 will be returned to him within a relatively few years, with some annual reward for parting with the $1.00 in the first place. Therefore, users of substantial amounts of money, such as the directors of a chemical company, recognize that a provision must be made in the original financing plans for the eventual return of the money to the investor.

In practically all cases in the chemical industry the original investor is the chemical company itself. New capital expenditures come from the surplus funds of the company, from funds raised by borrowing, or from the sale of new preferred and/or common stock. In recent years chemical companies have not sold much stock. The rate of recovery of these expenditures is set by the company's financial officers and is equal to the term eI in Eq. (2.7) and Fig. 2.1. In the venture worth treatment emphasized here e is assumed equal to the compound interest sinking fund deposit factor: $i/[(1 + i)^n - 1]$. This number is dependent on both i and n, and both of these numbers are dependent on the general level of industry interest rates, and on the estimated life of the project. As mentioned in Section 6.15, the estimated life of the project has an upper limit which is severely bounded by the allowable times for the depreciation tax allowance.

There are two primary problems connected with the recovery of a capital investment.

1. The ability of the project to repay all of the initial capital investment. The uncertainty connected with this repayment is a large part of the risk discussed in Section 6.01.

2. The length of time required for the project to recover the original investment.

Both of these problems may be examined together by studying the payout time. This time is defined by Eqs. (2.8a) and (2.8b). Different values are obtained depending upon which definition is used.

Equation (2.8a) is not really a practical or realistic definition because it makes no allowance for taxation and the government-allowed depreciation charges which accompany this. However, the result is useful for comparing one investment possibility with another because no ambiguity due to various ways of computing taxes and depreciation has been introduced. The smaller the payout time, the better. In this book T_m, the maximum acceptable payout time, is used; see Eqs. (3.43) and (3.45). Despite the utility of a payout time taken without regard to taxation and depreciation, such a number has no real meaning in connection with the recovery of invested capital.

The second definition of payout time—Eq. (2.8b)—is much more practical. It assumes that all of the project's revenues, after the payment of taxes and the taking of the depreciation tax credit, will be applied to the payment of the original capital investment. This means that no funds will be diverted from the retirement of the original capital investment. Here, $T = I/[R(1 - t) + dIt]$; this is the shortest possible investment recovery period. All revenues from the project are applied immediately to the repayment of the original capital investment. No money is assigned to the payment of common-stock dividends or to the redistribution of the earned surplus.

At the present time the chemical industry has an annual cash flow of about 13% of its gross plant, property, and equipment. If all of this money were to be employed for capital recovery, the time T, in Eq. (2.8b), would be about 8 years. This information has an interesting connotation. If the average payout time, as defined by Eq. (2.8b), is about 8 years, it must follow that a project deemed attractive enough for new investment must have a much shorter payout time—say 5 years.

The situation wherein a company pays no common-stock dividends and distributes no part of its surplus, but concentrates all of its after-tax earnings on capital recovery, does occur. Some corporations pay no dividends and distribute none of the surplus. All of the after-tax profits of the company are reinvested in the company. This practice makes for rapid growth and an increase in the stockholder's equity, but it does not pay any dividends or distribute any surplus. Nonetheless, it is conservative financial practice, and some companies follow it, always providing that they do not have a large number of common-stock holders that must be satisfied. In this connection, it must be remembered that a majority of common-stock holders purchase the stock for the purpose of participating in the growth of the company. The company grows fastest when no common-stock dividends are paid. Dividend payment is not a major consideration; growth is more important. A policy of this type is favored by many investors.

The net profit of the project is defined as $P = R(1-t) + dIt - eI$, that is, all the money remaining after the taxes have been paid and an allowance has been made for the recovery of the capital investment. If e is set equal to the sinking fund deposit factor, e will always be less than $1/n$ because the deposits in the sinking fund are earning compound interest at the rate i. In actual practice this comes about because the cash flow eI is put back into the general pool of corporate funds where it earns interest at the company's general level of earnings rate, i.

Management may set n equal to any value it wishes, but the lower the value for n the greater the portion of after-tax profits that will be credited to capital recovery, and the smaller the amount available for dividends and additions to earned surplus.

In recent years the chemical industry has been paying out about 50% of its net income in dividends. The remainder of the net profit is reinvested in various ways. This means that $0.5[R(1-t) + dIt + eI]$ is available for the retirement of capital investment, addition to net surplus, and further investment in company projects.

The combination of high taxation, high dividend payments, and strong competition (low R) has had an adverse effect on the chemical industry. At present, it is widely believed that the chemical companies are not spending enough on new facilities to meet the demands of the 1970s. That is, the value of the amount of new funds now being generated by the company's activities is not sufficient to provide the money necessary to expand existing facilities and create new ones, so that the future profitability of the company can be maintained. This gap between what is required and what is available can be bridged by borrowing, but only for a short period of time and at a considerable risk. It is always possible to reduce the common-stock dividend to provide more money for capital recovery and future growth. This can be a treacherous path because of the effect on the market price of the common stock and on the feelings of the common-stock holders, who may lose both income and capital value. The annual reports and the stock market behavior of many chemical companies support this.

Several trends are interesting

1. The amount of borrowing, by bonds or by loans, is increasing, and the capital structure of many chemical companies is now becoming appreciably higher in funded debt than in the years 1952-1960. In the case of the Dow Chemical Company the 1968 financial report revealed that the amount of funded debt was 40.4% of the capital structure. This is a substantial increase since 1960. However, by mid-1973 an unprecedented increase in the value of Dow common stock had brought this percentage back to about 20%. Nonetheless, the company is continuing to borrow large sums.

2. The percentage of the gross plant, property, and equipment that has been depreciated has decreased from about 55% to about 45% in the last 10 years. This may be interpreted as meaning that the companies are deliberately maintaining their common-stock dividend at a fairly high level in the face of declining profits with correspondingly smaller sums to be spent on capital recovery. This cannot be carried far before the common-stock dividend must be reduced.

3. The cost of new plant construction has been rising faster than any other cost to the chemical industry. New plants and equipment are now becoming so expensive that many worthwhile projects must be abandoned. Only projects with great promise, or with great certainty, are being undertaken. New process development is becoming severely handicapped because of the great cost of installing new plants. Even the replacement or repair of existing profitable facilities is being questioned because of the high cost of new construction.

REFERENCES

1. Moody's Industrials, Moody's Investor Services, Inc., 99 Church Street, New York.
2. B. Graham, D.L. Dodd, and S. Cottle, Security Analysis, 4th ed, McGraw-Hill Book Co., New York, 1962.
3. R.K. Jaedicke and R.T. Sprouse, Accounting Flows: Income, Funds, and Cash, Prentice-Hall, Inc., Englewood Cliffs, New Jersey, 1965.
4. Moody's Handbook of Common Stocks, Moody's Investor Services, Inc., 99 Church Street, New York.
5. D.G. Jordan, Chemical Process Development, Interscience Publishers, John Wiley and Sons, Inc., New York, 1968, p. 57.
6. D.G. Jordan, in Kirk-Othmer Encyclopedia of Chemical Technology, 2nd ed, Vol. 15, John Wiley and Sons, Inc., New York, 1968, p. 605.

PROBLEMS

6.1. A chemical company is contemplating the production of a chemical at a rate of 50,000,000 lb/yr. Let

I = \$5,000,000 c = \$0.15/lb i_m = 0.14 S_a = 0

I_w = \$500,000 i = 0.09 d_k = 0.10 (straight line)

What is the lowest sales price, s, that the company would deem acceptable if n is set at 2, 5, and 10 years?

6.2. Consider a chemical company whose balance sheet and profit and loss statement for the year 1970 are as follows:

	$
Assets	
Cash and receivables	147,000,000
Investments	8,000,000
Plant and equipment	
Cost	180,000,000
Less depreciation	74,500,000
Undepreciated plant and equipment	105,500,000
Total assets	260,500,000
Liabilities	
Current liabilities	46,000,000
Long-term debt	69,200,000
Total liabilities	115,200,000
Stockholders equity	
Common stock: 3,200,000 shs @ $10	32,000,000
Preferred stock: 100,000 shs $3.75 @ 100	10,000,000
Earned surplus	103,300,000
Liabilities plus net worth	260,500,000

The present price of the common stock is $60 a share.

Income statement

Sales		302,000,000
Operating costs and expenses		267,500,000
Interest paid		3,000,000
(Taxable income	31,500,000)	
Income taxes @ 50%		15,700,000
Total expenses		286,200,000
Net income		15,800,000

Compute: (a) the earnings per share of common stock, (b) the price-earnings ratio of the common stock, (c) the value of the term i.

6.3. Consider the company whose balance sheet and income statement are given in Problem 6.2.

The company wishes to purchase the U.S. rights to a German process for the manufacture of a certain chemical. The price is $2,000,000 plus a royalty of 1.0¢/lb of the product.

It is estimated that the total investment in the plant to manufacture this chemical would be $10,000,000, the sales price of the product would be 30¢/lb, the sales rate would be 30,000,000 lb/yr, and the manufacturing cost will be 20¢/lb (including the royalty payment and the depreciation charge).

Cost of process	$ 2,000,000
Cost of new plant	$10,000,000
Capacity of new plant	30,000,000 lb/yr
Royalty payment	$300,000/yr
Sales return	$9,000,000/yr
Cost of manufacture (including royalty)	$6,000,000/yr

Compute the consequences of financing the new plant by (a) using company funds from its earned surplus account, (b) borrowing money from banks or insurance companies at 6.5% interest.

6.4. In Problem 6.2 the company considered was in a strong financial position. It had ample reserve funds and its debt load was not heavy. Such a company could afford the rather small risk that the process would not succeed. Also, it could raise the necessary capital from its own funds or by borrowing.

Consider now a small company which is not in good financial condition. The financial statistics below show that the company is just barely profitable, pays its interest charges with little to spare, pays no common-stock dividend, owes quite a lot of money, and cannot pay its debts. Obviously, something should be done to improve the company's profitability and to strengthen its financial position.

Financial Statement

	$
Assets	
Cash and receivables	800,000
Investments	0
Plant and equipment	
Cost	6,000,000
Less depreciation	1,800,000
Undepreciated plant	4,200,000
Net assets	5,000,000
Liabilities	
Current liabilities	1,000,000
Long-term debt	2,000,000
Stockholders' equity	
Common stock: 100,000 shs @ $20/sh	2,000,000
Earned surplus	0
Liabilities plus net worth	5,000,000

Income Statement

	$/yr
Sales	4,500,000
Operating costs and expenses	4,270,000
Gross revenue	230,000
Interest paid	130,000
Taxable income	100,000
Income taxes	50,000
Net income	50,000
Earnings/share	$0.50
Market value of stock	$4.00/share

One possibility for improving the company's financial position lies in the company's discovery of a new process for the manufacture of an important chemical. This is believed to be a good process which should replace those existing and provide profitable business for years to come.

It is estimated that a 50,000,000 lb/yr plant would have the following economics:

New capital investment	$5,000,000
Working capital	$500,000
Factory cost	12¢/lb
Sales price	22¢/lb

Some of the criteria for economic attractiveness of new projects can be applied here. For instance,

T_m = 5,000,000/5,000,000 = 1 yr; very favorable

P = R - (R - dI)t - eI

P = 5,000,000 - (5,000,000 - 500,000)(0.5)
    - (0.069)(5,000,000)

P = $2,405,000; here $e = i/[(1 + i)^n - 1]$, if i = 0.08 and

n = 10; e - 0.069

p = 2,405,000/5,500,000 = 0.44; where p = rate of return

V = $(R - dI)(1 - t) - i_m(I + I_w)$; if i_m = 0.25

V = $845,000 (the venture worth)

From these rather quickly obtained results it can be seen that the proposed process does have a great deal of economic promise.

The company has three options open: (1) Commercialize the process itself. (2) Sell the process to another company for a fairly high price. (3) Give the process to another company in return for a royalty on each pound of material made.

Which of these options appears best for the company? If the process is to be sold, what price should be asked? If the process is to be licensed, what royalty should be asked?

6.5. A large oil company operates a producing department. In the exploitation of oil leases, one well out of every eight drilled is a successful producer. The average cost of bringing in each well is $50,000. Each producing well earns a net income at the start of $200,000/yr, and the earning rate declines on the average of 10%/yr. At the end of 10 years it is no longer profitable to operate a well owing to rapidly increasing producing costs.

If the depletion allowance is computed at a rate of 27.5%/yr on net income, what is the net rate of return on invested capital in such operations?

6.6. A chemical company has been doing laboratory research work on a new process for the manufacture of a well-known chemical. To date, the expenditures have been negligible. It is estimated that $200,000 more and 2 years of time will be needed to bring the process to the point where construction work might begin. From then, 2 more years will be required before production and sales can begin. During the 2-year construction period $100,000 will be spent each year for further development work.

The economic analysis of the project indicates the following:

$I = \$5,000,000$ $Q = 50,000,000 \text{ lb/yr}$ $i = 0.09$ $d_k = 0.1$

$I_w = \$ 500,000$ $c = 12¢/\text{lb}$ $i_m = 0.12$

                        $s = 18¢/\text{lb}$ $S_a = 0$

At the present time it is possible to purchase the rights to a process for the same chemical and so by-pass the development time, and not spend the development money. The economic parameters above still apply, but there is a purchase price for the new process of $400,000 plus a royalty of 0.5¢/lb on the product.

Which path should the company follow: (a) continue with its own development, or (b) purchase the process rights and pay the royalty?

6.7. Consider the company with the balance sheet described in Problem 6.2. What would be a reasonable value for the payout time, calculated from Eq. (2.8b), for any new project undertaken by this company?

Chapter 7

OVERALL CONSIDERATIONS
IN PROJECT ANALYSIS

7.00 NEW PROJECTS

In the analysis of any proposed project, two things should be done. First, a basic criterion, involving pertinent variables, must be established to measure project attractiveness. Second, this criterion must be developed from data on capital investment, operating costs, return anticipated, and, by assumption, as to risk and desired minimum profitability.

The equations developed at the beginning of this book dealt with the first problem: that of judging overall economic attractiveness. Information regarding suitable numbers to implement these equations was presented in later chapters; see particularly Chapters 5 and 6. A simple numerical illustration of how this material might be employed in a typical preliminary project study is given in Example 7.1.

It is usual for a series of economic and technical analyses to be made before a project is completed. The first will normally occur as soon as enough data are available to make a rough analysis. Additional studies will be made through pilot plant and design stages, involving greater elaboration from both the technical and economic standpoints; see Jordan [1, pp. 26-57].

The preparation of a large number of technical and economic evaluations can be laborious, time-consuming, and expensive.

Various approximations and shortcuts may be used to lighten this load.

1. A comparison with other projects which have points of similarity is helpful. This has been designated as "design by analogy."

2. It is usually assumed that certain portions of the subject project may be studied independently of other portions of the same project. For preliminary investigations, the effects of small changes in the design of one portion of the plant on another part are ignored.

3. If the proposal is an addition to an existing plant, the addition can be considered as incremental. No new problems of sales, raw material costs, equipment design, or utilities consumption will arise.

Once the rates of production and consumption are established, it is usually possible to subdivide a plant physically into component parts or operations for detailed economic balances to ensure optimum sizing and consumption of utilities.

Advanced mathematical techniques, such as those described in Chapter 4, may be useful in the final stages of the process analysis, but very often the actual design of a chemical-process plant is too complex to admit of exact calculations. Usually a preliminary design must first be generated using reasonable approximations based on past experience or pilot plant data. Once the initial design has been established by the application of these approximate calculations based on rather rough or incomplete data and knowledge, it will be found that certain items will be responsible for the major profit or loss elements in a process. If it is thought desirable, more complex calculations may be made on these items in an effort to refine their design and to approach an optimum more closely.

Appendix C lists a collection of approximation methods of calculation. Methods for developing them further are given in Chapter 8, which also provides techniques for more refined economic balances on individual plant components, once a base design has been established.

The present chapter will explore the first class of problems: those having to do for the most part with a substantial part of the entire plant rather than its component unit operations.

Overall Considerations in Project Analysis 291

Example 7.1

In previous illustrative examples two preliminary economic calculations have been shown. In Example 3.2 the detailed year-by-year calculation of the venture worth was presented, and a capital investment and a manufacturing cost were assumed. The details of the estimation of these two numbers were not shown. In Example 5.1 the details of the estimation of the new capital investment were discussed, but no effort was made to estimate operating cost, pay-out time, or return on investment. In any real situation detailed calculations for new capital investment, operating costs, selling price, and the values of the various profitability indices must be made.

The example that follows is typical of a preliminary approach to plant design problems.

Isopropyl alcohol is produced in the United States in very large quantities; in 1969, about 2.1 billion lb/yr. However, the growth rate of such production has been slow during the last 5 years, and the selling price has changed very little from about 6¢/lb. The recent rapid growth of the polypropylene industry has put something of a premium on sources of propylene, so that consideration might well be given to producing pure propylene instead of isopropanol.

The company desires to investigate the economics of manufacturing isopropyl alcohol from propylene. The source of the propylene is a fuel gas which contains other light hydrocarbons besides propylene. Ordinarily this gas would be used as a fuel and so does not have a very high monetary value. The product isopropanol would be marketed as an antifreeze and the company estimates that it can sell 5,000,000 gal/yr.

The process involves absorption of the propylene in 70 wt% sulfuric acid, hydrolysis of the isopropyl acid sulfate, and purification of the product by distillation. The product will contain 88 wt% isopropyl alcohol and 12 wt% water because of the existence of an isopropanol-water azeotrope at this composition. Purification of the isopropanol cannot be carried beyond this composition by ordinary fractional distillation. The specific gravity of the isopropanol-water azeotrope is 0.815 at 20°C. This is 6.8 lb/gal.

Figure 7.1 is a flow sheet of the process. All of the important process lines are numbered and the analyses, flow rates, temperatures, and pressures of each of these process streams are given in Table 7-1.

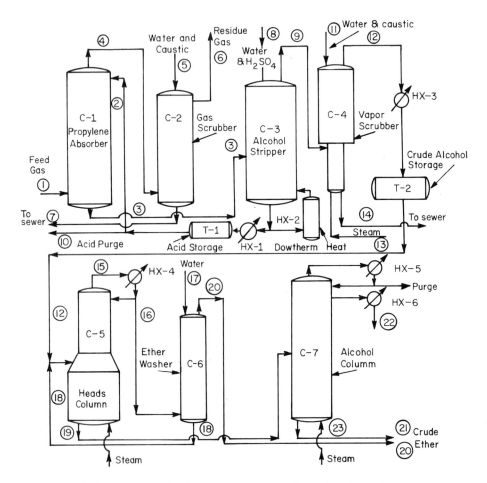

Fig. 7.1. Isopropyl alcohol plant: a schematic flow diagram.

The absorption of the propylene from the fuel gas may be accomplished by direct contact with the 70 wt% sulfuric acid in a plate type absorption tower. The propylene dissolves in the acid, but the other gases do not, and consequently the raw material charge to the process is only that of the heating value of the dissolved propylene. The exit gas is washed with water and caustic to remove entrained acid and is then available for further use in the plant. The propylene absorber exit liquid stream containing the isopropyl acid sulfate is pumped to a third tower where the acid sulfate is hydrolyzed by dilution of the acid solution with water to the equivalent of about 40 wt% sulfuric acid. The alcohol is stripped out of the acid by heating. At the same time, excess water is removed so that the sulfuric acid is reconcentrated to 70 wt%. Some sulfuric acid is destroyed in the hydrolysis reaction and new acid must be added to the top of the hydrolysis tower. Heavy tars and other high molecular weight compounds must be purged from the system before the reconcentrated acid is sent back to the absorption tower. A net acid consumption of 0.224 lb H_2SO_4 per pound of finished product is allowed. The alcohol and water vapors stripped off are neutralized and condensed as crude intermediate.

The first purification step removes the low-boiling ether, in column C-5, the "heads column." The ether-containing overhead from the distillation column is then washed with water in a liquid-liquid extraction column, C-6, in order to recover isopropyl alcohol which is returned to column C-5. The raffinate is a crude grade of isopropyl ether.

The main product is distilled in a rectifying tower to concentrate it to 88 wt%. Best results are obtained when the alcohol product is taken as a side stream. Impurities lighter than isopropanol must be purged from the system at intervals.

For preliminary design and cost estimation purposes the feedstock to the process may be regarded as being a refinery fuel gas of the following composition.

	Mole %
Ethane and lighter	78
Propylene	10
Propane	10
Butanes	2

An extract solution from the bottom of C-1 will contain all the absorbed propylene as $(C_3H_7)HSO_4$.

The conversion of absorbed propylene will yield the following final products:

	lb/100 lb propylene absorbed	Wt% of total products
Isopropyl alcohol	126	90.0
Isopropyl ether	12	8.6
Polymers	2	1.4
	140	100.0

Solution. The material balance shown in Table 7-1 was constructed from the above data. The line numbers may be obtained from Fig. 7.1.

An energy balance similar to the material balance was also prepared. From these balances, the various key items of equipment were designed, at least to the extent that physical dimensions and materials of construction could be determined with fairly good precision. At the same time, the various utilities consumptions were established: cooling water, steam, and electrical power. These items can be displayed in various lists which summarize the results of the preliminary design calculations; for instance, Table 7-2, which lists the size and the specifications for the columns. Similar lists are made for other pieces of equipment such as vessels, heat exchangers, pumps, compressors, and instruments.

Lists of equipment such as described above may then be used as a basis for preparing cost estimates, the results being summarized in Table 7-3. The costs of the equipment shown in Table 7-3 were based on the costs given by Happel [2] in the first edition of this book. Those costs were adjusted to the year 1970 by the following method:

1. Material costs were multiplied by the ratio of the Chemical Engineering Plant Cost Index for 1970—taken as 110—to the same index for 1955—taken as 90; see Fig. 5.1. This factor is 110/90 = 1.218. Thus, the new material, for a plant built in 1970, would be 21.8% more expensive than that used for a new plant built in 1955.

2. The labor costs for construction of the plant were multiplied by the ratio of the Engineering News Record Indices for 1970—taken

Overall Considerations in Project Analysis 295

as 172—and that for 1955—taken as 90; see Fig. 5.1. This ratio is 172/90 = 1.91.

Whether this procedure is exactly correct is not known—nothing in cost estimation is exactly correct. However, it does account for two well-known characteristics of present-day chemical plant construction costs: (a) The cost of equipment is higher than in 1955, but not greatly higher. (b) The cost of construction labor has increased by a much larger ratio. Thus, the estimated cost for this plant in 1970 is quite a bit higher than the 1955 estimated cost.

The various profitability indices (see Tables 7-4 and 7-5) may be computed in the following way:

$$\text{Gross return rate} = p = \frac{\text{net profit}}{I + I_w} = \frac{395,500}{4,394,808} = 9.0\%$$

$$\text{Payout time} = \frac{4,190,808}{1,300,500} = 3.2 \text{ years}$$

Venture profit may be obtained by subtracting from the net profit, $395,500, the minimum acceptable return on the total investment, $I + I_w$.

Minimum acceptable return - $(0.15)(4,394,808) = \$660,000$

Venture profit = -$264,500

Equation (2.10) gives the same answer for the venture profit:

$$V = \left[R - dI - \frac{i_m(I + I_w)}{1 - t} \right](1 - t)$$

$$V = \left[1,300,500 - 420,000 - \frac{(0.15)(4,394,808)}{0.45} \right](0.45)$$

$$V = -\$264,500$$

Quite obviously, the project will not come close to providing the minimum acceptable rate of return.

It is instructive to compare the results of this calculation with that made by Happel in the first edition of this book [2, pp. 140-147]. See also Table 7-5 of this book. There are several important points:

1. The sales price of the isopropanol has not changed in 15 years. The gross income in the two cases is the same.

Table 7-1 Material Balance: Isopropyl Alcohol Plant Production of 5,000,000 gal/yr[a]

Line number	Ethane and lighter	Propylene	Propane	Butanes	Sulfuric acid (100%)	Isopropyl acid sulfate
1	10,840	3,640	3,820	981	-	-
2	-	-	-	-	13,900	-
3	-	-	-	-	6,960	9,890
4	10,840	670	3,820	981	-	-
5	-	-	-	-	-	-
6	10,840	670	3,820	981	-	-
7	-	-	-	-	-	-
8	-	-	-	-	952	-
9	-	-	-	-	-	-
10	-	-	-	-	952	-
11	-	-	-	-	-	-
12	-	-	-	-	-	-
13	-	-	-	-	-	-
14	-	-	-	-	-	-
15	-	-	-	-	-	-
16	-	-	-	-	-	-
17	-	-	-	-	-	-
18	-	-	-	-	-	-
19	-	-	-	-	-	-
20	-	-	-	-	-	-
21	-	-	-	-	-	-
22	-	-	-	-	-	-
23	-	-	-	-	-	-

[a] Product is 88 wt% isopropyl alcohol and 12 wt% water. Rates: lb/hr.
[b] Will include some sodium sulfate, sodium bisulfate, and tar.

Overall Considerations in Project Analysis 297

Water	Isopropyl alcohol	Isopropyl ether	Tar	Sodium hydroxide	Temperature (°F)	Pressure (psia)
-	-	-	-	-	170	140
5,940	-	-	31	-	170	140
5,940	-	-	31	-	170	140
-	-	-	-	-	170	140
8,800	-	-	-	176	100	67
-	-	-	-	-	100	67
8,800	-	-	-	176^b	100	67
20,400	-	-	-	-	100	45
20,000	3,771	362	25	-	370	45
409	-	-	2	-	100	45
33,200	-	-	-	332	100	40
19,600	3,771	362	22	-	225	40
4,000	-	-	-	-	270	40
17,600	-	-	-	332^b	270	40
-	747	2,650	-	-	152	14.7
-	90	362	-	-	152	14.7
161	-	-	-	-	100	14.7
171	84	-	-	-	100	14.7
20,200	3,765	-	21	-	187	14.7
5	6	362	-	-	240	25
33,100	25	-	21	-	225	14.7
510	3,740	-	-	-	100	14.7
13,400	-	-	-	-	40	30

Table 7-2 Column Schedule, Isopropyl Alcohol Plant

	Temperature (°F)	Design pressure (psig)	Size	Specifications
C-1 Propylene absorber	170	124	48 in. × 70 ft	A285; lead-lined 30 Karbate trays
C-2 Gas scrubber	100	52	60 in. × 50 ft	A285
C-3 Alcohol stripper	370	30	75 in. × 50 ft	A285; lead-lined 20 Karbate trays
C-4 Vapor scrubber	225	25	75 in. × 50 ft	A285; Monel clad
C-5 Heads column	200	0	Top section 24 in. × 23 ft	A285; 10 trays
			Bottom section 48 in. × 33 ft	A285; 15 trays
C-6 Ether washer	100	0	12 in. × 30 ft	A106; 20-ft packing 1-in. steel Pall rings
C-7 Alcohol column	200	0	48 in. × 90 ft	A285; 40 trays

Table 7-3 Summary of Isopropyl Alcohol Plant Costs, 1970[a]

Item	Material ($)	Labor ($)	Total ($)
Columns (see Table 7-2)	234,000	47,900	281,900
Vessels	79,200	13,400	92,600
Exchangers	141,500	23,000	164,500
Boiler (Dowtherm)	67,000	38,200	105,200
Pumps and compressors	97,500	15,300	112,800
Tanks	34,200	15,300	49,500
Instruments	61,000	9,560	70,560
Total key accounts	714,400	162,660	877,060
Insulation	72,000	168,000	240,000
Piping	288,000	451,000	739,000
Foundations	36,600	86,000	122,600
Buildings	29,200	32,600	61,800
Structures	29,200	9,560	38,760
Fireproofing	3,660	28,700	32,360
Painting and clean-up	3,660	28,700	32,360
Electrical	42,700	95,700	138,400
Total	505,020	900,260	1,405,280
Sum of materials and labor			2,282,340
Overheads			685,000
Total erected cost			2,967,340
Engineering fee			296,734
Contingency			296,734
Total plant cost			3,560,808
Offsite construction			630,000
Total investment			4,190,808

[a] Production rate: 5,000,000 gal/yr. Operating time: 8,000 hr/yr.

Table 7-4 Summary of Isopropyl Alcohol Economics[a]

	$/yr	¢/gal
Income		
Isopropyl alcohol; 88 wt% @ 40¢/gal	2,000,000	
Isopropyl ether (crude) @ 4¢/gal	116,300	
Total income	2,116,300	42.3
Expense		
Raw materials: Refinery gas, absorbed Propylene only; 2,970 lb/hr; Heating value 2,400 Btu/cuft @ $0.10/MM Btu	52,500	1.05
Sulfuric acid @ 1.2¢/lb	91,500	1.83
Sodium hydroxide @ 1.5¢/lb	61,000	1.22
Total raw material cost	205,000	4.10
Operating cost		
Labor 4 men/shift @ $4.00/hr	128,000	2.56
Supervision @ 10% of DL	12,800	0.26
Utilities	100,000	2.00
Total operating cost	240,800	4.82
Indirect costs @ 125% of DL	160,000	3.20
Maintenance @ 5% of plant investment/yr	210,000	4.20
Total expense (not including depreciation)	815,800	16.32
Depreciation @ 10%/yr	420,000	8.40
Total expense (including depreciation)	1,235,800	24.72
Total profit (not including depreciation)	1,300,500	26.0
Net profit (including depreciation)	880,500	17.58
Gross profit	1,300,500	
Depreciation	420,000	
Net profit	880,500	
Taxes @ 55%	485,000	
Net profit (after taxes)	395,500	
Working capital @ 25% of total expense	204,000	

[a] Production of 5,000,000 gal/yr.

Table 7-5   Comparison of the Economics of Isopropanol Production
1955-1970

	1955 ($)	1970 ($)
Income	2,116,300	2,116,300
Expense		
Raw materials	53,600	205,000
Operating	358,000	240,800
Indirect costs	465,000	160,000
Maintenance		210,000
Total expense	876,600	815,800
Gross profit	1,239,700	1,300,500
Depreciation	244,000	420,000
Net profit	995,700	880,500
Taxes @ 55%	548,000	484,000
Profit after taxes	447,700	396,500
Working capital, I_w	219,000	204,000
Invested capital, I	2,434,000	4,190,808
Gross return rate	16.9%	9.0%
Payout time	2.0 years	3.2 years
Minimum acceptable return	398,000	650,000
Venture profit	49,700	-244,500

2. The two values of the gross profit are within 10% of one another. This is so close as to be essentially the same.

3. The 1970 new capital investment is about 75% higher than in 1955. This, of course, makes the depreciation charge higher in the 1970 case, and, as a result, the profit after taxes in the 1970 case is somewhat less (about 9%) than that in 1955, despite the higher gross profit in 1970.

4. The economic profitability indices are strongly affected by the much larger capital investment. The return rate has been reduced from 16.9% to 9.0%; the payout time has been increased from 2 years to 3.2 years, and the venture profit has been made strongly negative rather than positive as in 1955.

5. In actuality, the 1970 case is really not undesirable. The return rate and the payout time are quite reasonable by 1970 standards. The problem is that a minimum return rate of 15% is not attainable in 1970, while it was in 1955.

This particular problem shows the impact of a high capital investment on the economics of a project. The result obtained helps to explain why management is always so concerned with the dollar magnitude of the new capital investment. As explained in Section 6.12, the new capital investment is always a major business risk.

There are several courses of action that might be taken in an effort to improve this situation:

1. A careful redesign of the plant could be made. Every effort would be expended to bring the capital investment down. For instance, columns C-1 and C-3 might use acid brick lining and ceramic packing instead of lead lining and Karbate trays. The new design must pass minimum design criteria, but no more than the minimum.

2. Special methods of construction, deliberately used to reduce labor costs, could be employed. The design of foundations, piping, buildings, and electrical connectors could be simplified so that construction labor is reduced to as low a value as possible. Elegance in design and conservation of materials would be sacrificed to simplicity and reduction of construction labor costs.

3. The redesign of the plant would also attempt to reduce such items as utilities, labor, and supervision. Also, a redesign might reduce maintenance from the 5% a year assumed here to, say, 3%.

All of these efforts might help, but it is unlikely that much improvement would result. The situation prevailing here is quite

Overall Considerations in Project Analysis 303

common to the chemical industry today. Higher costs for construction, utilities, labor, and maintenance are forcing production costs upward, but competition and overcapacity keep the sales price constant, or move it downward. The result is a serious profit "squeeze" on the companies, and many formerly profitable processes are now not as attractive as they once were.

7.01 EXISTING PLANTS

In addition to the evaluation and selection of new ventures for development, the engineer must be concerned with maintaining existing investments in the best possible competitive condition. Because of the uncertainties accompanying many new ventures, it is not an exaggeration to say that the maintenance of an existing process in a profitable condition is more important than the development of a new process.

A company's investment is subject to continual turnover; no process and no plant can remain changeless. There will be a continual improvement in existing processes, a steady addition of new processes, and a discontinuance of those that are no longer satisfactory.

In Chapter 2, Sections 2.6 and 2.7, the problem of the termination, or discontinuance, of an operation was discussed, and a criterion for the desirability (or not) of termination was derived. The principle involved was, when the change in the venture worth of a project with one year of time, $\Delta W/\Delta k$, becomes equal to zero, the project was nearing its termination point. The equations resulting, (2.6) and (2.32), show this criterion in quantitative form. Basically, the equations propose that termination occur when the earnings from the operations are not quite as large as the earnings from the recovered capital—which has been reinvested. Because of the difficulty of realizing anything more than a pittance from recovered capital the criterion essentially requires that the project be kept in operation as long as there are even small earnings from sales. Of course, once there are losses, the project would be suspended or terminated.

Another situation frequently encountered is that in which discontinuance of the investment does not involve discontinuance of the function that it accomplishes. The investment is simply removed and replaced by another one which performs essentially the same function. As, for instance, when a worn reciprocating pump is replaced by a new and more efficient centrifugal pump. At the other extreme of values, the same basic idea applies when a plant which is producing a chemical by one process is replaced by one which manufactures the same product by a more efficient process. Fairly recently, the older plants for the manufacture of acrylonitrile from acetylene and hydrogen cyanide have all been terminated and replaced by plants producing acrylonitrile from ammonia and propylene. This replacement took place before the end of the depreciable life of the acetylene-hydrogen cyanide plants.

Example 7.2 illustrates a frequently encountered case in which it is desired to combine an existing operation with new facilities in order to meet a demand for increased plant capacity. It relates to the problem of the rate of operation of existing boiler capacity when a considerable increase in steam requirement is to be met. The same situation would apply in operating a chemical plant where alternatively the product could be purchased from outside sources or produced in a new unit. The criterion to be employed in all cases is the same: namely, that the entire operation yield a maximum total profit above an acceptable minimum rate of return on new investment proposed.

Example 7.2

A boilerhouse contains five coal-fired boilers, each with a nominal rating of 10,300 lb steam/hr (300 boiler hp). Owing to the growth of manufacturing facilities, it is necessary for additional boilers to be installed in the plant in which these boilers are operating. From the data below, determine the percent of nominal capacity rating at which these boilers should be operated.

The cost of coal, including the cost of handling and removing cinders is $7/ton, and the coal has a heating value of 14,000 Btu/lb. The overall efficiency of the old boilers, from coal to steam, has been determined from tests at various ratings and is given in Table 7-6.

Annual fixed charges C_A on each boiler are given by the equation

$$C_A = 14,000 + 0.04N^2$$

Assume 8,550 hr of operation per year.

In order to apply the methods presented here, it is necessary to have additional information regarding the characteristics of the

Overall Considerations in Project Analysis

Table 7-6 Boiler Efficiency

N, % of nominal rating	E, % overall thermal efficiency
100	75
150	76
200	74
225	72
250	69
275	65
300	61

proposed new boiler installation. Assume that the capital requirement for installation of new boiler capacity will be approximately $6/(lb)/(hr) of steam generated. All direct operating costs for steam generation with new facilities will be assumed constant at $0.30/M lb of steam. It will be assumed that a maximum payout time $T_M = 3.5$ years will be required on new capital for a boiler installation.

Solution. The assumptions made regarding the new installation, for simplicity, consider that new production will be made at a constant cost. For the new boilers

$$\text{Fixed charge} = \frac{(6)(1,000)}{(3.5)(8,550)} = \$0.20/\text{M lb steam}$$

Since the operating cost is $0.30/M lb, the total cost of producing steam in the new boilers is $0.50/M lb. This is also the incremental cost of producing steam in the new boilers.

A solution is found by observing three facts:

1. The total demand for steam is so high that it cannot be met entirely from the old boilers. Therefore some steam from the new boilers must be used.

2. The cost of producing steam from the new boilers is constant at 50¢/M lb. No matter what amount of steam is produced in the new boilers the cost is the same. This means that the

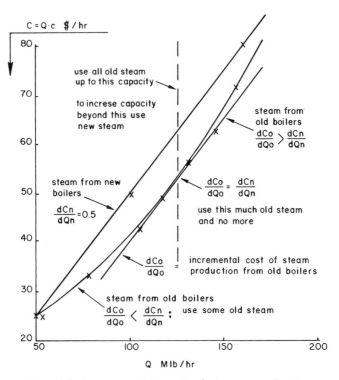

Fig. 7.2. Incremental cost of steam production.

incremental cost for the production of steam from the new boilers is constant at 50¢/M lb.

3. The cost of producing steam from the old boilers is variable, but in every case the steam from the old boilers costs less than the steam from the new boilers. Therefore, it is advantageous to use as much of the old boiler capacity as possible.

Figure 7.2 shows the situation graphically. Here the steam production rate, Q (for either old or new boilers, but not both together), is plotted against the product Qc, the cost of steam production in dollars per hour for either new or old boilers. The lower curve, for the steam from the old boilers, is taken from Table 7-7. The shape of the line is concave upward and the slope of the line is the incremental cost of producing steam in the old boilers. This slope starts at a low value at low steam rates and increases steadily as the steam rate increases.

Overall Considerations in Project Analysis

The upper curve shows the slope of the cost-production rate curve for the steam from the new boilers. It is constant at 0.50. The exact vertical location of the line is not important, and not known. The important thing to know is the slope; the incremental cost of producing steam in the new boilers.

It should be apparent that since some steam from the new boilers at an incremental cost of 50¢/M lb must be used, the proper amount of steam from the old boilers is that amount which has an incremental cost exactly equal to the incremental cost of the steam from the new boilers. If more steam from the old boilers is used the last increment of this steam added to the system will be more expensive than the new steam; this would be undesirable. Also, if less steam from the old boilers is used the deficiency must be made up by the more expensive new steam; this too would be undesirable.

Obviously, the solution to the problem is found at the point where the slope of the old steam cost-capacity curve is equal to the slope of the new steam cost-capacity curve. In this case this is at an old boiler rating of about 240%, or 125,000 lb/hr.

At the present writing the cost of coal is rising and is now well above the $7/ton figure used in the problem above. However, a change in the price of the coal will have little effect on the result of the calculation. The cost of coal is the same for both old and new boilers, and, although the absolute value of the steam cost will change, the incremental value will not. Hence, the answer will be the same even though the price of the coal increases substantially.

Table 7-7 may be constructed by taking 1 boiler horsepower equal to 34.5 lb/hr of steam (= 33,500 Btu/hr) and designating C_F as the hourly cost of fuel at $7/ton. Then

$$C_F = \frac{(5)(300)(N/100)(33,500)(7)}{(E/100)(14,000)(2,000)} = 12.56 \, (N/E)$$

and

$$C_A = \frac{(5)(14,000 + 0.04N^2)}{8,550} = 8.18 + 0.0000234N^2$$

Inspection of Table 7-7 shows that the incremental steam cost using the old boilers matches the incremental steam cost using the new boilers at a rating of between 225 and 250%. This is the same answer as obtained using Fig. 7.2.

Table 7-7 Cost Analysis of Boiler Installation

(1) N	(2) E	(3) Steam rate (M lb/hr)	(4) Increase in steam rate (M lb/hr)	(5) Fuel cost (C_F)	(6) Fixed charges (C_A)	(7) C_F+C_A	(8) Average steam cost ($/M lb) (7)/(3)	(9) Increase on C_F+C_A ($/hr)	(10) Incremental steam cost ($/M lb), (9)/(4)
100	75	51.8	–	16.7	8.4	25.1	0.49	–	–
150	76	77.8	26.0	24.8	8.7	33.5	0.43	–	–
175	75	90.8	13.0	29.3	8.9	38.2	0.42	4.7	0.36
200	74	103.8	13.0	33.9	9.0	42.9	0.41	4.7	0.36
225	72	116.8	13.0	39.2	9.4	48.6	0.42	5.7	0.44
250	69	129.8	13.0	46.5	9.6	56.1	0.43	7.5	0.57
275	65	143.8	13.0	53.2	10.0	63.2	0.44	7.1	0.55
300	61	155.8	13.0	61.8	10.3	72.1	0.46	8.9	0.68

Overall Considerations in Project Analysis 309

7.02 REPLACEMENT OF PLANT COMPONENTS

In the chemical process industry it is quite common for a plant to operate continuously (except for short planned shutdowns for cleaning and maintenance) for a period of time that is long in comparison to the plant payout time. In all this time the service, or function, performed by the plant does not change. For instance, the phenol plant of the Durez Plastics and Chemicals Company (now a division of Occidental Petroleum) operated almost without stopping for a period of 25 years. In cases of this type, even though the operation continues for an indefinite period there will be a succession of replacements of those components which wear out or become obsolete.

There are many different types of process equipment, installed in long-lived plants, that are subject to replacement by newer and/or more efficient machines. Several examples are ball mills used for the crushing of solid feed to metallurgical operations, shaker screens used in vegetable seed oil extraction plants, or centrifuges used in the production of pure solid crystals. Furthermore, parts of larger pieces of process equipment are constantly being renewed or replaced with something different; for instance, heat exchanger tubes or plates in a distillation tower. In addition, plant practices such as cleaning, painting, and lubrication are also subject to change.

However, despite these instances it must be remembered that in many cases, especially in the synthetic organic chemistry field, obsolescence often limits plant life to less than 10 years, so that no succession of replacements occurs.

Terborgh [3,4] has made a detailed analysis of the equipment replacement problem with special reference to machine tools. However, the analysis is applicable to all process equipment.

Terborgh assumes that an existing machine will accumulate an operating inferiority at a constant rate. This operating inferiority with respect to a proposed replacement, to accomplish essentially the same function, is a combination of two elements, deterioration and obsolescence: the one reflecting internal change caused by gradual wear and tear, and the other reflecting external change, resulting from product obsolescence, the development of inadequacy, and the improvement of available mechanical alternatives. This is equivalent to the assumption of a linear variation in the annual return rate R_k as discussed in the development of formulas (3.29) to (3.33). If either the inferiority gradient or the service life of a proposed new piece of equipment is assumed, it is possible to arrive at the corresponding minimum average annual cost. This "adverse minimum" is used as a model or basis for comparing the advantages of an existing machine with the possible disadvantages of a proposed replacement. For this purpose, the "adverse minimum" of the "challenger" is compared with the anticipated disadvantage of operation of the existing machine for the next year. In principle this is the same as employing Eq. (3.29) for comparing alternatives.

The problem of calculating the "challenger's" adverse minimum first involves the development of a simple formula for the equivalent operating inferiority for n years in terms of a gradient g (the rate, in dollars per year, at which inferiority is accumulated) and the interest rate, i. The operating inferiority figures constitute an end-of-year series of $0, g, 2g, 3g, \ldots, (n-2)g, (n-1)g$ over the period of n years.

This series is not uniform since each term is somewhat larger, by the gradient g, than the preceding term. If the sum of the amounts accruing at the end of n years is desired, the summation

$$(1+i)^n g \sum_{2}^{n} \frac{k-1}{(1+i)^k}$$

$$= (1+i)^n g \sum_{2}^{n} \frac{k}{(1+i)^k} - (1+i)^n g \sum_{2}^{n} \frac{1}{(1+i)^k} \qquad (7.1)$$

must be evaluated. This sum is the total inferiority at the end of n years. Note that, in contrast to the summations in Chapter 3, following Eq. (3.29), no inferiority is assumed during the first year of operation. Consequently, the summation index begins at $k = 2$ instead of $k = 1$.

The first and second terms on the right-hand side of Eq. (7.1) may be evaluated by means of Eqs. (A.8) and (A.4), respectively.

$$(1+i)^n g \sum_{2}^{n} \frac{k-1}{(1+i)^k} = (1+i)^n g \left[-\frac{1}{i(1+i)^n} \left(n+1+\frac{1}{i} \right) \right.$$

$$\left. + \frac{1}{i(1+i)} \left(2+\frac{1}{i} \right) - \frac{(1+i)^{n-1}-1}{i(1+i)^n} \right] \qquad (7.2)$$

$$= \frac{g}{i} \left[\frac{(1+i)^n - 1}{i} \right] - \frac{ng}{i} \qquad (7.3)$$

Overall Considerations in Project Analysis

The equivalent uniform annual figure for n years may then be found by multiplying this sum by the sinking fund deposit factor [see discussion following Eq.(A.7)] to obtain

Equivalent annual operating inferiority =
$$\frac{g}{i} - \frac{ng}{i}\left[\frac{i}{(1+i)^n - 1}\right] \qquad (7.4)$$

Thus, the exact formula for total equivalent annual operating inferiority will be the sum of the above and the annual cost of capital

$$A = I\frac{i(1+i)^n}{(1+i)^n - 1} + \frac{g}{i} - \frac{ng}{i}\left[\frac{i}{(1+i)^n - 1}\right] \qquad (7.5)$$

The value of A_{min}, the adverse minimum, is determined by setting n at a value that will make $(\Delta A)/(\Delta n) = A_{n+1} - A_n = 0$. This can be accomplished by numerical evaluation or by the use of approximation formulas developed by Terborgh [3].

However, usually as a practical matter, the value of the gradient g is not known explicitly, so that the determination of A_{min} in effect simply involves the assumption of n equal to an appropriate service life for the new machine. In this case it is clear that, since the old machine is expected to have an inferiority of ng for the year following the n-th year, the value of A for a new machine must be set equal to this figure to warrant replacement. For the same range of n and i as Terborgh uses, a better approximation for capital recovery factor is

$$\frac{i(1+i)^n}{(1+i)^n - 1} \cong \frac{1}{n} + 0.65i$$

A further improvement in the derivation may be made by noting that

$$\frac{i}{(1+i)^n - 1} = \frac{i(1+i)^n}{(1+i)^n - 1} - i \cong \frac{1}{n} + 0.65i - i = \frac{1}{n} - 0.35i$$

Substitution of these expressions and A = ng in Eq. (7.5) gives

$$A_{min} + A_{min}\left(\frac{1}{ni} - 0.35\right) - \frac{A_{min}}{ni} \cong I\left(\frac{1}{n} + 0.65i\right) \qquad (7.6)$$

and

$$A_{min} \cong I\left(\frac{1}{0.65n} + i\right) \cong I\left(\frac{1.54}{n} + i\right) \qquad (7.7)$$

Equation (7.7) should be just as accurate as more complex ones for the employment of Terborgh's method and is substantially easier to use. The values of n and i which were used in the derivation of the approximation for the capital recovery factor were n, 5 to 20 years, and i, 0.05 to 0.15. These numbers would seem to cover the range of values commonly encountered.

It is suggested, however, that a further modification might be made to allow for taxes and a minimum acceptable return rate, i_m. To accomplish this Eq. (3.42) may be employed, using in place of C, a constant annual cost, the equivalent annual operating inferiority, Eq. (7.4). The following result is then obtained, if m is taken equal to zero since maintenance is included in the inferiority gradient:

$$A = \left\{ \frac{i_m - i + \dfrac{i(1+i)^n}{(1+i)^n - 1}\left[1 - \sum_{k=1}^{k=n} \dfrac{d_k I}{(1+i)^k}\right]}{1 - t} \right\} I$$

$$+ \frac{g}{i} - \frac{ng}{i}\left[\frac{i}{(1+i)^n - 1}\right] \qquad (7.8)$$

For simplicity it may be assumed that straight-line depreciation is employed and the tax period corresponds to the plant life; $d_k = d = 1/n$. Employment of the approximations used in developing Eq. (7.7) gives

$$A_{min} \cong \frac{A_{min}}{ni} = A_{min}\left(\frac{1}{ni} - 0.35\right) + \left(\frac{i_m - 0.35i}{1 - t} + \frac{1}{n}\right)I \qquad (7.9)$$

and finally

$$A_{min} \cong \left(\frac{1.54}{n} + \frac{1.54i_m - 0.54i}{1 - t}\right)I \qquad (7.10)$$

Thus, as the tax rate is raised, the challenger will be discouraged. With zero tax rate and $i_m = i$, this equation reduces to Eq. (7.7). Note also that this development differs from Eq. (3.42) in that m is not considered separately but is included in the cost C.

A simple example will be used to illustrate the numerical application of the above formulas.

Overall Considerations in Project Analysis

Example 7.3

Assume that a machine is in operation in an existing plant and the question has arisen as to whether it should be replaced by a newer model. The new machine costs $27,000. It is estimated that the new machine should have a service life of 11 years as judged by its wearing out and by newer models being developed to accomplish the same purpose. The plant in which this machine is installed is, however, assumed to have an indefinitely longer life. Current cost of capital is 12%, and the terminal salvage value of the equipment is zero.

The existing machine has a next year operating inferiority of $5,000. This includes in addition to direct labor saving such items as cost saving through lower maintenance, decreased production of off-specification product, reduced overhead, lower power consumption, reduced toolroom costs, increased convenience and flexibility, greater reliability, better worker morale.

It is required to determine the desirability of installing the new machine.

Solution. First the "challenger's" adverse minimum is computed using Eq. (7.7).

$$A_{min} = 27,000 \left(\frac{1.54}{11} + 0.12\right) = \$7,020$$

The Machinery and Allied Products Institute Replacement Manual has a collection of charts showing plots of adverse minimum as a percentage of investment versus service life anticipated, with separate curves for various salvage values. These may be used in place of Eq. (7.7). A few of these charts may be seen in texts such as Taylor [5]. However, since the salvage value is usually quite low—say zero—Eq. (7.7) is almost as fast and simple as reading a chart.

Eq. (7.10), which makes an allowance for taxes and a minimum rate of return, may also be used to compute the challenger's adverse minimum. If $t = 0.55$ and $i_m = 0.20$ the following results:

$$A_{min} = 27,000 \left[\frac{1.54}{11} + \frac{1.54(0.2) - 0.54(0.12)}{0.45}\right] = \$18,400$$

If it had been assumed that $i_m = i = 0.12$, A_{min} would have been $11,000.

Here, $7,020, $11,000, and $18,400 are well above the "defender's adverse minimum" of $5,000, so that the proposed new installation will not be attractive.

Usually in chemical-process plants replacement is judged on the basis that equipment to be replaced, such as pumps, is performing a standardized operation, so that, over the course of a few years obsolescence will not occur, as regards the service being performed, though the overall process may become obsolete. Maintainence is taken as a separate constant percent per year of investment. These and other appropriate assumptions lead to the development of Eqs. (3.42) and (3.46).

With this approach, the function $U = S - V/(1 - t)$ must be minimized to ensure maximum venture profit V. If V is assumed to be the same each year, and for simplicity we again assume $d_k = d = 1/n$, then Eq. (3.41) reduces to the following sample form:

$$U = C + \left[m + \frac{\frac{i(1+i)^n}{(1+i)^n - 1} - \frac{t}{n} + i_m - i}{1 - t} \right] I \qquad (7.11)$$

If it is desired to compare costs on the basis of present worth, it is simply necessary to convert the uniform annual series represented by Eq. (7.11) by multiplying both sides by the present worth factor $[(1 + i)^n - 1]/i(1 + i)^n$.

Example 7.4

A piece of equipment in a highly corrosive atmosphere lasts 5 years and costs $10,000. How much can be spent per year on painting if the life is extended to 10 years? Assume that $t = 0.55$ and $d = 1/n$; i.e., depreciation for tax purposes is a straight line for the plant life. Cost of capital is 10.0% and maintenance $m = 0$.

Solution. The value of U without painting corresponds to a life of 5 years.

$$U_1 = \left[\frac{\frac{0.10(1 + 0.10)^5}{(1 + 0.10)^5 - 1} - \frac{0.55}{5}}{1 - 0.55} \right] 10,000 = \$3,420$$

For the case with painting and life of 10 years, if x is the unknown annual payment for painting,

Overall Considerations in Project Analysis   315

$$U_2 = x + \left[\frac{\frac{0.10(1 + 0.10)^{10}}{(1 + 0.10)^{10} - 1} - \frac{0.55}{10}}{1 - 0.55}\right] 10{,}000 = x + 2{,}400$$

By equating $U_1 = U_2$,

$$3{,}420 = x + 2{,}400 \; ; \qquad x = \$1{,}020$$

Therefore, \$1,020/yr can be spent on painting.

For examples like this, where the alternatives considered do not involve any change in investment, considerable simplification is possible. Thus, with U_k constant and $m = 0$, Eq. (3.46) becomes

$$U = C + \left(\frac{1}{n} + \frac{i_m - 0.35i}{1 - t}\right) I \qquad (7.12)$$

In the above example,

$$U_1 = \left(\frac{1}{5} + \frac{i_m - 0.35i}{1 - t}\right) I$$

and

$$U_2 = x + \left(\frac{1}{10} + \frac{i_m - 0.35i}{1 - t}\right) I$$

Setting $U_1 = U_2$ as before gives

$$x = \frac{1}{10} I$$

so that, with $I = \$10{,}000$, $x = \$1{,}000$, a value that checks the above closely. Note that values are not needed for either i or t in order to obtain this result. i_m was not specified in the problem and is not needed.

Example 7.5

The tubes of an evaporator cost \$8,000, last 8 years, and require a labor cost of \$2,000/yr for cleaning. Tubes can be operated under essentially nonscaling conditions. This will shorten the life of the tubes but will reduce the labor cost for cleaning to $500/yr and result in an estimated additional saving of $1,200/yr because of better heat transfer. If current cost of capital is 10%, how long must the tubes last under the second set of conditions to permit this method of operation?

Solution. For the first set of conditions, use of formula (7.12) gives

$$U_1 = 8,000 \left(\frac{1}{8} + \frac{0.65i}{1-t} \right) + 2,000$$

For the second set of conditions

$$U_2 = 8,000 \left(\frac{1}{x} + \frac{0.65i}{1-t} \right) + 500 - 1,200$$

If we set $U_1 = U_2$,

$$1,000 + 2,000 = \frac{8,000}{x} - 700$$

whence

$$x = 2.2 \text{ years}$$

Thus the answer is obtained readily without assuming values of i, i_m, or t. Jelen [6] solves the same problem, using $i = 0.04$ and $t = 0$ to arrive at the result that the tubes must last between 2 and 3 years.

Jelen [6] has also given a number of examples for situations where expenses vary, following the general procedure discussed by Terborgh [3].

It should be emphasized, in connection with the use of formulas similar to those developed above, that the problem of replacement in chemical plants differs from that in other industries. The "inferiority" due to obsolescence is often reflected in gradual reduction in the sales price of the product, rather than the possibility of new replacements of a given part of the plant operating at systematically higher efficiencies. Replacements are likely to be entirely different; as, for example, using more corrosion-resistant material instead of a new design. Furthermore, replacements in process plants must be keyed to the life of the plant, of which the replacement is only a part. Thus, toward the end of useful plant life, a different replacement policy may prevail from the one that existed when the plant still had many years of useful life ahead. Also, if the replacement can be considered capital equipment, it may be possible to obtain tax deductions that will be larger than on old equipment which has been depreciated for tax purposes.

7.03 TIME EFFICIENCY

In Chapter 3 and Appendix A, equations are developed for the general case where various factors entering into overall profitability may vary. It is assumed that the effect of variations with time can be evaluated by calculations based on a periodic series of interest periods. The factors themselves are assumed to remain constant or to vary with time in some simple fashion as, for example, the variation in allowable depreciation for tax purposes from year to year. Where variations occur over a long period of time, it is usually convenient to tabulate anticipated annual figures (assuming a yearly interest period); see Example 3.2. Allowance for the effect of time is then made by discounting each item to a present worth contribution, as discussed in Chapter 3.

Often, however, unsteady operations occur over a relatively short period of time, so that differences in interest rates need not be considered. Often, too, operations may be cyclic in nature and thus can be handled analytically. Unsteady operations either arise as a result of some intermittent feature in supply and demand or are due to shutdowns in the processing operation. Thus they may be generally classified as the application of economic balances to processes requiring inventory or cyclic operations as has been done in considerable detail by Schweyer [7].

In many process plants raw materials are received intermittently by car or tanker and stored. The processing operation itself may run continuously 24 hr per day except when terminated because of an emergency or for inspection and maintenance. The final product again goes to storage preliminary to outgoing shipment. Thus, depending on the shipping schedules involved, whether the product is sold on a seasonal basis, whether it is perishable or unstable, and the nature of the processing step, there will be variations in the inventory of raw materials and finished products. The in-process inventory is usually a fixed quantity and depends on plant design and capacity. In the design of new plants it can be selected by an economic balance, the cost for its use including interest on working capital as well as fixed charges on the storage facilities necessary. The alternative to increased inventory in economic balances is usually enlarged production facilities.

The general procedure [7] for considering alternatives involving different inventories entails plotting or tabulating a series of production and withdrawal schedules to obtain average inventory. A separate calculation of costs for each series can then be made and tabulated to determine which combination is the most economical, taking all factors of the economic balance into consideration.

In the mechanical industries and in process operations involving solids or materials handling, inventory problems are frequent. Grant and Ireson [4] and Thuesen [8] treat a number of such problems in detail. The problem of determining the economic size of a manufacturing lot arises whenever a machine or group of machines is capable of being shifted from one part or product to another. In the process industries, an excellent example is in the use of expensive glass-lined or stainless steel vessels for the production of seasonal products such as insecticides and other agricultural and food products. Should these materials be produced and stored (produced against inventory) or produced only on order? Another problem [8] involves balancing the cost of carrying feed inventory against the cost of reordering at frequent intervals.

All processes are more or less cyclic. Even primarily continuous processing operations require shutdowns for repairs, inspection, cleaning, and catalyst changing. In addition, batch processes are characterized by production itself being unsteady during normal operation. Usually laboratory and small scale operations are conducted in batch systems. As size increases, a point is reached with most operations (as a rough rule of thumb, 10 million lb/yr) where continuous production is most economical. The important timing consideration in continuous operation is the optimum length of run between shutdowns.

Bertetti [9] has presented an analysis of the economics of run-length determination for processing units. He notes that, with the assumption of no excess capacity, the termination of runs occurs for essentially three reasons: namely, economics, safety, and emergency. Considering economics alone, a typical on-stream period of a processing unit is characterized by operation for a considerable length of time at a comparatively high rate of earnings, followed by a declining-rate period in which fouling or damage to equipment has forced a reduction in feed rate, in severity of processing conditions, or in conversion efficiency.

The general principles are the same as those involved in considering the discontinuance of existing operations as discussed in the preceding section of this chapter, or in Chapter 3. In this case, since no incremental investment is involved, it is simply necessary to operate so as to ensure that gross earnings are maintained at a maximum. The different variables involved are expressed as a function of the run length which it is desired to determine, and in turn the overall profit for the entire period (both off and on stream) is

Overall Considerations in Project Analysis 319

expressed as a function of run length. Differencing (or approximately differentiating) gives the condition when the earning level will be reduced by an additional day on stream.

Example 7.6

A rather heavy organic molecule is to be catalytically cracked (essentially in half), with a conversion of 25 to 50%, to make the desired product (a vinyl monomer similar to styrene) and one of the original reactant molecules. The latter can be separated by distillation and recycled to the first reactor; not the cracking reactor. The unconverted heavy molecule may also be recovered by distillation and recycled to the cracking reactor.

The reaction occurs in the vapor phase at 1,000°F in the presence of substantial quantities of steam. Some carbonaceous material is deposited on the catalyst during the cracking operation so that the activity of the catalyst declines during the period of exposure to the process vapors. After a period of exposure of 30 min the catalyst must be regenerated by passing an air-steam mixture through it. The entire period of chemical reaction, regeneration, and purging with steam takes 60 min.

Despite regeneration, the catalyst gradually loses both activity and selectivity. The average conversion per pass declines and the amount of undesirable by-products such as polymers, tars, and the saturated form of the vinyl monomer increase. After some period of time the catalyst must be removed and new catalyst installed.

The problem is to estimate the optimum time between catalyst changes given certain charges for the cost of the new catalyst and the catalyst change, the loss of yield, the loss of production, and the values of the product and the by-products.

Solution. Two catalytic reactors will be used. When one is on stream, the other will be undergoing regeneration. At the end of the on-stream period the valves will be changed and the roles of the catalytic reactors will be reversed. In this way a steady production is maintained and the catalyst is regenerated at proper intervals.

The effect of the deterioration of the catalyst's activity can be expressed by following the changes in the composition of the reactor effluent gases as the number of passes of the gases through the catalyst increases.

Laboratory and pilot plant work have established that the effluent gases will have the following average compositions (on a steam-free basis) during a period of N exposures of the catalyst:

Desired product:	$0.25 - 0.000029N$ wt. frac.
Recovered by-product:	$0.23 - 0.000029N$ wt. frac.
Recovered reactant:	$0.50 + 0.0000486N$ wt. frac.
Undesirable by-products:	$0.02 + 0.0000139N$ wt. frac.

It should be noted that, as the number of exposures increases, the effluent gases contain less of the desired product and more of the uncracked reactant. This means a decrease in the productivity of the plant and a decrease in plant earnings. If N is allowed to increase to a large number the plant productivity will become uneconomically low.

It is estimated that the catalyst in one reactor can be changed in 24 hr and that the cost of changing the catalyst in one reactor will be $8,000. This amount will include $5,600 for the cost of the new catalyst plus a labor charge of $2,400 for the physical work involved in the change. A charge and a time as low as these indicate that the catalyst bed is of the bulk dumped small particulate solid type; see Jordan [1, p. 253].

During the time of changing the catalyst the plant will continue to run, but using only one reactor. While that reactor is undergoing regeneration the reactor feed stream will be stopped and no reaction will occur.

The economic value of the various compounds involved is as follows:

Desired product:	30¢/lb
Reactant (both new and recycled; see Note):	10¢/lb
Recovered by-product:	2.5¢/lb

(Note: Actually the recovered reactant will cost a small amount more than the new reactant because of the cost of recovering and purifying the recycled material. A detailed economic study would take this difference into consideration, but for present purposes it can be neglected.)

A solution to this problem can be found by writing out an algebraic expression for the earnings of the plant for an average on-stream day. These earnings are to be maximized.

Overall Considerations in Project Analysis

It is convenient to sum the total earnings over one entire cycle of plant operations. A cycle of plant operations includes the changing of the catalyst in both reactors and then steady operation for N passes of the reactant vapors through each catalyst bed.

The earnings per cycle will be

$E = 10{,}000\,N[(0.25 - 0.000029\,N)(0.30)]$

(credit for the sale of the desired product)

$+\ 10{,}000\,N[(0.23 - 0.000029\,N)(0.025)]$

(credit for the recovered by-product)

$+\ 10{,}000\,N[(0.50 + 0.0000486\,N)(0.10)]$

(credit for the recovered reactant)

$-\ 10{,}000\,N[(0.10)]$

(debit for the cost of the feed to the reactor)

$-\ 16{,}000$

(debit for changing catalyst in both reactors)

$-\ 240{,}000\,\{[(0.25 - 0.000058\,N)(0.30)] + [(0.23 - 0.000058\,N)(0.025)]\}$

(debit because of no production during 24 hr of catalyst change)

$+\ 240{,}000(0.1)$

(credit because of no reactant feed during 24 hr of catalyst change)

Summing: $E = 312\,N - 0.04565\,N^2 - 11{,}400$ \$

The total earnings per average operating day are

$E/(N/24) = 7{,}500 - 1.1\,N - 274{,}000/N$ \$/day

Differentiating with respect to N and setting equal to zero gives

$-1.1 + 274{,}000/N^2 = 0$

$N = 507$ runs

At one run per hour this result says that the catalyst in each reactor should be changed every 21 days. This seems to be fairly frequent, but it could be done. Actually, an examination of the expression for the total average earnings per operating day shows that the maximum in the curve is very flat. The earnings are quite insensitive to the value of N. If N were made as high as 1,000 runs, or almost 42 days between catalyst changes, the resulting earnings would only be 4.4% below the maximum.

An analysis of the various factors requiring shutdowns of petroleum refinery equipment has been reported by Shannon [10]. Although Shannon's work is specifically directed at petroleum refineries, much of his discussion is applicable to chemical plants.

In addition to direct economic factors such as the deterioration of catalyst treated in Example 7.6, other items such as fouling, mechanical deterioration, and corrosion are listed, as well as planned inspection periods and emergency considerations. Means for increasing run length are discussed, and it is suggested that the following four deserve special attention:

1. Preventive maintenance
2. Thorough inspection
3. Improving the quality of maintenance
4. Use of spare equipment

Many technical as well as economic aspects enter into the detailed implementation of items like those listed above, so that exhaustive discussion here will be impractical.

The problem of the reliability of the component parts of a system has been thoroughly investigated [11, p. 381; 12]. The theory of probability has been used to analyze the problem of providing redundant equipment to improve the reliability of the system. There is an extensive development, and there have been many applications to modern high-technology systems: computers, rockets, and nuclear reactors. As is apparent, any kind of useful answer must rest on the statistical data regarding the probability of failure of the equipment. Accurate data of this kind are hard to obtain.

Two aspects will be briefly discussed because they often enter into economic calculations: namely, the problems of specification of corrosion-resistant materials and of providing adequate spares.

The problem of economics of installation of spares or spending extra money on corrosion-resistant materials is essentially one of

Overall Considerations in Project Analysis 323

providing an initial capital investment to avoid future down time. In order to evaluate the economic advantage of such investments, it is desirable to have the following information:

1. Statistical data regarding the probability of failure of equipment with and without the additional investment. In the case of spares it is sometimes possible to accumulate such information from plant operating data, though frequently nothing more than an educated guess is possible. Corrosion data are often not available except by actual experience with the process installation involved.

2. An estimate of the loss resulting from shutdown caused by equipment failure.

3. The cost of the proposed additional equipment or material.

It is then possible to weigh the economic advantage of the extra investment, assuming an appropriate payment time, against the loss of product due to forced shutdowns.

One point that should not be overlooked in considering problems of this type is the case where operating facilities are not always running at peak capacity to meet market demands. If a particular facility operates at less than peak capacity, it is possible to catch up after a forced shutdown, and, since market demands will be met at all times, there is not much financial loss from the failure of a facility to operate continuously. This situation applies very often in chemical plants where batch operations and multipurpose units are employed. In the larger continuous units used in refining and petrochemical manufacture, down time is more serious since little intermediate process storage capacity is available.

Bell [13] analyzing the economics of spare, or standby (redundant), equipment in oil refineries discusses the above matters in some detail. He also presents an interesting development for the calculation of the number of units to install in order that capital costs will be a minimum, while providing one idle standby unit.

Let I equal total cost of a group of identical installed units each having a fraction f of the total design capacity. With one unit as an idle standby the number of such units required will be $1 + 1/f$. If I_1 is the cost of a single unit with capacity equal to the normal design operating capacity of the plant; then the cost of one small unit having the fraction f of the capacity of the large unit will be

$$I_1 f^a$$

This comes from an application of the cost-size exponential

relationship discussed in Chapter 5. From this it can be seen that the total cost I will be

$$I = (1 + 1/f)I_1 f^a$$

A minimum value for I may be found by differentiating the above equation with respect to f and setting the derivative equal to zero. Whence,

$$f_{optimum} = (1 - a)/a \tag{7.13}$$

Since the number of units to be used is $1 + 1/f$, and a is frequently between 0.6 and 0.7, the number of units required will be 3.

It is important to note that the true economic criterion for installation of spares is not minimum investment, however, but maximum overall profitability. This will result in application of Eq. (7.13), only when the size of the single facility I_1 is fixed at the optimum; for example, if a pump is involved, it must be operating at such an efficiency that pumping costs for power and fixed charges are at a minimum. In addition, the various fractional-sized equipment must then also operate at the same efficiency. If the efficiency of the smaller equipment is less, this item would have to be considered in making an economic balance.

Another consideration in specifying spares is that it is not always desirable to have the standby equipment idle. Thus, with an installation involving four compressors it might be desirable to run all four at a reduced load during normal operation, and, in the event of failure of one, to operate those compressors at a reduced efficiency. In this way maintenance costs might be less, and it would be possible to employ a lower capital investment than if the spare were to remain idle during normal operation.

7.04 OPTIMUM MATERIAL BALANCE

In many industrial chemical processes the major portion of the manufactured cost of the final product comes from the costs of the raw materials. In some other processes, such as those involving large amounts of electrical power, the costs of the raw materials may not be so important, but, quite generally—especially in the synthetic organic and petrochemical industries—the cost of the raw

Overall Considerations in Project Analysis

materials is from 60 to 90% of the cost of the final product; see Jordan [1, p. 13].

A consequence of this fact is that industrial chemists and engineers must pay very careful attention to several considerations: (1) the cost of the raw materials; every effort should be expended to purchase raw materials for the minimum price. (2) The chemical reaction should be made as efficient as possible; the production of unusable by-products should be minimized. (3) The plant should be operated in the optimum manner. The last, of course, means that a combination of process variables must be found that will cause the plant to operate at minimum cost, or maximum profitability.

For industrial chemical reactions there are two properties of the reaction that are all-important and are interrelated: (1) The conversion of the reaction(s)

$$\text{Fractional conversion} = \frac{\text{lb raw material in - lb raw material out}}{\text{lb raw material in}}$$

There may be different conversions for each of the reactants. The fractional conversion shows how much of the raw material(s) is converted into something else, not necessarily into what is desired, but into something besides the original material. (2) The yield of the reaction. This is defined by

$$\text{Fractional yield} = \frac{(\text{lb desired product})(\text{raw material/product Stoichiometric ratio})}{(\text{lb raw material in - lb raw material out})}$$

The yield shows how much of the raw material which has been converted into something else has been converted into the desired product; 1 - yield represents the fraction of material which has been converted to by-product or waste material. These by-products may have little or no value, but the quantity produced must be known so that proper plans may be made for their disposal. If there are two or more reactants there may be a different yield for each reactant.

It is frequently true that there is an inverse relationship between conversion and yield. This cannot be predicted or calculated, but must be measured in the laboratory and/or pilot plant. It is a matter of experience that a higher conversion, caused by more severe reaction conditions, will almost always result in a lower yield. The proper value of the conversion to be used (the optimum material balance) can be found by an economic balance between

(a) high conversion with lower capital investment and operating costs (smaller recycle of unconverted raw material), but with higher raw material costs (lower yield), (b) lower conversion (with higher yield) and higher capital investment and operating costs (greater recycle of unconverted raw material), but lower raw material costs.

In some cases, such as the manufacture of monochlorobenzene where a high conversion will lead to the production of higher chlorobenzenes, a low conversion (say 6%) may be used. In other cases, such as the production of phthalic anhydride from naphthalene, a conversion of 100% is best.

It is apparent from these considerations that the entire plant may be involved with each alternative studied. Lower yields, caused by a more severe reaction or by the rejection of raw materials or product to waste streams, will increase the amount of feed material which must be fed to produce a pound of product. This will raise the manufactured cost of the product and perhaps increase the size and expense of the feed preparation unit. A higher yield will reduce the raw material charge to the product factory cost, but will increase the size, capital investment, and operating costs of the reactor and the separation equipment.

Every time the reaction conditions are changed to give a new conversion-yield relationship the separation and recycle conditions must also be changed so as to give an optimum operation.

In practical operations it is frequently true that the optimum is rather flat. Fairly large changes in process conditions may not make significant changes in the cost of the plant or the product. In such circumstances the various sections of the plant may be treated independently of one another. The assumption here is that slight variations in the amount and composition of the steams flowing between sections will not affect the designs of the units.

There are, of course, other processes where the entire plant should be considered as a unit. This is not often done because of the cost of carrying out the extensive computations and because of the lack of accurate data. However, such problems are now yielding to an attack based on a combination of greater amounts of data and experience (such as for ammonia and sulfuric acid), modern high-speed computing machines, and recent advances in optimization and control theory.

In many chemical plants the chemical reaction section may be the only section which is optimized, because the consumption of raw materials, which is so important to the factory cost of the product,

Overall Considerations in Project Analysis 327

is largely controlled by the operation of the chemical reactor. The remainder of the plant is then designed to fit the amounts and compositions of the streams entering and leaving the reaction section. These amounts and compositions are determined from an optimum material balance on the chemical reactor.

Example 7.7 illustrates the situation of a low conversion-high yield plant where an economic balance around the reaction section may be used to determine the amounts and compositions of the other plant streams.

Example 7.7

A preliminary design has been prepared for a unit to produce 10,000,000 gal/yr (67,000,000 lb/yr) of 95 wt% ethanol by catalytic vapor-phase hydration, using 97 wt% ethylene as the feed material. This process is described by Nelson and Courter [14]. A schematic flow diagram is shown in Fig. 7.3.

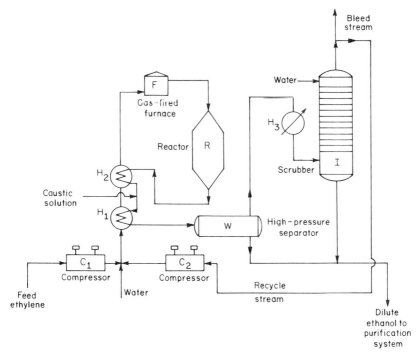

Fig. 7.3. Reaction section—ethylene hydration unit.

In this unit, ethylene is compressed to 1000 psig and fed into the recycle stream. Water is combined with the ethylene to give a feed containing 0.6 mole of water per mole of ethylene at a temperature of 180°F. This feed is heated first by the reactor products, in exchangers H_1 and H_2, and then by the gas-fired furnace, F, to the reaction temperature of 570°F. From the top of the reactor, the feed moves downward through a phosphoric acid catalyst supported on Celite.

The reaction product is partially condensed in exchanger H_2 by the incoming reactor feed stream. Phosphoric acid carried from the reactor is neutralized with caustic soda. Further condensation then occurs by more heat exchange in exchanger H_1. The condensed liquid is separated at high pressure in vessel W, from the vapor stream, which is cooled and scrubbed with water in tower I to remove alcohol. It must be noted that the entire system is kept at high pressure. Compressor C-2 simply supplies the pressure drop incurred by the gas during its passage around the system. In order to prevent accumulation of inerts (ethane), a small part of the vapor is removed while the remainder is recycled.

In the purification section, the reaction product is fed to a stripping column to concentrate the ethanol. The overhead is hydrogenated to convert aldehydes to alcohols. The crude product is then sent to a column where diethyl ether and other light products are removed, and then to a fractionating column where the ethyl alcohol is concentrated to a final product.

The preliminary evaluation indicates that the process is attractive economically, but there are a number of items in the design which should be more precisely specified for optimum performance. One of these involves an economic balance to determine the optimum material balance for the reaction section.

The object of this example is to detail this economic balance. For this purpose the reaction section can be considered separately from the purification system. The design of the purification system will not depend to any great extent on the small variations in feed composition produced by variations in the vented stream.

The problem of the quantity and quality of the vented stream from a chemical process is a recurring one, particularly with those processes that utilize gaseous feed materials. Any chemical reaction system that recycles unreacted feed material to the reactor will accumulate inert material. No reactor feed stream is 100% pure; the chemical reaction will remove the active portion of the feed stream and the inert portion will remain. On recycle, the unreacted portion will accumulate. This inert material must not be allowed to accumulate for more than a short period of time because it will eventually seriously reduce the efficiency of the reaction. In the limit, the system would be full of inerts and no reaction at all would occur.

The concentration of inerts may be kept low by venting, or "bleeding" a small amount of the recycle stream out of the system. When this is done an economic balance problem arises which centers around the quantity of material to be vented. If a large amount is vented the accumulation of inerts is kept small, the size and power of the recycle compressor are low, and the size and cost of other equipment (such as heat exchangers) is less. Heat exchangers, reactor feed preheaters, compressors, pumps, and the chemical reactor itself can be smaller and cheaper. In addition, the costs for fuel and cooling water are reduced. However, valuable raw material will be removed with the vented stream and this must be recovered, or replaced, and this will increase the cost. Also, the new feed stream must now be larger and its processing equipment, such as pumps, compressors, and preheaters must also be larger, thus making the cost greater.

If the quantity of vented material is reduced the reverse of the above situation occurs. Thus, there is an optimum quantity of recycled material to be vented.

In the subject problem, decreasing the amount of the bleed stream increases the quantity of recycle. This increases the size and power consumption of compressor C_2. However, there would be a decrease in the make-up ethylene required, and a reduction in size and power consumption of compressor C_1. Because of the accumulation of inerts, the sizes of heat exchangers H_1, H_2, and H_3 would increase, and a larger furnace would also be required. In addition, the cost of fuel and cooling water would rise. To keep the space velocity in the reactor constant, a larger reactor would be required. It is assumed that small changes in ethylene concentration would not influence the conversion per pass through the reactor. Also, changes in the separator W and the scrubber I are assumed to be small.

The schematic flow diagram, Fig. 7.3, shows the pertinent equipment involved in this economic balance.

The design is based on the following information:

Reaction temperature	570°F
Reaction pressure	1000 psig
Reaction feed ethylene concentration	85% (water-free)

Ethylene make-up concentration	97 wt%
Molal ratio of water to ethylene in feed	0.6
Space velocity in reactor (vol of gas @ 60°F and 1 atm/min/vol catalyst)	30
Ethylene conversion per pass	4.2%
Time on stream	95%

Ethylene in the bleed stream, which must be reconcentrated elsewhere, is assumed to have a value of $0.01/lb, as compared with a value of $0.04/lb in the concentrated feed.

Utilities costs are assumed as follows:

Cooling water	$0.08/M gal
Gas	$0.30/MM Btu
Power	$0.02/kw-hr

Maintenance costs on incremental investment are taken at 7%/yr, and the maximum acceptable payout time is assumed to be 2 years. In computing relative investments for different alternatives, it will be assumed that piping and instrument costs will remain unchanged as well as various overheads which would enter into erection of a complete new plant.

The original article by Nelson and Courter [14] says nothing about the yield of the hydration reaction. It cannot be 100% because the authors state that some diethyl ether and some heavy polymeric material is formed. On the other hand, the yield must be fairly high because the authors state that only small quantities of by-products are made. In industrial organic chemistry a yield of greater than 95% is unusually high, but for present purposes a yield of 98% will be assumed. Then,

The specific gravity of 95 wt% ethanol-water @ 20°C = 0.804

$(10,000,000)(8.35)(0.804) = 67,200,000$ lb/yr of 95 wt% ethanol made

Operating 95% time on stream

$(67,200,000)(0.95)/(0.95)(365)(24) = 7,680$ lb/hr 100% ethanol made

$0.042 = $ (lb ethylene in - lb ethylene out)/(lb ethylene in)

Overall Considerations in Project Analysis

$$0.98 = \frac{(7{,}680)/(28/46)}{(0.042)(\text{lb ethylene in})}$$

Lb ethylene in = 113,300 lb/hr

Lb ethylene converted = (0.042)(113,300) = 4,750 lb/hr

Lb ethylene converted to ethanol = (0.98)(4,750) = 4,655 lb/hr

Lb ethylene converted to by-products = (0.02)(4,750) = 95 lb/hr

Solution. For the present study the reactor feed concentration is to be varied from the 85% design concentration. It is assumed that the reaction is sufficiently far from equilibrium so that the effect of this change on the fractional conversion can be neglected.

Fig. 7.4 shows a flow plan for the reactor. The number of pounds per hour of the bleed stream, z, is to be varied. The other streams in the reactor system—x, the pounds per hour of fresh 97 wt% ethylene; y, the number of pounds per hour of the recycle stream (which has a composition p weight fraction ethylene, and the products 7,680 lb/hr of ethanol and 95 lb/hr of waste—are connected by means of a series of material balances.

Ethylene balance around feed mixing point

0.97x + py = 113,300

Inerts balance around all reactor

0.03x = (1 - p)z

Ethylene balance around all reactor

0.97x = pz + 4750

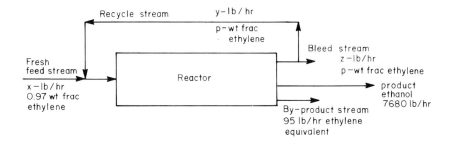

Fig. 7.4. Flow plan for ethylene hydration reactor.

Table 7-7 Material Balance on Reactor

p Wt frac ethylene in recycle and bleed streams	x Fresh feed stream (lb/hr) 0.97 wt frac ethylene	y Recycle stream (lb/hr)	z Bleed stream (lb/hr)
0.85	6,140	126,000	1,225
0.80	5,769	134,000	863
0.75	5,567	156,400	667
0.70	5,438	153,800	542
0.65	5,350	165,400	457
0.60	5,290	179,000	398

These equations may be solved simultaneously for several possible values of p. These results are tabulated in Table 7-7.

With these quantities fixed it is then possible to design the plant. That is, the various pieces of equipment can be sized, the materials of construction chosen, the quantities of utilities to be consumed calculated, and then both fixed and operating costs for the plant found. The sum of the fixed and operating costs plus the cost of the fresh ethylene feed minus the credit for the ethylene in the bleed stream will give the total operating cost for the plant for each value of p. This cost will be expressed in dollars per year.

It was stated previously that the maximum acceptable payout time will be 2 years and that the maintenance cost will be 7% of the installed cost of the equipment per year. Thus the fixed charges on each piece of equipment will be (0.5 + 0.07)(installed cost) $/yr.

The raw material cost to the plant may be calculated by multiplying the quantity x by $0.04/lb, and the credit to the project for the ethylene in bleed stream will be (0.01)(z). A typical set of these numbers is given in Table 7-8 for the single case of p = 0.80.

Figure 7.5 and Table 7-9 show this summary. Apparently the sum goes through a minimum at a value of p of about 0.64. However, as has been encountered before, the minimum is flat and low-cost operation can be achieved at any value of p lower than 0.75. In fact, the "design" value used by Nelson and Courter of 0.85 is

Overall Considerations in Project Analysis

Table 7-8 Total Annual Cost, p = 0.80

	$/yr
Fixed charges	
C_1	64,100
R	166,000
H_1	32,750
H_2	40,500
H_3	36,000
F	16,820
C_2	27,200
Total fixed charges	380,170
Operating expenses	
C_1, power	49,600
H_3, water	17,200
F, fuel	19,700
C_2, power	27,000
Total operating expense	113,500
Raw material cost (0.04)(5769)(8760)(0.95)	1,920,000
Credit, dilute ethylene	-57,500
Total operating cost	2,362,170

only about 5% higher than the minimum. In view of the uncertainties involved in the cost data and in the design procedures it cannot be said that operation at a p of 0.64 is significantly more economical than operation at a p of 0.85.

If values of p as high as 0.95 are used the cost does increase rather strongly as p increases. High values of p lead to excessive raw material costs and feed stream compression (C_1) charges. Lower values of p become uneconomical largely due to high costs for the reactor.

Table 7-9 Total Annual Cost ($/yr)

	\multicolumn{6}{c}{p}					
	0.85	0.80	0.75	0.70	0.65	0.60
Fixed charges	373,230	380,170	395,450	409,700	419,000	434,000
Operating expenses	112,300	113,500	115,500	120,000	124,800	130,500
Raw material cost	2,038,000	1,920,000	1,850,000	1,808,000	1,780,000	1,760,000
Credit	86,900	57,500	41,600	31,600	24,750	19,900
Total operating cost	2,436,630	2,362,170	2,319,350	2,306,100	2,299,050	2,304,600

Overall Considerations in Project Analysis 335

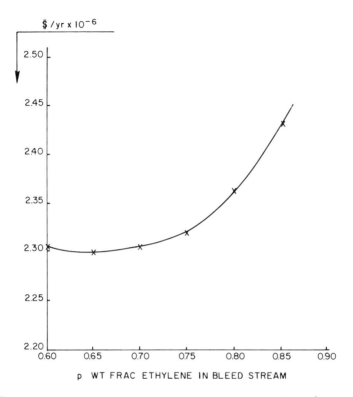

Fig. 7.5. Total annual cost of ethanol by hydration of ethylene.

It is interesting to compare the results of this calculation, done in 1970, to that in the previous edition, done in 1958. In the 1970 case the costs for the equipment, the installation, and construction, and for some of the utilities were quite a bit higher than the comparable costs in 1958. However, the principal cost to this process is that of the raw material, and this did not change from 1958 to 1970. Thus, despite higher equipment and utility charges, the total operating cost of this process is only 7 to 8% greater in 1970 than in 1958. It is typical of large-scale synthetic organic chemical processes such as this one that the cost of the raw material is the major part of the final product cost.

Note that, since the amount of ethanol to be produced is fixed by the design specified, the value of ethanol does not enter into the economic balance. If the problem had been stated as a design problem involving the production of the economic amount of ethanol from a fixed available quantity of ethylene, it would have been necessary to take the design of the purification unit into consideration in order to establish the value of dilute ethanol from the reaction section. Similarly, if the unit were an operating unit of fixed design and it were desired to determine the economic level of operation with no limitation on feedstock or product quantity, economic balances would involve the entire unit. It is always important that the restrictions necessary to establish the solution be carefully specified.

7.05 THE USE OF COMPUTERS IN PROJECT ANALYSIS

In Section 7.00 it was stated that "a large number of technical and economic evaluations can be laborious, time-consuming, and expensive." As described in Chapter 4, the amount of computation required to find the optimum point can be quite large and prohibitively expensive.

This fact has been known for some time, and much ingenuity has been applied in an effort to reduce the burden of such work.

At present, the electronic digital computer is being employed to do the calculations necessary for economic and technical evaluations. This work includes the calculation of heat and material balances, the sizes of the pieces of equipment, the cost of the equipment, and a value for the venture worth.

Much high-level work has gone into the development of such calculation procedures. It would not be appropriate to discuss these methods here because of their complexity, the lack of space in this rather small book, and because the book attempts to focus attention on simple fundamental matters without undue complication and extensive calculation.

However, in industry, where greater amounts of time and money are available, and where a minor change in operating conditions or design specifications can result in substantial savings, these computer programs for design and economic evaluation are now widely used.

Those interested in such matters may consult Refs. [15] through [19].

As an illustration of computer application, the following is a description of a computer program which was developed to solve Example 7.7. The computer program includes a Main Program and

Overall Considerations in Project Analysis 337

five Subprograms. The structure of the main program and the functions of these subprograms are as follows:

Program	Function
Main	(a) Reads input data such as operating conditions, utility costs, raw material cost, by-product cost.
	(b) Prints output data such as fixed charges, operating expenses, cost of raw material, credit from by-product, and total operating cost.
Subroutine 1	Solve the simultaneous equations for specified value of p to obtain the values of x, y, and z.
Subroutine 2	Compute the fixed charge of each of equipment items for various values of p, x, y, and z, which are obtained in Subroutine 1 and sum up these fixed charges to calculate the total fixed charges for each value of p.
Subroutine 3	Compute the operating expense of each of equipment items for the values of p, x, y, and z, which are obtained in Subroutine 1 and sum up these operating expenses to calculate the total operating expenses of all the equipment for each value of p.
Subroutine 4	Calculate the cost of raw material and the credit from dilute ethylene for the various values of p.
Subroutine 5	Compute the total annual cost by summing up the costs obtained in Subroutines 1 through 4.

For the computation of both the fixed charge and the operating expense of each piece of equipment we first chose the operating conditions for p = 0.80 as a standard condition and calculated the fixed charges and the operating expenses of all the equipment for the standard condition. For the cost estimation of the equipment items a manual entitled Chemical Engineering Costs by C.E. Dryden and R.H. Furlow (Columbus, Ohio: Ohio State University, 1966) was used. For the costs of equipment items for operating conditions other than standard state, the 0.6 power rule was employed. For the costs of utilities we assumed that the cost is proportional to the corresponding gas flow rate. For example, the cost of furnace and the cost of gas when p = 0.82 were calculated as follows:

$$\begin{pmatrix} \text{Fixed charge of furnace} \\ \text{for } p = 0.82 \end{pmatrix} = \begin{pmatrix} \text{fixed charge of furnace} \\ \text{for } p = 0.80 \end{pmatrix} \times \left(\frac{x \text{ for } 0.82}{x \text{ for } 0.80} \right)^{0.6}$$

$$\text{Cost of gas for } p = 0.82 = (\text{cost of gas for } p = 0.80) \times \left(\frac{x \text{ for } 0.82}{x \text{ for } 0.80} \right)$$

In this computation the values of p were changed from 0.5 to 0.9, incrementing with 0.01. Needless to say the total costs printed should be plotted against p to find the optimum operating conditions.

Execution time (UNIVAC 1108)	4.1 sec
Computer program (to write)	3 hr
Search for cost data	5 hr

Results obtained were similar to those shown in Fig. 7.5.

SUMMARY

In this chapter the analysis of projects was discussed from an overall point of view. That is, concentrating on a complete material balance for the entire operation, instead of performing an economic balance, or detailed design, on each item of equipment.

Economic balances were used to fix yields, conversions, and recoveries of important steps in the processing sequence.

Once the optimum material balance was established a more detailed design could be attempted.

REFERENCES

1. D.G. Jordan, Chemical Process Development, Interscience Publishers, John Wiley and Sons, Inc., New York, 1968.
2. J. Happel, Chemical Process Economics, John Wiley and Sons, Inc., New York, 1958.

3. G. Terborgh, Dynamic Equipment Policy, McGraw-Hill Book Co., New York, 1949.

4. E.L. Grant and W.G. Ireson, Principles of Engineering Economy, 4th ed., The Ronald Press Co., New York, 1960.

5. G.A. Taylor, Managerial and Engineering Economy, D. Van Nostrand Co., Princeton, New Jersey, 1964.

6. F.C. Jelen, Chemical Engineering, 61 (No. 2), 199 (1954); 62 (No. 8), 181 and (No. 12), 183 (1954).

7. H.E. Schweyer, Process Engineering Economics, McGraw-Hill Book Co., New York, 1955.

8. H.G. Thuesen, Engineering Economy, 2nd ed., Prentice-Hall, Englewood Cliffs, New Jersey, 1957.

9. J.W. Bertetti, Petroleum Refiner, 34, No. 3, 134 (1955).

10. R. Shannon, Petroleum Refiner, 34, No. 3, 132 (1955).

11. D.F. Rudd and C.C. Watson, Strategy of Process Engineering, John Wiley and Sons, Inc., New York, 1968.

12. I. Bazovsky, Reliability: Theory and Practice, Prentice-Hall, Englewood Cliffs, New Jersey, 1961.

13. J.E. Bell, Oil and Gas Journal, 51, 244 (1953).

14. C.R. Nelson and M.L. Courter, Chemical Engineering Progress, 50, 527 (1954).

15. I.V. Klumpar, Chemical Engineering, September 22, 1969, pp. 114-122; January 12, 1970, pp. 107-116; June 29, 1970, pp. 76-84.

16. I.V. Klumpar, Chemical Engineering Progress, 67, No. 4, 74 (1971).

17. C.M. Crowe, P.T. Shannon, et al., Chemical Process Simulation, John Wiley and Sons, Inc., New York, 1970.

18. E.J. Henley and E.M. Rosen, Material and Energy Balance Computations, pp. 337-437, John Wiley and Sons, Inc., New York, 1969.

19. S.A. Bresler and M.T. Kuo, Chemical Engineering, June 26, 1972.

PROBLEMS

7.1. Styrene (vinylbenzene) is made in very large quantities by the vapor-phase dehydrogenation of ethylbenzene. The ethylbenzene may be made by one of two methods: (a) the ethylation of benzene by the reaction of ethylene and benzene, and (b) by the fractional distillation of the ethylbenzene-mixed xylenes stream which is made during the reforming of certain selected petroleum naphtha streams.

The approximate analysis and the boiling points of a typical refinery stream is as follows:

	Boiling point 760 mm Hg (°C)	Weight%
Ethylbenzene	136.2	20
o-Xylene	144	21
m-Xylene	139.3	42
p-Xylene	138.5	17

The difference in boiling points between ethylbenzene (EB) and the next most volatile xylene, p-xylene (pX), is only 2.3°C, but this is sufficient to make possible a clean separation between the EB and the pX by fractional distillation.

However, it must be recognized that there are two other factors that also apply here. If these special characteristics of this system were different, the distillation could not be made. These special characteristics are (1) the xylenes are thermally stable at temperatures near their boiling points, and can be boiled for long periods without suffering degradation; (2) the xylenes are essentially noncorrosive, so that ordinary low-carbon steel can be used in the fabrication of the distillation system.

The combination of these three characteristics makes it possible, at least in principle, to produce pure ethylbenzene by fractional distillation.

What would be the capital investment and the operating cost for a distillation system which would produce 99.5% EB as product,

Overall Considerations in Project Analysis 341

and recover 90% of the EB in the feed, at the rate of 50,000,000 lb/yr of EB?

The following data apply:

1. The system may be regarded as the binary, ethylbenzene-p-xylene, with ethylbenzene as the more volatile component.

2. The relative volatility of this ideal binary system is equal to the vapor pressure ratio. This value is 1.060 and may be regarded as constant over the temperature range involved.

3. The latent heat of vaporization for both compounds is 15,500 Btu/lb mole or 146 Btu/lb. The liquid-phase specific heat of the compounds is 0.4 Btu/(lb)(°F).

7.2. A company is contemplating the production of styrene by the ethylene-benzene-ethylbenzene route. The company is a large one, used to dealing with costly endeavors, and is willing to invest the capital necessary for a styrene plant to produce 1×10^9 lb/yr of product.

There are two approaches to the problem of the design of very large chemical plants.

1. The "single-train" design. Here, there is one primary flow line and only one set of equipment. The operations proceed sequentially and all of the process flow stream contacts all of the equipment. This design makes use of the "economy of scale" concept as discussed in Chapter 1.

The disadvantage of the single-train concept is that any shutdown due to power failure or equipment malfunction will cause a shutdown of all operations with accompanying losses of great magnitude.

2. The "multiple-train" design. Here, two or more production units are connected in parallel, and production goes on in the parallel connected units. If there is an equipment failure in one train, the other train can continue to produce. The rate of production is halved for a while, but it is not stopped. Thus, the plant is less susceptible to unplanned shutdowns. The initial capital investment will be greater, and the cost of operation will be somewhat higher, but the greater reliability of operation may more than offset these drawbacks.

(a) For a styrene plant of 1×10^9 lb/yr capacity list the advantages and disadvantages of the single-train design philosophy.

(b) According to Guthrie [K.M. Guthrie, Chemical Engineering, June 15, 1970] a 5×10^8 lb/yr styrene plant will cost $22,000,000 and a 1×10^9 lb/yr styrene plant will cost $33,500,000. Calculate the depreciation charge for (i) a plant producing 1×10^9 lb/yr of styrene using one single-train unit costing $33,500,00; (ii) a plant producing 1×10^9 lb/yr of styrene using two trains each of 5×10^8 lb/yr capacity and each costing $22,000,000. Let $d_k = 0.10$.

(c) Consider the comparison between a single-train unit of 1×10^9 lb/yr and a double-train unit where each unit has a capacity of 5×10^8 lb/yr. Derive an expression for the yearly profit of both types of plants allowing d days per year for unplanned shutdowns. Assume that only one of the trains of the double train unit will be shut down.

For various values of the sales price s calculate how many days of shutdown can be tolerated before it would be more economical to build a double-train unit than a single-train unit.

7.3. In recent years the chemical industry has seen the productive capacity for many pure chemicals increase to quite high levels. At the same time, the sales price of these chemicals has dropped in an astonishing manner. It is stated that some chemicals are sold in the billions of pounds a year for little more than it costs to produce them.

Make an approximation calculation showing the effect of sales volume on the profit margin for a high-volume chemical. Just how small can a profit margin be?

7.4. A process for the sulfonation of phenol requires the use of a 3,000-gal kettle. It is desired to determine the most suitable material of construction for this vessel.

The following data and assumptions are applicable. If the reactor is replaced, the process will be liable for the following expenses:

Idle manpower	$13,500
Loss in production	63,000
Plant burden (overhead) for cost of replacement	8,500
	$85,000

Overall Considerations in Project Analysis 343

The time value of money is taken into account by use of an interest rate i = 0.15. Minimum acceptable return rate is i_m = 0.20. The total income-tax rate is to be taken at t = 0.55.

The life of a kettle n is calculated by dividing the corrosion allowance 0.15 in. by the estimated corrosion rate. The equipment is assumed to have no salvage value at the end of n years.

For the case in question, corrosion data indicate that only a few corrosion-resistant alloys will be suitable [G.A. Nelson, Corrosion Data Survey, Shell Development Co., 1954].

Vessel type	Installed cost	Average corrosion rate (in./yr)	Average life (years)
Monel clad	$ 95,000	0.01	15
Nickel clad	80,000	0.02	7.5
Hastelloy B[a]	250,000	0.002	25

[a] A maximum life of 25 years is arbitrarily set to allow for obsolescence.

7.5. As was discussed in Example 4.6, natural gas is pumped over quite long distances through steel pipelines at pressures near 300 psia. In recent years, other gases, such as ethylene and oxygen, are also being transported in a similar manner, although over shorter distances.

At present, it is contemplated that pure elemental hydrogen be used as a source of energy, in a manner similar to natural gas, that is, to be burned with air, as a fuel, and that the hydrogen be transported through pipelines over long distances.

Using Example 4.6 as a reference, discuss the general problem of transporting hydrogen gas through a pipeline and estimate the cost of this operation.

7.6. A small sulfite pulp mill produces 200 tons/day of air-dried wood pulp by means of the traditional calcium bisulfite digestion process. There is no recovery process for either heat or chemicals, and the stream pollution problem is severe.

The managers of the mill have been told by the public health authorities that the stream pollution must stop.

The company wishes to estimate the economics of modernizing its wood digestion practice and, at the same time, complying with the ruling of the health authorities.

The steps that have been decided on are as follows:

1. A change from calcium bisulfite to magnesium bisulfite cooking liquor. This change should have several advantages, among them, an increased capacity for the same digester volume.

2. The installation of a recovery system to conserve heat and chemicals and to abate stream pollution.

J.J. Voci and F.D. Iannazzi [Chemical Engineering Progress, 61, 110 (1965)] give the following data:

	Existing mill	Modernized mill
Production (tons/day)	200	250
New capital investment ($)	0	5,200,000
Operating costs ($/ton)		
Wood	50.6	45.7
Pulping chemicals	5.2	3.6
Bleaching chemicals	4.1	6.1
Operating labor	7.3	7.0
Maintenance	4.0	5.2
Fuel oil	4.2	0
Electrical power	1.6	0.6
Overhead expenses	8.0	7.7
Depreciation, local taxes, and insurance	2.4	10.5
Total operating cost	87.4	86.4
Credits (steam)	0	1.4
Net operating cost ($/ton)	87.4	85.0

Overall Considerations in Project Analysis 345

Wood pulp can be considered to sell for $140/ton.

Calculate: (a) the incremental profit after income taxes for the new installation; (b) the incremental percent return on the new investment; (c) the payout time for the new investment after taxes.

7.7. Vinyl chloride is the principal vinyl monomer now manufactured in the world. In the United States in 1971, 4.19 billion pounds were sold, and the capacity to produce this chemical is even greater than that. The production and sale of vinyl chloride monomer is very big business indeed.

Until approximately 1968 vinyl chloride was manufactured by the vapor-phase reaction of acetylene and hydrogen chloride. Now, because of the low price and ready availability of high-purity ethylene, a new process has been introduced that uses ethylene, air, and chlorine to produce vinyl chloride according to the reactions:

$$C_2H_4 + Cl_2 \rightarrow C_2H_4Cl_2 \tag{1}$$

$$C_2H_4 + 2HCl + 1/2\, O_2 \rightarrow C_2H_4Cl_2 + H_2O \tag{2}$$

$$C_2H_4Cl_2 \rightarrow C_2H_3Cl + HCl \tag{3}$$

The two chlorination reactions are balanced so that all the by-product HCl is consumed, and the required amount of $C_2H_4Cl_2$ is made by a combination of reactions (1) and (2). The result is

$$C_2H_4 + 1/2\, Cl_2 + 1/4\, O_2 \rightarrow C_2H_3Cl + 1/2\, H_2O$$

Some trichloroethylene appears as a by-product.

This process has been described in British Chemical Engineering for May 1969. The following data were given:

Production rate	100,000,000 lb/yr VC
Capital investment	$5,000,000
Raw materials	$/ton VC
Ethylene @ 3.5¢/lb	34.20
Chlorine @ 2.75¢/lb	36.30
Catalysts and chemicals	1.00
Total raw materials	71.50
Total utilities	5.40

            Direct costs including maintenance   7.90
            Fixed costs   2.20
            Depreciation   9.80
                Total manufacturing cost   96.80 (say 97.0)
                    or 4.85¢/lb

    Calculate: (a) the percentage of the manufacturing cost represented by each of the cost items in the list above; (b) the effect of increasing production to 200,000,000 lb/yr and to 500,000,000 lb/yr on the manufacturing cost and on the percentages calculated in (a) above; (c) the return on investment for each of the above production rates if all the production could be sold at a price of 4.75¢/lb; (d) the break-even point (see Chapter 2) for each of the productive capacities listed above; (e) suggest several research and development projects that might be undertaken with the goal of reducing the manufacturing cost of vinyl chloride.

7.8. Contaminated water is often purified by a three-stage process wherein the final stage is the adsorption of certain organic compounds on activated carbon. In such a process the activated carbon is contained in steel vessels, the water is pumped downward through the bed of activated carbon, the organic compounds are adsorbed, and acceptably pure water leaves the adsorbers. After the passage of a rather large quantity of water the carbon becomes saturated with impurities and must be regenerated. This may be accomplished by removing the carbon, conveying it to a furnace, and there regenerating it by strong heating in the presence of steam. The regenerated carbon is then returned to the adsorbers. In practice, this is done by having the carbon in at least two vessels connected in series. The carbon in the lead vessel becomes exhausted first, is drained, the valving is changed, and the water then flows through the second vessel first.

    This process has been described in some detail by Cover and Wood [A.E. Cover and C.D. Wood, Chemical Engineering Progress, Symposium Series, Vol. 67, American Institute of Chemical Engineering, New York, 1971, pp. 135-146], and by English et al. [J.M. English et al., loc. cit., pp. 147-153]. These authors have presented data, flow sheets, and cost information. Their work may be consulted for more details.

    In a process of this type several interesting questions arise: (a) How much carbon must be used? (b) What part of the original capital investment in the process is represented by the carbon?

Overall Considerations in Project Analysis 347

(c) It is almost certain that some part of the carbon will disappear because of attrition, mechanical losses, and process difficulties. What effect will a loss of carbon have on the economics of the operation? (d) The regeneration of the carbon is an expense to the process. It may be asked, Is regeneration necessary? Perhaps the used carbon could be discarded instead of regenerated. Obviously, if the price of the carbon exceeds a certain value regeneration is necessary, but if it is cheap enough it could be discarded and new carbon brought in. At what price for the carbon does regeneration become economically necessary?

Data:

Water flow rate: 10,000,000 gal/day

Inlet water concentration: 1.72×10^{-4} lb organic matter/gal

Outlet water concentration not to exceed 0.47×10^{-4} lb organic matter/gal

Retention time in adsorbers (based on empty vessel): 50 min

Bulk density of carbon: 30 lb m/cu ft

Cost of carbon: 25¢/lb

Cost of adsorption process
10,000,000 gal/day

New capital required (includes $32,000 for regeneration system and $450,000 for carbon)	$1,660,000	
Operating costs (¢/M gal)	With regeneration	Without regeneration
Make-up carbon @ 25¢/lb (5% loss of carbon/yr)	0.49	0
Power @ 1¢/kw-hr	1.00	0.99
Backwash water	0.31	0.31
Fuel for regenerator	0.15	0
Labor	1.05	0.21
Overhead	1.02	0.58
Depreciation @ 7.4%/yr	2.45	2.38
Maintenance @ 6%/yr	1.99	1.94
Insurance @ 1%/yr	0.44	0.32
Total	8.89	6.73

7.9. A chemical plant produces 30,000,000 lb/yr of a dry solid to be used as a pigment. One of the processing steps is a filtration where a wet cake is produced from a slurry containing 25 wt% solid and 75 wt% water. The filtration is now being done in four plate-and-frame presses having 20 frames in each press. The filtration cycle is 0.5 hr filtration, 0.5 hr wash, and 1 hr clean. The four filters can maintain the production rate on the above schedule, but any increase in capacity would require additional filtration equipment. The plant is 15 years old and is still operating efficiently. The depreciation tax allowance on the filter presses ceased to apply 5 years ago.

It is now desired to increase the capacity of the plant by 25%. This can be done with only minor alterations in the reaction section of the plant and any additional cost there can be ignored. However, there will have to be an expansion of the filtration system. This can be done in one of two ways:

(a) A new plate-and-frame press can be purchased and installed. This would give the plant five plate-and-frame presses of similar design—four older, and fully depreciated, and one new.

(b) It has been found that two used rotary vacuum filters can be purchased for 50% of their new value, and that these filters can process 75% of the total new expanded capacity. Because of their lower manpower charges the rotary filters, combined with two of the older plate-and-frame presses, might provide a cheaper operating unit than the five plate-and-frame presses.

Using the cost data in Chapter 5 calculate the cost of operation of each proposal and find the most economical combination.

Chapter 8

PROCESS PLANT COMPONENTS

8.00 INTRODUCTION

When the overall energy and material balances of a process have been established, the individual components of a chemical plant, such as heat exchangers, distillation towers, and dryers can often be designed for optimum economic performance without reference to each other. This effects a substantial simplification in the calculations. Often such calculations will be similar in nature, regardless of plant type, so that standard methods can be developed for the economic optimum design. Sometimes the relationships are simple enough to be expressed as approximations of the type discussed in Appendix C. In cases where the equipment involves a substantial cost, more detailed calculations are justified. In the past these calculations were difficult and time-consuming, but, at present, the electronic computer has greatly reduced the time and labor for such calculations.

It is the purpose of this chapter to detail a few situations of this nature commonly encountered in chemical process plants. The simplest problem occurs when the operating cost of the equipment in question does not vary. Then the optimum solution consists in finding the minimum capital cost for the equipment to accomplish the desired function. Such a situation exists in choosing the optimum diameter for tanks and vessels. More commonly, the optimum results from balancing incremental investment against operating savings. Thus, problems in fluid flow involve balancing the cost of larger diameter pipe against the saving in pumping costs, and problems in heat transfer involve balancing the cost of extra exchanger surface against the cost of the heat saved.

In studying problems of this type it is usually convenient to assume constant values for all the variables involved except one. The total cost function is then set up in terms of this variable. Differentiation with respect to this variable and setting the derivative equal to zero give the required optimum value. Where more elaborate treatment is desired, the techniques described in Chapter 4 may be employed.

8.01 THE OPTIMUM DESIGN OF PROCESS PIPING

The capital investment in the process piping in a chemical plant can be as high as 40% of the total plant investment; very seldom will it be less than 25%. It is apparent that careful attention should be paid to the selection of piping of the optimum size and type.

In cases of fluid transport over long distances in fairly large diameter pipelines, the piping and the pumps represent essentially the entire process plant. The case of the optimum design of gas transport lines has been considered in some detail in Example 4.6. Also, Bodman [1] and Sherwood [2] have discussed the optimum design of transport pipelines for liquid sulfur and natural gas.

Process-plant piping is specified with two objects in view: (1) The piping must have sufficient strength and corrosion resistance to withstand the physical and chemical stresses produced by the process streams. (2) The piping must be sufficiently large to enable the process streams to pass through without excessive pressure drop and high pumping costs; yet the piping must not be so large that its cost becomes too high.

The first consideration determines part of the capital cost of the pipe, the fittings, the valves, and the auxiliaries such as piping supports (some pipe, such as stainless steel at relatively low temperatures, has great strength, but other pipe, such as glass, has relatively little). Many different kinds of pipe and piping auxiliaries have been developed to solve such problems, and the selection of the proper kind requires the careful attention of a specialist. In process-plant piping many difficult problems must be solved: thermal expansion, mechanical vibration, and mechanical stress. Despite all this, the economic considerations can all be treated in the same fundamental manner.

Process Plant Components

The second consideration determines part of the capital cost and all of the operating charges. From the well-established principles of fluid flow it is known that a small-diameter pipe will have a high fluid velocity and a high fluid pressure drop through the pipe, valves, and fittings. This will require a pump delivering a high discharge pressure and a motor having large energy consumption; these may be fairly expensive. These higher charges will be counterbalanced by the lower costs for the smaller pipe, valves, and fittings. If the pipe diameter is made larger, the fluid velocity drops markedly and the pumping costs become substantially lower. Conversely, the capital investment into larger pipe, fittings, and valves becomes greater.

There will be a pipe diameter at which the total cost for installation and operation of the pipe and pumping system will be a minimum.

An equation for the total cost of the piping system may be obtained by summing the expressions for the yearly cost of piping, valves, insulation, pump equipment, and steam or electricity. All other indirect costs are assumed to remain constant. Differentiating the total cost equation with respect to the pipe diameter (while holding constant the density and the quantity of liquid flowing), and setting the derivative equal to zero yields a general equation for the economic pipe size. Bodman [1] and Sherwood [2] show how a similar equation may be expressed in two and three variables. Partial differentiation is used for solution. Example 4.6 considers four variables and uses the Lagrangian multiplier method of solution.

By calculating the liquid velocity for each pipe diameter which is an economic optimum size for a given density and quantity of liquid flowing, the "economic velocity" is obtained.

The total annual cost for the pumping system will consist of: (1) the annual cost of the piping and fittings; (2) the annual cost of the valves; (3) the annual cost of the pumps and motors; (4) the annual cost of the pumping energy required.

The installed costs of piping and fittings are known, or may be correlated, as functions of the pipe diameter. Such correlations take the form $C = KD^n$, where K and n are empirical constants which change as the kind of pipe changes. D is the pipe diameter in inches and C is the cost in dollars per foot of pipe.

For standard weight carbon steel process piping the data of Guthrie [3] show that the installed cost may be expressed well enough by the equation: Cost = 1.75 D $/ft for pipe between 3 and 8 in. in diameter.

For gate and check valves the data of Guthrie [3] show that the installed cost of one valve is Cost = $77 D^{0.7}$ $.

It must be understood that the cost of piping and valves can change substantially as the overall economic price level, the competitive nature of business, and the relationship between consumer and supplier changes. However, the numbers given above are useful enough at present.

The cost of pumps and motors is usually correlated in terms of their required horsepower. Here also, prices change rapidly with time, volume purchase, or by negotiation. These cost relationships are usually of the form: Cost = $F(Hp_H)^m$, where Hp_H is the hydraulic horsepower required of the system, and F and m are empirical constants which include the efficiency of the pump and motor. For fairly large installations steam turbines are also used as pump drivers; see Braca and Happel [4].

The horsepower required to pump the fluid from origin to destination may be calculated from

$$Hp_H = (\Delta p_s + 8.63 \times 10^{-4} f L_e Q^2 \rho/D^5)(Q/1{,}714) \qquad (8.1)$$

where Δp_s = differential pressure head on pump, exclusive of pipe friction (lb f/sq in.)

f = friction factor; may be taken as 0.0045

L_e = equivalent length of pipe (ft) the resistance to fluid flow of pipe fittings such as angled bends and valves is accounted for in terms of an equivalent length of straight pipe. Frequently the "equivalent length" will exceed the nominal length. See Foust et al. [5, p. 401 and Appendix C-2].

ρ = density of liquid (lb m/cu ft)

D = pipe diameter (in.)

Q = rate of flow (gal/min).

The numerical value of Δp_s will be equal to the net static head and the pressure drop through the process equipment and control valves. (The pressure drop through the control valves is usually taken as 30% of the total pressure drop.)

The total cost of the piping, fittings, valves, pumps, and motors will be expressed as a fixed amount of dollars. This is converted to an annual cost by multiplying by an annual "fixed charge factor." Equations (2.13) and (3.42) show this in some detail. Usually the fixed charge factor is evaluated by assuming

Process Plant Components   353

some reasonable payout time, T_m, and some reasonable maintenance factor, m. These multiplying factors usually range between 0.4 and 0.6.

With this number, and a cost for the pump motor energy charge, the annual capital charges and the annual operating charges may be added to give the total annual cost of the pumping system. This equation may be differentiated with respect to the pipe diameter, set equal to zero, and the resulting expression solved for the diameter. As explained before, this results in an economic velocity for a fluid of a given density flowing at a given volumetric rate through the pipe of the economic diameter.

Example 8.1.

Consider a typical pumping problem. Water, 400 gal/min, is to be pumped from one point to another. The length of the pipeline is 250 ft and there are two gate valves (open), one check valve (open), and four right-angle bends.

The static pressure drop (exclusive of pipe friction resistance) is 100 psi.

The horsepower required is given by Eq. (8.1). The various quantities in the equation are as follows:

$f = 0.0045$ $\qquad \rho = 62$ lb m/cu ft

$Q = 400$ gal/min $\qquad \Delta p_s = 100$ psi

From Appendix C-2 of Foust et al. [5]

$\qquad$ 90° bends $\quad L_e/D = 30$

Gate valve (open) $\qquad = 13$

Check valve (open) $\qquad = 50$

Total equivalent length; $L_e = 250 + 4[30/(D/12)] + 2[13/(D/12)]$
$\qquad\qquad + 50(D/12)$

$\qquad\qquad L_e = 250 + 196(D/12)$

The pump horsepower is

$$Hp_H = 100 + (8.63 \times 10^{-4})(0.0045)\left(250 + \frac{196\,D}{12}\right)\left(\frac{Q^2 \rho}{D^5}\right)\left(\frac{Q}{1714}\right)$$

The installed cost of the pump and motor may be estimated (see Chapter 5) as $1{,}500(Hp_H)^{0.6}$, or

$$1,500\left\{100 + (8.63 \times 10^{-4})(0.0045)\left(250 + \frac{196\,D}{12}\right)\left(\frac{Q^2\rho}{D^5}\right)\left(\frac{Q}{1714}\right)\right\}^{0.6}$$

The cost of the power to pump the fluid will be $8,000(1.5)(Hp_H)(0.01)$. The term 1.5 includes efficiencies of 83% for the electric motor and 60% for the pump; electrical power is taken at \$0.01/kw-hr. The installed cost of the piping and valves is $(250)(1.75\,D) + 3(77\,D^{0.7})$.

The total annual cost of the piping system is

$$C_T = 0.6\left\{250(1.75\,D) + 231\,D^{0.7} + 1,500\left[5.4 \times 10^{-2}Q \right.\right.$$
$$\left.\left. + 2.26 \times 10^{-9}\left(250 + \frac{196\,D}{12}\right)\frac{Q^3\rho}{D^5}\right]^{0.6}\right\}$$
$$+ 120\left[5.4 \times 10^{-2}Q + 2.26 \times 10^{-9}\left(250 + \frac{196\,D}{12}\right)\frac{Q^3\rho}{D^5}\right]$$

Fixed charges on pipe and valves = $0.6\left[250(1.75\,D) + 231\,D^{0.7}\right]$

Fixed charges on pump and motor = $900\left\{5.4 \times 10^{-2}Q + 2.26 \times 10^{-9}\right.$
$$\left. \times \left(250 + \frac{196\,D}{12}\right)\frac{Q^3\rho}{D^5}\right\}^{0.6}$$

Pumping costs = $120\left[5.4 \times 10^{-2}Q + 2.26 \times 10^{-9}\left(250 + \frac{196D}{12}\right)\frac{Q^3\rho}{D^5}\right]$

There are two ways to find the optimum diameter for a given Q and ρ.

1. Assume various values of D, substitute them into the equation for C_T above, and find the minimum value of C_T by plotting C_T against D. This is simple and straightforward. It has the advantage of showing the shape of the curve (sharp or flat minimum) and the actual value of the cost in dollars per year.

2. The equation for C_T may be differentiated with respect to D, the resulting equation set equal to zero, and the optimum value of D obtained by algebraic manipulation. Quite frequently this operation leads to complex equations and a trial-and-error solution. Method 1 is simpler and easier.

In the present case both methods were used. Method 2 proved to be quite complex, while Method 1 was simple. Table 8-1

Process Plant Components

Table 8-1   Pumping Costs Versus Pipe Diameter

D (in.)	Fixed charges on pipe and valves ($/yr)	Fixed charges on pump and motor ($/yr)	Pumping cost ($/yr)	Total cost ($/yr)
3	1,090	7,290	3,930	12,310
4	1,418	6,110	2,930	10,458
6	2,060	5,750	2,640	10,450
8	2,700	5,670	2,600	10,970

shows the values of the total cost and of the component parts of the total cost as a function of the pipe diameter.

When these results are plotted there is a long flat minimum between 4 and 6 in. There is essentially no difference in total pumping cost in the region where D is between 4 and 6 in.

The very sharp drop in cost between 3 and 4 in. diameter pipe followed by the rather gradual rise in cost after D exceeds 6 in. should be noted. When pumping fluids through circular pipe, it is apparent that the pipe should be greater in diameter than a certain value, but that still larger diameters have only a small effect on the system economics. There is a sharp drop to the minimum value and then a gradual rise away from it.

For the present assumption of Q = 400 gal/min and ρ = 62 lb m/cu ft, the economic velocity in the 4 in. diameter pipe is 10 ft/sec. Other values of Q and ρ may be assumed and other optimum diameters and velocities obtained. These are shown in Table 8-2.

The length of the pipe, L or L_e, is not an important variable except in cases where Δp_s is low and L is high. Such cases are not usually encountered in ordinary plant design problems.

Steam turbines may also be used for driving pumps. A cost analysis similar to that shown above, with the cost of a turbine and the cost of steam replacing the costs of an electric motor and electricity gives the results shown in Table 8-3.

Process engineers frequently encounter situations wherein the applicable costs are different from those used above. Calculations of the sort shown, but using other values for steam and electrical power costs, payout times, and equipment costs were made. The results show that the economic velocity will not change

Table 8-2 Economic Pipe Velocities, Motor Drive[a]

Nominal pipe diameter (in.)	Specific gravity of liquid		
	0.50	0.75	1.0
2	10.7	9.2	8.4
3	12.0	10.3	9.4
4	12.9	11.1	10.1
6	13.9	12.0	10.9
8	14.6	12.6	11.4
10	15.0	13.0	11.7
12	15.2	13.2	11.9

[a] Liquid flow in feet per second, using motor-driven centrifugal pumps.

Table 8-3 Economic Pipe Velocities, Turbine Drive[a]

Nominal pipe diameter (in.)	Specific gravity of liquid		
	0.50	0.75	1.0
2	7.0	6.0	5.5
3	7.6	6.6	5.9
4	8.0	7.0	6.3
6	8.5	7.4	6.7
8	8.8	7.7	7.0
10	8.9	7.8	7.1
12	9.0	7.9	7.1

[a] Liquid flow in feet per second, using 3,600-rpm turbine-driven pumps.

by more than 10-15% for fairly large changes in the values assumed for the solution of Example 8.1.

The importance of proper piping design and selection to the plant design and construction effort is so great that many firms have attempted to devise a computer system to do the design work. These systems can calculate optimum sizes, select the most suitable material of construction, and make piping diagrams. The devisers of such systems claim that piping costs (design, fabrication, and erection) can be reduced by as much as 30% through the use of computer-aided techniques. The duPont company now offers such a computer-aided piping design system [6].

8.02 THE OPTIMUM DESIGN OF HEAT-TRANSFER APPARATUS

Of the various types of problems involving heat transfer in process plants, two have been studied in detail by many investigators: (1) the design of tubular equipment such as is employed in heat exchangers, coolers, and condensers; (2) the selection of the proper thickness of insulation required on equipment operating at elevated temperatures. These problems lend themselves to the application of standard methods and so are considered in detail here and in Section 8.03.

Tubular Exchangers

Heat transfer problems in tubular exchangers can be divided into two types:

1. Those wherein process streams exchange heat so as to conserve heat within the process, for example, preheating the feed to a distillation column with the effluent from the column reboiler. In such cases the economic balance involves the cost of the incremental heat-exchange area with the cost of the incremental heat that would have to be supplied if the exchange did not take place.

2. Those processes wherein the heat is supplied or taken away by a utility stream. For example, the cooling of a distillation column product with water or air before storage. In this case the economic balance involves the cost of the utility with the cost of the incremental heat exchange area.

Frequently, these two types are combined and a more complicated analysis is required.

In both cases there are three different situations which may be characterized by the temperature changes in the two streams.

1. Those wherein both streams change temperature. The hot stream is cooled and the cool stream is warmed. This is probably the most commonly encountered form of heat exchange.

2. Another type frequently met is that wherein one stream stays at a constant temperature and the other stream heats or cools. The condenser, the waste heat boiler, and the steam-heated preheater are good examples.

3. The type wherein neither stream changes in temperature. Only latent heat is transferred. The reboiler and the evaporator are of this type.

For process-stream heat exchangers, where heat is conserved that would otherwise be wasted (and then would have to be supplied by burning fuel in a furnace), the economic balance results in the determination of the optimum terminal temperature differences (and consequent amount of heat transferred) that will result in in maximum earnings.

In order to calculate the optimum terminal temperature difference the designer must specify the overall heat-transfer coefficient and the mechanical construction of the exchanger. By varying the number of passes (both shell and tube), the diameter of the shell, the diameter and length of the tubes, the baffling, and the velocity of fluid flow through both tubes and shell, it is possible to obtain substantial variations in the overall heat-transfer coefficient and in the pressure drop.

Since pressure drop and heat transfer coefficient are closely related, the economic balance should involve the determination of the optimum pressure drop. This could be found by balancing pumping costs against exchanger costs. In practice, this is seldom done because the optimum pressure drop is high. If such high pressure drops (with their attendant high flow rates) were used, the fouling which would normally occur would cause a rapid rise in pressure drop and a premature shutdown. Usually, engineers use lower pressure drops and larger exchangers in order to get longer operating times between shutdowns for cleaning and tube replacement. The pressure drop is usually specified and is not considered a design variable.

Process Plant Components

What frequently happens in modern heat-transfer design work is that the process engineer specifies flow rates, temperatures, pressure drops, and physical and chemical properties for a certain heat-transfer problem. Every engineering office has a heat-exchanger specification sheet that must be filled out by the process engineer. When this has been done almost all of the pertinent questions have been answered, at least to a first approximation. These numbers are then studied by a heat-transfer specialist who suggests the mechanical design which will accomplish the task. The specialist is often assisted by an electronic computer which has been programmed for heat-exchanger design and can calculate the details of the heat exchanger in a few seconds. Thus, the consequences of many different ideas may be calculated quickly.

The ideas of the process engineer and the practical knowledge of the heat-transfer specialist are frequently in conflict. There then occurs a period of friction between the two, involving a sequence of give and take, which finally results in a design acceptable to both.

In addition to this, the heat-exchanger manufacturer must also be consulted. His ideas, based on years of manufacturing heat exchangers which must be sold at a profit, may be different from the other two.

Out of all this argument and consultation a practical design will result. It must always be remembered that a heat exchanger is a piece of mechanical equipment that must be forged, rolled, welded, chipped, installed, serviced, and have a very long life. All the sophisticated heat-transfer science arguments in the world will not change the fact that a heat exchanger is a mechanical device that can cost a great deal of money to design, build, ship, install, and maintain.

The heat-transfer coefficient that is used may be the product of extensive calculations, or of practical experience, or experiment, or all together. By changing the geometry of the exchanger it is possible to make significant changes in the coefficient. The proper relationship between all these is a matter for specialists. In addition to the above-mentioned complications, there is the matter of fouling. For industrial work it is important to realize that the coefficient will become smaller as time passes due to the fouling of the heat-exchange surface by dirt and scale. The coefficient that is used for clean surfaces (which is what will be obtained from the use of heat-transfer correlations) will exist for a short time after the start of operations, but will decrease steadily as time passes until it may become only half as large as at the beginning. Industrial designers allow for this by incorporating fouling factors in the various resistances to heat transfer [7]. This steady decrease in the coefficient will have an important effect on the temperatures of the flowing streams.

It is apparent that a calculation based on the use of an assumed or calculated overall heat-transfer coefficient (U) will give a result which will be correct for only a rather short period of time. The rate of heat transfer must become progressively lower as time passes.

When all the complications involving the calculation of the overall heat-transfer coefficient are considered, it is obvious that, for preliminary considerations, a practical solution is to assume a value which long experience has proven to be suitable (See Ref. [7], p. 840, and Appendix C of this book) and to assume that this value does not change. It must also be recognized that whatever U is used, the outlet temperatures will change as time passes and suitable controls must be installed to deal with this [7, p. 221].

The analysis of tubular heat exchangers will start with the simplest case, that in which the medium on one side does not change temperature (such as a steam-heated preheater or a waste heat boiler).

If the heating medium is at a constant temperature, T, and the stream being heated increases in temperature from t_1 to t_2 without vaporizing, the operating earnings, O, in dollars per year above the minimum acceptable return on investment, will be

$$O = Q H_t\, 8{,}760\, Y\, (10^{-6}) - r(C + E A) \qquad (8.2)$$

$\underbrace{\phantom{Q H_t\, 8{,}760\, Y\, (10^{-6})}}_{\text{value of heat recovered}} \quad \underbrace{}_{\text{cost of recovering heat}}$

where   Q = heat-transfer rate (Btu/hr)

H_t = total cost of supplying incremental heat ($/$10^6$ Btu)

8,760 = number of hours in 1 year (hr/yr)

Y = fraction of year equipment is in operation

r = total annual capital-related fixed charges factor

Equation (3.46) gives $r = m + 1/n + (i_m - 0.35i)/(1 - t)$ where

i_m = minimum acceptable return rate on invested capital (frac/yr)

n = allowable life for depreciation of equipment; 1/n = straight-line depreciation factor

Process Plant Components

i = average rate of return on invested capital (frac/yr)

t = income tax rate, Federal plus state (frac/yr)

m = maintenance charge (frac/yr)

When r is multiplied by the capital investment in the heat exchanger the result is the annual capital-related charges, the so-called "fixed charges," of the cost of the heat-exchanger equipment.

The cost of the exchanger is represented by an equation of the form $C + EA$, a straight line when cost is plotted against area. C is a constant and E is the slope of the line. Thus, E is the incremental cost of heat-exchange area, $d(cost)/dA = E$. This method of representing the cost-area relationship gives quite a good correlation over ranges of A that are not more than 10-fold. In Chapter 5 and Appendix C the relationship, $C = C_o A^n$, a straight line on double logarithmic coordinates, is used to correlate the cost-area relationship for heat exchangers. In this case, the incremental cost of the heat-exchange area is $C_o n A^{n-1}$. This is a more complex expression than the straight line used above.

The double logarithmic coordinate form of cost-size correlation is widely used for all manner of process equipment, but, in fact, the lines are rarely straight, except over size ranges of less than 10-fold. As a matter of practical usage, where the range in sizes is usually much less than 10 times, a straight line of the $C + EA$ form is adequate. In fact, it makes little difference which form is used, because the actual number employed in the calculation will be assumed independent of A in order to reduce the complexity of the derivation. Since the range of areas considered in any one optimization problem is small, it is correct enough to use a constant value for E.

When it is known that E is not independent of A, a simple first-try calculation will be made to establish the approximate value of A. From this a more accurate number for the incremental cost may be obtained.

The optimum will be determined by expressing all variables in terms of a single one, differentiating, and setting equal to zero. In this case A will be selected as the independent variable. It should be noted that this optimum takes into consideration the incremental rate of return, as in Fig. 2.5, because of the use of r in terms of $(i_m - 0.35i)/(1 - t)$.

Before differentiation, it is necessary to express Q in terms of A. This is accomplished by a heat balance and the heat-transfer equation

$$Q = cw(t_2 - t_1) \tag{8.3}$$

where c is the specific heat of the fluid being heated, and w is its flow rate in pounds per hour.

Whence

$$t_2 = \frac{Q + cwt_1}{cw} \tag{8.4}$$

$$Q = UA \left[\frac{\dfrac{Q + cwt_1}{cw} - t_1}{\ln \dfrac{T - t_1}{T - \dfrac{Q + cwt_1}{cw}}} \right] \tag{8.5}$$

where U is the overall heat-transfer coefficient. If the value of dQ/dA from Eq. (8.5) is inserted into Eq. (8.2) after the differentiation with respect to A is performed, there is obtained

$$dO/dA = U(T - t_2)H_t(10)^{-6}(8,760)Y - rE \tag{8.6}$$

Note that in the differentiation the constant term in the cost equation disappeared. In subsequent derivations the constant term will be omitted. However, it must be remembered that the incremental cost is used only in an equation that has already been differentiated; see Section 2.15.

If dO/dA is set equal to zero,

$$(T - t_2)_{opt} = \frac{114 \, rE}{UYH_t} = \frac{Z}{H_t} \tag{8.7}$$

where Z is defined equal to $114 \, rE/(UY)$.

The same result can be more simply obtained by using t_2 as the independent variable. The method used here is employed for consistency with the rest of the derivations. Example 2.5 shows the solution of a heat-transfer problem of the type described here. In Example 2.5 a different method of solution was used, but an identical answer may be obtained using Eq. (8.7). It is of interest that for the case where one fluid does not change in temperature, the equivalent temperature difference for each incremental square foot is equal to the terminal temperature difference.

Where the temperature of both fluids changes, Eq. (8.2) is still applicable, the hot fluid being cooled from T_1 to T_2 instead of remaining at the constant temperature T. In this case, for countercurrent flow

Process Plant Components

$$Q = cw(t_2 - t_1) = CW(T_1 - T_2) = UA \frac{(T_2 - t_1) - (T_1 - t_2)}{\ln[(T_2 - t_1)/(T_1 - t_2)]} \quad (8.8)$$

The value of Q from Eq. (8.8) may be inserted into Eq. (8.2) and, if the differentiation with respect to A is performed,

$$\frac{dO}{dA} = \frac{UD(1 - P)(1 - RP)H_t Y}{114} - rE \quad (8.9)$$

where $D = T_1 - t_1$, difference in entering temperatures between hot and cold fluids

$R = \dfrac{WC}{wc}$, the ratio of the heat capacities of the two streams

$P = \dfrac{T_1 - T_2}{T_1 - t_1}$, fractional approach of hot fluid temperature difference to difference in entering temperatures

Upon setting dO/dA equal to zero and collecting terms, there results

$$(1 - P)(1 - RP)_{opt} = Z/H_t D \quad (8.10)$$

For any given problem the term $Z/H_t D$ will be a constant, and R will be fixed. The problem is solved by calculating P, and from P and Eq. (8.8), T_2 and t_2 may be calculated.

For a multipass exchanger of the 1 - 2 type, a more complicated expression must be employed in place of the log mean temperature difference which was used in the derivation of Eq. (8.8). It is then possible to arrive at the following expression for the optimum [8]:

$$1 - P\left(1 + R - \frac{RP}{2}\right)_{opt} = \frac{Z}{H_t D} \quad (8.11)$$

Ten Broeck [8] has presented a convenient nomograph for evaluation of P for all three cases of countercurrent, 1 - 2 multipass, and 2 - 4 multipass exchangers. In order to use this nomograph the terms $Z/H_t D$ and R must be known. This nomograph is reproduced in Fig. 8.1.

Once P has been calculated from Fig. 8.1, values of T_2 and t_2 are easily computed.

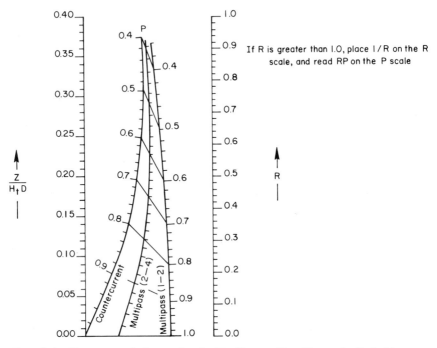

Fig. 8.1. Nomograph to evaluate P. From Ten Broeck, Ind. Eng. Chem., 36, 64 (January 1944). Copyright 1944 by the American Chemical Society and reprinted by permission of the copyright owner.

Although every problem is different from every other problem it is possible to make two useful generalizations.

1. When the cost of the heat H_t is high, the cost of the heat-exchanger area E is low, and the overall heat-transfer coefficient U is high it is apparent that the circumstances favor the use of a large amount of heat-exchange area and the recovery of a large fraction of the heat. The term $Z/H_t D$ will be near 0.1 and P will be near 0.75 to 0.90. This is a common occurrence in process heat exchange involving organic fluids at temperatures of 200°F and higher. For low-temperature streams the quantity of heat to be recovered is so small that it is not worth recovering. For ordinary organic fluids, or water, at temperatures near their boiling points, it is economically feasible to recover 75 to 80% of the recoverable heat in the hot stream.

Process Plant Components 365

2. The reverse situation occurs when the cost of the heat is low, the cost of the heat-exchange area is high, and the overall heat-transfer coefficient is poor. Under these conditions it is apparent that the situation favors the use of only a small amount of the heat-exchange area and the recovery of 50% or less of the recoverable heat. Gas-gas heat exchange at temperatures near 1,000° F comes in this category. Here it is economically feasible to recover about 50% of the heat in the hot stream.

Many other conditions may apply. Each problem should be solved on its own merits, but the complexity of the solution is greatly reduced by the use of Fig. 8.1.

In process plants it commonly occurs that a heat exchanger which cools a process stream may be followed by a water-cooled exchanger which will cool the effluent process stream from the first exchanger to a still lower temperature. In many instances the water-cooled exchanger is a necessity because the process stream cannot be sent to storage at temperatures above, say, 150°F. If the cool process stream entering the first heat exchanger is fairly warm, say 200°F, then it is apparent that a water-cooled exchanger is necessary.

It can be considered that an economic balance will exist in the exchanger-cooler situation. It is reasonable to assume that an exchanger effluent temperature T_2 will exist which will maximize the earnings of the exchanger-cooler combination. Not only must the costs of exchanger area, cooler area, and cooling water be considered, but it must also be recognized that transferring heat in the process heat exchanger effects a large saving in the costs of the water-cooled exchanger. If the heat were not removed from the process stream by heat exchange, which is a saving, it would have to be removed in the cooler, which is an expense.

Figure 8.2 shows a typical exchanger-cooler combination. In this case a hot stream at temperature T_1 is cooled to T_2 in preheating a cold stream from temperature t_1 to t_2. The temperature T_2 is still above the desired storage temperature, and the original stream is cooled from T_2 to T_3 in a cooler by means of water which rises in temperature from t_{w1} to t_{w2}. The previous derivation, applying to heat exchanger 1, was based on given quantities of hot and cold fluids and given inlet temperatures T_1 and t_1. Its object was to specify the outlet temperatures of both fluids. The problem now is essentially the same: that is, to specify t_2 and T_2. However, it is assumed that cooler-related cost items will enter the picture as auxiliary dependent variables. If the cooling water temperature

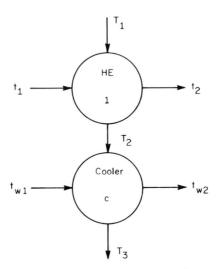

Fig. 8.2. Typical exchanger-cooler flow (definition sketch).

t_{w1} is specified and also the rise in temperature is fixed, t_{w2} will be specified. If it is assumed that T_3, the temperature of the cooled stream to storage, is also fixed, the problem can be set up as a function of one independent variable as previously. The following derivation, after Ten Broeck [8], may be applied to this situation.

The operating earnings O will be equal to: value of the heat recovered - cost of cooling - heat-exchanger surface cost - cooler cost + cost of cooling water in the absence of the heat exchanger + cost of cooling surface in the absence of the heat exchanger

$$O = Q_1 H_f (10)^{-6} 8,760Y - Q_c H_w (10)^{-6} 8,760Y - rE_1 A_1 - rE_c A_c$$
$$+ (Q_1 + Q_c) H_w (10)^{-6} 8,760Y + rE_c A'_c \qquad (8.12)$$

where Q_1 = heat transferred in the heat exchanger (Btu/hr)

Q_c = heat transferred in the cooler (Btu/hr)

H_f = value of incremental heat ($/MM Btu)

H_w = value of cooling water ($/MM Btu); this can be known only if the temperature rise of the water is known

E_1 = incremental installed cost of heat exchanger ($/sq ft)

E_c = incremental installed cost of water cooler ($/sq ft)

Process Plant Components

r = fraction of total charges on exchanger or cooler surface per year to allow for maintenance, depreciation, and an acceptable minimum profit = $m + 1/T_m$

A_1 = heat-exchanger surface (sq ft)

A_c = cooler surface (sq ft)

A'_c = cooler surface in absence of exchanger (sq ft)

Y = fraction of year equipment will be in operation

H_f includes fixed charges on furnace equipment, but this item is usually not dependent to any extent on the temperature of the exchanging streams. Heat transfer in a furnace occurs at a sufficiently high temperature so that small differences in temperatures of streams being heated will not affect the economics of heat transfer.

Equation (8.12) can be simplified to

$$O = 8{,}760 Y Q_1 (H_f + H_w)(10)^{-6} - rE_1 A_1 - rE_c A_c + rE_c A'_c \quad (8.13)$$

Maximum operating earnings will result when $dO/dA_1 = 0$.

$$\frac{dO}{dA_1} = 0 = 8{,}760 Y (H_f + H_w)(10)^{-6} \left(\frac{dQ_1}{dA_1}\right) - rE_1 - rE_c \frac{dA_c}{dA_1} \quad (8.14)$$

There are several points of interest about Eqs. (8.13) and (8.14):

1. The algebraic manipulation resulting in Eq. (8.13) eliminates the quantity Q_c, the amount of heat transferred in the cooler. However, this term is reintroduced later in order to evaluate the term dA_c/dA_1.

2. The differentiation with respect to A_1 results in the elimination of the term $rE_c A'_c$. This term is a constant and not dependent on A_1.

3. The term dA_c/dA_1, the change in the area of the cooler with the change in the area of the heat exchanger, is formed. A way must be found to evaluate this.

The development of Eq. (8.14) can proceed as follows:

$$\frac{dA_c}{dA_1} = \left(\frac{dA_c}{dQ_c}\right)\left(\frac{dQ_c}{dA_1}\right) \quad (8.15)$$

and

$$Q_1 + Q_c = WC(T_1 - T_3) = \text{constant} \tag{8.16}$$

$$dQ_1 + dQ_c = 0 \tag{8.17}$$

$$dQ_c = -dQ_1 \tag{8.18}$$

Therefore,

$$\frac{dA_c}{dA_1} = -\left(\frac{dA_c}{dQ_c}\right)\left(\frac{dQ_1}{dA_1}\right) \tag{8.19}$$

Substituting this expression in (8.14) gives

$$0 = 8{,}760(10)^{-6} Y(H_f + H_w)\left(\frac{dQ_1}{dA_1}\right) - rE_1 + rE_c\left(\frac{dQ_1}{dA_1}\right)\left(\frac{dA_c}{dQ_c}\right) \tag{8.20}$$

Solving for dQ_1/dA_1

$$\frac{dQ_1}{dA_1} = \frac{114 rE_1/Y}{H_f + H_w + \dfrac{114 rE_c}{Y\, dQ_c/dA_c}} \tag{8.21}$$

$$\left(\frac{dQ_1}{U_1 dA_1}\right)_{opt} = \frac{Z_1}{H_f + H_w + \dfrac{Z_c}{dQ_c/(U_c\, dA_c)}} = \frac{Z_1}{H_t} \tag{8.22}$$

where Z_1 is defined as $114 rE_1/U_1 Y$, and Z_c is defined as $114 rE_c/U_c Y$. It can be shown that $Z_1/H_t = D(1-P)(1-RP)_{opt}$ for countercurrent exchangers, and that $Z_1/H_t = D - DP(1+R-RP/2)_{opt}$ for 1 - 2 multipass exchangers. The only new element involved is the evaluation of H_t, which now includes H_f, H_w, and $Z_c/(dQ_c/U_c dA_c)$. The last item is evaluated as follows for countercurrent flow in the cooler:

$$Q_c = WC(T_2 - T_3) \tag{8.23}$$

$$T_2 = T_3 + Q_c/WC \tag{8.24}$$

$$Q_c = U_c A_c \Delta t_{lm} \tag{8.25}$$

Process Plant Components

$$Q_c \ln\left(\frac{T_2 + Q_c/(WC) - t_{w2}}{T_3 - t_{w1}}\right) = U_c A_c \left[\frac{Q_c}{WC} - (t_{w2} - t_{w1})\right] \quad (8.26)$$

$$\frac{dQ_c}{U_c dA_c}\left[\ln\left(\frac{T_3 + Q_c/(WC) - t_{w2}}{T_3 - t_{w1}}\right) + \frac{Q_c/WC}{T_3 + Q_c/(WC) - t_{w2}} - \frac{U_c A_c}{WC}\right]$$

$$= \frac{Q_c}{WC} - (t_{w2} - t_{w1}) \quad (8.27)$$

Let

$$T_2 - t_{w2} = \Delta t_1$$

and

$$T_3 - t_{w1} = \Delta t_2$$

After simplification Eq. (8.27) becomes

$$\frac{dQ_c}{U_c dA_c} = \frac{\Delta t_1 - \Delta t_2}{\left[\ln\left(\frac{\Delta t_1}{\Delta t_2}\right)\right]\left(1 - \frac{T_2 - T_3}{\Delta t_1 - \Delta t_2}\right) + \frac{T_2 - T_3}{\Delta t_1}} \quad (8.28)$$

The equations presented above are solved by assuming a value for T_2. This allows the calculation of $dQ_c/U_c dA_c$ from Eq. (8.28). This, in turn, allows a value for Z_1/H_t to be calculated from Eq. (8.22), and from this result a value of P may be computed. However, the assumption of T_2 leads directly to P--from the definition of P. The two values of P must be equal. If they are not, a new value for T_2 must be assumed. The trial-and-error computation goes on until a value of T_2 is found that results in the two values of P being equal.

In practice, the trial-and-error calculation is not necessary because the contribution of the T_2-dependent term, $dQ_c/U_c dA_c$, is small; see following discussion. It is entirely practical to omit the $\frac{Z_c}{dQ_c/U_c dA_c}$ term from Eq. (8.22) and treat the heat exchanger as though it existed independently of the cooler. This result appears to make ludicrous the preceding elaborate development. However, this could not have been known until the analysis had been forced to a conclusion.

McAdams [9] considers the problem where two hot streams and one cold stream are involved. This is in contrast to the above treatment where one hot stream and two cold streams enter into the economic balance. In McAdams' problem a stream is partially heated by exhaust steam, the remainder of the desired preheat being furnished by more expensive high-pressure steam. The problem is to determine the optimum amount of heating in the first heater, in which the low-pressure steam is used. Since the vapors condense at constant temperature, some simplification is possible in the final equation derived.

More complicated examples may be encountered in practice where additional hot and cold streams are involved in the same process unit. Ten Broeck [8] has considered the situation for a stream being preheated in a battery of exchangers. The stream flows through a number of units in series and engages in heat transfer with a number of hot side streams from a fractionating column. A series of simultaneous partial differential equations is set up and solved by trial and error for the optimum area for each exchanger. The same problem can also be solved by trial and error by systematically choosing a series of intermediate temperatures for the stream being heated and determining the optimum design for each condition.

The calculations and the effects of the variables are shown in Example 8.2 for a countercurrent exchanger followed by a countercurrent cooler.

Example 8.2

Consider Fig. 8.2.

Assumptions:

1. Average incremental installed cost of heat exchanger and cooler, approximately 1,000 sq ft = \$16/sq ft. The same for both exchanger and cooler.

2. Average incremental furnace cost = 9.0 (\$)(hr)/1,000 Btu absorbed.

3. $Y = 0.95$, or total annual operating time = 8,320 hr.

4. U and U_c = 50 Btu/(hr)(sq ft)($^\circ$F). This is probably a conservative estimate. For many kinds of fluids the heat-transfer coefficient could be substantially higher. However, a value of 50 is within reason and is adequate for an illustrative example.

Process Plant Components 371

5. Cooling water cost = $0.06/1,000 gal.
6. Cooling water enters at 80°F (t_{w1}) and leaves at 115°F (t_{w2}).
7. Maintenance charges on heat-transfer equipment = 10%/yr.
8. Fuel cost = $0.40/$10^6$ Btu.
9. T_3, the temperature of the hot stream leaving the cooler = 120°F.
10. The furnace efficiency = 75%. This number, combined with the fuel cost and the fixed charges on the furnace, will give a value for the cost of the heat.
11. The maximum allowable payout time is set at 2 years. Together with m at 0.10 this makes the annual fixed charges cost factor r = 0.60. The annual fixed charges on furnace, heat exchanger, and cooler are found by multiplying their installed cost by r.

a. The cost of the fuel, H_f:

$$H_f = (9.0)(0.60)(10^3)/(8,760) + (0.40)/(0.75)$$

$$= 0.617 + 0.533 = \$1.15/10^6 \text{ Btu}$$

b. The cost of the cooling water, H_w:

$$H_w = (0.06)(10^3)/(8.35)(35) = \$0.205/10^6 \text{ Btu}$$

Note that the cost of the cooling water depends upon the rise in temperature that it experiences during its passage through the cooler.

c. $Z_1 = Z_c = \dfrac{114rE}{UY} = \dfrac{(114)(0.6)(16)}{(0.95)(50)} = 23 \dfrac{(\$)(°F)}{10^6 \text{ Btu}}$

In this particular example $U_1 = U_c$ and $E_1 = E_c$, therefore $Z_1 = Z_c$

There are two ways to solve the problem. The first method is as follows:

1. Specify T_2, calculate $dQ_c/U_c dA_c$ from Eq. (8.28), and calculate P from Fig. 8.1.

2. Choose a series of values of R and D and for each R and D calculate P from Fig. 8.1.

3. From these values of R, D, T_2, and P, the values of T_1, t_1, and t_2 may be computed. These temperatures, combined with spec-

ified T_2, t_{w1}, and t_{w2} will give a solution to the problem for each value of R and D.

In the present case

$$\frac{dQ_c}{U_c \, dA_c} = \frac{\Delta t_1 - \Delta t_2}{\left[\ln\left(\frac{\Delta t_1}{\Delta t_2}\right)\right]\left(1 - \frac{T_2 - T_3}{\Delta t_1 - \Delta t_2}\right) + \frac{T_2 - T_3}{\Delta t_1}}$$

$T_2 = 300$ $\quad t_{w2} = 115$ $\quad \Delta t_1 = 185$

$T_3 = 120$ $\quad t_{w1} = 80$ $\quad \Delta t_2 = 40$

$$\frac{dQ_c}{U_c \, dA_c} = \frac{185 - 40}{\ln\frac{185}{40}\left(1 - \frac{180}{145}\right) + \frac{180}{185}} = 239$$

$H_t = 1.15 + 0.205 + 23/239 = 1.46$

$Z_1/H_t = 23/1.46 = 15.7\,(\$)(°F)/(10^6\,\text{Btu})$

The values of R and D given in Table 8.4 were chosen, and the corresponding values of P and $T_1 - t_2$ were calculated.

Table 8-4 Calculated Values of P and $T_1 - t_2$

| | | | | R = | |
$Z_1/H_t D$	D	R	0.5	1.0	2.0
0.326	50	$T_1 - t_2$	P = 0.55 37	P = 0.43 29	P = 0.28 22
0.161	100	$T_1 - t_2$	P = 0.75 62	P = 0.60 40	P = 0.38 24
0.082	200	$T_1 - t_2$	P = 0.86 114	P = 0.73 54	P = 0.43 28

Process Plant Components 373

If D = 200 and R = 1.0, the following solution is obtained:

$T_1 = 446°F$; $t_1 = 246°F$; $t_2 = 392°F$; $T_2 = 300°F$

```
                            ↓ T₁ = 446°
       t₁ = 246° → | Heat Exchanger | → t₂ = 392°
                   ↓ T₂ = 300°
       t_w2 = 115° ← | Water Cooler | ← t_w1 = 80°
                   ↓ T₃ = 120°
```

Using this method of solution a trial-and-error calculation is avoided.

The second method of solution is to specify T_1, t_1, and T_3 (the usual industrial case), and to calculate the economic optimum value of T_2.

Let $T_1 = 446°F$; $t_1 = 246°F$; $T_3 = 120°F$; $t_{w2} = 115°F$; $t_{w1} = 80°F$; $R = 1.0$.

Assume $T_2 = 300°F$; then $P = (446 - 300)/(446 - 246) = 0.73$. From the previous calculation: $dQ_c/U_c\, dA_c = 239$; $H_t = 1.46$; $Z_1/H_t = 15.7$; $D = T_1 - t_1 = 200$, then $Z_1/H_t D = 0.0785$. If $R = 1.0$ and $Z_1/H_t D = 0.079$, then from Fig. 8.1, $P = 0.73$, which is exactly in agreement with the value of P resulting from the assumption of $T_2 = 300°F$. This is a correct solution.

It can be seen that if some temperature other than 300°F had been assumed for T_2: (1) the numerical value of H_t would not have been affected significantly, and (2) the value of P resulting from the assumption would have been different from that resulting from the calculation, which would still have been 0.8. There would then have been no alternative but to have assumed another value of T_2 until both values of P agreed. Also, since the numerical value of T_2 has such a small effect on the calculated value of P, through $Z_1/H_t D$, a solution can be readily obtained without trial and error by neglecting the $Z_c/U_c\, dA_c$ term and calculating T_2 from the value of P read from Fig. 8.1 as a consequence of $Z_1/H_t D$ and R.

The term $dQ_c/U_c\, dA_c$ is obviously dependent on the temperatures of the streams entering and leaving the water cooler. Under almost all conditions the water stream temperatures will be in the range: $t_{w1} = 80°$ to $90°F$ and $t_{w2} = 100°$ to $120°F$. The exit cooled process stream temperature, T_3, will almost always be near $120°F$. Consequently, the value of $dQ_C/U_C dA_C$ depends most strongly on the entering process stream temperature, T_2, which may have values ranging from $500°$ to $150°F$. Under these conditions the following values of $dQ_C/U_C dA_C$ are typical.

T_2 (°F)	$dQ_c/U_c dA_c$
150	50
200	125
300	240
400	345
500	450

From this it can be seen that the term $Z_C/(dQ_C/U_C dA_C)$ will nearly always be quite small. The value of Z_C will depend on the cost of the heat-exchange area, which will be different for different materials, and upon the overall heat-transfer coefficient. Even at a high value, the contribution of the term $Z_C/(dQ_C/U_C dA_C)$ will not be more than 10% of the total value of H_t.

The consequence of all this is that the presence of the water-cooled exchanger has almost no effect on the design of the heat exchanger. A trial-and-error solution to Eqs. (8.22) and (8.28) is not needed. The heat exchanger may be designed without reference to the water cooler, and the water cooler may be designed only to cool T_2 to T_3.

It is interesting to note that if t_1, the heat-exchanger inlet cool temperature, is near $100°F$, and this could often be true, the heat exchanger alone would be sufficient to cool T_2 to $150°F$ or lower. No water cooler would be needed. A water cooler should be used only when the inlet temperature, t_1, is not low enough to bring T_2 to a satisfactorily low level, say $120°F$.

In addition to the effect of changes in heat-exchanger cost and overall heat-transfer coefficient, there can be the effect due to changes in maximum acceptable payout time. In general, this effect is small because furnace charges tend to counterbalance exchanger fixed charges. At $R = 1$, the approach temperatures $T_1 - t_2$ indicated in Table 8-4 are reduced only 20% by employing a payout time of 5 years (instead of 2 years), and correspondingly increased by about 20% by using a payout time of 0.5 year. For the 1 - 2 multipass exchangers the same range of payout time changes causes very little variation in optimum approach temperatures.

Process Plant Components 375

Coolers

As discussed in the previous section, it is usually assumed that the cost of cooling water is directly proportional to the amount used, line pressure being sufficient to provide the desired flow rate. Also it is assumed that the rise in cooling water temperature is fixed and is governed by the scaling tendencies of the water. Sometimes process conditions involved will not permit the entire allowable water temperature rise to be employed economically. Even though the amount of heat to be transferred is fixed, the optimum amount of cooling water is not established. Under such circumstances this optimum corresponds to the minimum annual sum of the cooling water costs and the fixed charges on the condenser or cooler employed. This problem has been treated by McAdams [9].

Cooling by Air Using Extended Surface Heat Exchange. The use of ambient air instead of water for the cooling of process fluids has been practiced for many years. The present-day shortage of plentiful supplies of good quality cooling water and the concern for the thermal pollution of natural water sources have led to an increased interest in the use of air cooling. It is interesting to calculate the costs of air cooling and to compare them with the costs of water cooling. This work was influenced by that of Williams and Damron [10]. Kern [7, Chapter 16] gives a lengthy discussion on extended surface heat exchangers.

The cooling of process fluids by air is usually done by forcing the air past a bundle of extended surface heat-exchange tubes. The air passes at right angles over the tubes, usually in a vertical direction, with the tubes being horizontal; of course, there are other geometries. The process fluid to be cooled (sometimes condensed) is passed through the inside of the tubes. The air used for cooling is moved either by induced or forced draft by fairly large fans.

As is well known, the heat-transfer coefficient for air to a metal tube wall is quite low. In a system having an organic fluid flowing through the inside of a tube and air moving past the outside of the tube almost all of the resistance to the flow of heat will be in the gas film. The overall heat-transfer coefficient will be low--say 5 Btu/(hr)(sq ft)($^{\circ}$F). This poor heat-transfer coefficient will result in there being a very large heat-transfer area, many tubes, and a high cost for the exchanger.

The use of extended surface tubes helps to alleviate this problem by providing a much larger amount of heat-transfer area for each tube. For example, a commonly used extended surface tube has nine aluminum fins per inch of tube length, firmly secured to the outside of the tube. This increases the area available for heat transfer by a factor of almost 20. Thus, commercial air coolers use extended surface to compensate for the poor gas film heat-transfer coefficient.

The use of extended surface heat exchange poses several problems. Kern [7, Chapter 16, and, in particular, Example 16.5] and Williams and Damron [10] discuss these problems. The three most important problems, from the process-engineering standpoint, are as follows:

1. The evaluation of the overall heat-transfer coefficient. For preliminary estimation purposes a value of 90 Btu/(hr)(sq ft)(°F) may be used. This is based on the internal surface area of the tube. Kern gives a detailed example of the method of computing U, while Williams and Damron argue for the use of previously measured coefficients from existing similar devices. The above-mentioned value of 90 appears reasonable.

2. The pressure drop through the apparatus on the extended surface side. The pressure drop should be low because the volume of air to be used is large and the cost of moving it can be high. Kern gives a detailed description of the pressure drop across transverse finned tubes. This pressure drop is approximately 0.125 in. of water per layer of tubes. Usually, the total pressure drop will be 0.5 in. of water.

3. The evaluation of the proper overall temperature difference for heat transfer. Flow through a standard air-cooled exchanger is usually of the cross-flow type and a correction to the true countercurrent log mean temperature difference must be used; Kern [7, p. 549] shows these correction factors.

The annual cost of operating an air-cooled heat exchanger is composed of two parts: (a) the cost of the power required to move the air across the bank of finned tubes, and (b) the charges related to maintenance, depreciation, and return on investment. This fixed charge factor is given by the second term in Eq. (3.47): $r = m + 1/n + (i_m - 0.35i)/(1 - t)$.

Maintenance: Air coolers do not require a great deal of maintenance; let $m = 0.02$.

Depreciation: Use straight-line 10-year depreciation; $1/n = 0.10$.

Return on investment: Let $i_m = 0.12$; $i = 0.08$; $t = 0.55$; then, $(i_m - 0.35i)/(1 - t) = 0.204$.

Process Plant Components

Then r = 0.324, or a maximum acceptable payout time, T_m, of 3.1 years.

The power cost: An air-cooled heat exchanger usually requires a fan to move the air past the heat-exchanger tubes. Because large volumes of air must be used, the power cost can become substantial if the air side pressure drop is at all large. It is common practice to keep this pressure drop small by using a large area for air flow and a rather small number of tubes in line with the air flow.

Kern [7, pp. 554-559] describes a calculation method for computing such pressure drops. Williams and Damron [10] describe a standardized heat-exchanger bundle and give a figure of 138 kw-hr per square foot of inside tube area as the yearly power consumption. For present purposes it would be reasonable to use a figure of 150 kw-hr per square foot of inside tube area. The power cost will then be (0.01)(150)(A) \$/yr, where the cost of the electrical power is taken as \$0.01/kw-hr.

The capital investment-related charges: The installed cost of extended surface heat-exchange equipment is given in Chapter 5 as $180 A^{0.75}$ for carbon steel tubes with aluminum fins, where A is the inside tube area in square feet. Therefore, the capital investment-related charges are $(0.324)(180 A^{0.75})$ \$/yr.

Thus, the total annual operating cost for an extended surface air-cooled heat exchanger is

$$C_a = (0.01)(150)(A) + (0.324)(180 A^{0.75}) \text{ \$/yr}$$

$$A_a = Q/(U_a \Delta T_a) = Q/(90)(\Delta T_a)$$

$$C_a = (0.01)(150)[Q/(90)(\Delta T_a)] + (0.324)(180)[Q/(90)(\Delta T_a)]^{0.75}$$

where Q = heat transferred in cooler (Btu/hr)

T_a = corrected logarithmic mean temperature difference (°F).

Cooling by Water. The cost of a water-cooled heat exchanger which would remove the same amount of heat may be considered as composed of three parts.

1. The cost of the cooling water itself. This cost varies substantially from one plant to another depending upon the amount of make-up water used; on the necessary chemical treatment for the abatement of corrosion, algae growth, and scaling; and on the cost of cooling tower operation. Brooke [11] presents a discussion of

water costs and gives values for the costs of the various kinds of water used: raw water, cooling tower make-up water, and boiler feed water. For water coming from a cooling tower and used as a coolant in a heat exchanger, a value of $0.04/1000 gal circulated seems reasonable. This is within the range of costs given in Table 5-5.

If the heat to be absorbed by the cooling water is Q Btu/hr, and the water temperature rises 35°F during its passage through the cooler, the cost of the cooling water will be 1.1×10^{-3} Q $/yr.

2. The cost of the power required to force the cooling water through the heat exchangers, attendant piping, and valves. Williams and Damron [10] suggest a figure of 0.04 hp/gal per minute circulated. If the cost of the electrical power is $0.01/kw-hr, the cost of the electrical power for pumping the water will be 1.36×10^{-4} Q $/yr. The capital investment-related charges on the circulating pumps, the valves, the piping, and the motors are incorporated in the installed cost of the heat exchanger and in the cost of the cooling water.

3. The capital investment-related charges. Here again, the maintenance, depreciation, and rate of return factors may be combined and then the sum multiplied by the capital investment in the water-cooled exchanger. Here, m will be set equal to 0.06, because water-cooled exchangers require more maintenance, but the other terms will be the same as for the air-cooled exchanger. The capital investment in water-cooled countercurrent flow heat exchangers of carbon steel is given in Chapter 5 as $346 A^{0.62}$, where A is the square feet of needed heat-exchange area. Then,

$$I = 346(Q/U_w \Delta T_w)^{0.62} \text{ \$}$$

The total annual cost of cooling by water will then be

$$C_w = 1.1 \times 10^{-3} Q + 1.36 \times 10^{-4} Q + (0.364)346(Q/U_w \Delta T_w)^{0.62}$$
$$\$/yr$$

Air Cooling vs. Water Cooling. The total annual costs for both water and air cooling may be calculated from the equations above for a given Q and for certain values for the stream temperatures. Consider the two sketches below; the process stream outlet temperature is kept constant at 125°F and the inlet process stream temperature is varied from 260° to 200° to 150°F. The temperatures of the cooling air and water streams are kept as shown.

Process Plant Components

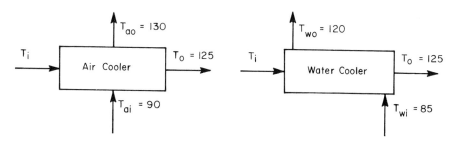

For the case where $Q = 10^6$ Btu/hr, the results of such a calculation are shown in the following table, and are also reported graphically in Fig. 8.3.

$$Q = 10^6 \text{ Btu/hr}$$

T_i (°F)	C_w ($/yr)	C_a ($/yr)
260	3,446	3,443
200	3,961	4,312
150	4,896	5,778

The curves of Fig. 8.3 show some interesting things about air and water cooling.

1. Increases in process stream inlet temperature favor air cooling over water cooling. In fact, in the subject case the two costs are exactly equal at T_i of 260°F. At higher temperatures air cooling will be less expensive.

2. For air cooling, the capital investment-related charges are a much greater proportion of the total charges than is the case for water cooling.

3. Water-cooling charges are sensitive to the cost of cooling water. It is difficult to obtain a reliable figure for this cost so that calculations of the kind shown above can be made with confidence.

4. The magnitude of the heat load, Q, has an effect on the economics of air versus water cooling. In the expression for the cost of water cooling Q appears to the 0.62 power, whereas in the similar expression for the cost of air cooling Q appears to the 0.75 power.

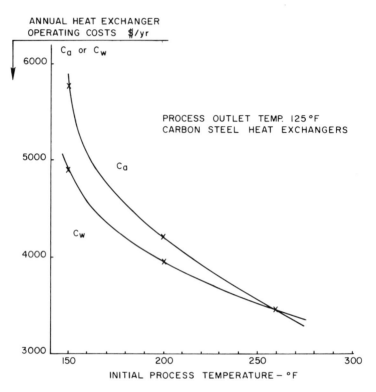

Fig. 8.3. Operating costs: air-cooled and water-cooled heat exchangers.

This will cause the capital investment-related charges for air cooling to increase faster, as Q increases, than those for water cooling. Thus, for heavy heat loads at lower temperatures water cooling may be preferred, whereas for smaller values of Q, air cooling might be better.

8.03 THE OPTIMUM DESIGN OF INSULATION

The thickness of insulation that it is desirable to employ to reduce heat losses from piping systems and vessels is one of the classic problems in process-plant economics. The problem is to minimize the sum of the annual cost of heat loss and the annual fixed charges on the insulation under given conditions. Both of these costs can be set up in terms of a single variable, the insulation thickness,

$$C = \frac{YH_t(10^{-6})\Delta T/\pi D_m}{\frac{1}{h_i \pi D_i} + \frac{x}{k\pi D_m} + \frac{1}{(h_c + h_r)\pi D_o}} + r(Bx + F) \qquad (8.29)$$

where C = total cost of insulation fixed charges and heat loss through the insulation [\$/(mean sq ft)(yr)]

x = insulation thickness (in.)

k = coefficient of thermal conductivity of insulation [Btu/(hr)(sq ft)(°F/in.)]

D_m = mean diameter of insulation (in.). For thick insulation around a small pipe it is necessary to employ the logarithmic mean, though usually the arithmetic mean of inside and outside diameters is satisfactory.

$h_c + h_r$ = surface film coefficient for transfer of heat by natural convection [Btu/(hr)(sq ft)(°F)]

h_i = film coefficient inside process equipment [Btu/(hr)(sq ft)(°F)]

D_o = outside diameter of insulation (in.)

B = incremental installed cost of insulation [\$/(in.)(mean sq ft)]

F = constant cost of insulation (\$/mean sq ft). Total insulation cost per mean square foot is equal to $F + Bx$.

r = annual fixed charges factor (1/yr)

Y = hours of operation per year (hr/yr)

H_t = value of incremental heat (\$/10^6 Btu)

ΔT = overall temperature difference (°F); process fluid temperature minus ambient temperature.

Equation (8.29) may be differentiated with respect to x, the result set equal to zero, and the expression solved for x, the optimum insulation thickness. A correct approach to this problem must take into account the fact that D_o and D_m are functions of x. If this is done, a complex expression results that can only be solved by trial and error. Dodge [12] discusses the solution of insulation problems by means of a computer which performs the trial-and-error calculation.

However, for preliminary evaluation purposes, some simplifications can be used which make the problem more tractable and do not introduce any great error.

1. The greater part of the resistance to heat flow is offered by the insulation. In many cases the portion of the total resistance due to the insulation is 85 to 90%. Therefore, simple approximations of the other resistances are allowable. For instance, the internal film resistance $1/h_i \pi D_i$ is almost negligible for a mobile liquid, and is not great for a gas. Therefore, it is allowable to neglect it. However, it must be recognized that it is no more logical to neglect the internal gas film resistance than to neglect the external resistance term.

2. The ratio D_m/D_o may be considered constant and independent of x. This would be particularly true for larger diameter pipes with thin insulation.

If these assumptions are made and Eq. (8.29) is differentiated, some terms combined, and the result set equal to zero, there is obtained

$$dC/dx = 0 = (-ak)/(x + Rk)^2 + b \qquad (8.30)$$

where $\quad a = YH_t(10^{-6})\Delta T$

$b = rB$

$R = [1/(h_c + h_r)](D_m/D_o)$

Then

$$x_{opt} = (ak/b)^{0.5} - Rk \qquad (8.31)$$

In order to establish approximate optimum values which reflect changes in the important variables involved, a brief tabulation has been prepared (Table 8-5). The following assumptions were used as a basis for the computations necessary:

1. Maximum acceptable payout time, 2 years.

2. Maintenance charges, 5%/yr.

3. Thermal conductivity of insulation, assumed to be 85% magnesia, taken roughly as 0.50 Btu/(hr)(sq ft)(°F/in.).

4. Values of $(h_c + h_r)$ taken as reported by McAdams [9]. For For the purpose of the present approximation $h_c + h_r = 2.0$

Table 8-5 Optimum Thickness of Pipe Insulation (In.)[a]

	Uninsulated pipe temperature (°F)		
	200	400	600
Base Case	1/2	1	1-1/4
Payout time, 1 year	1/4	1/2	3/4
Payout time, 5 years	3/4	1-1/2	2
Value of heat, $0.10/MM Btu	1/4	1/2	3/4
Value of heat, $0.40/MM Btu	1/2	1-1/4	1-1/2

[a] Base case assumptions are tabulated at top. Nominal pipe diameter 2 in. to 8 in.

Btu/(hr)(sq ft)(°F). These coefficients refer to ambient temperature of 80°F. For more accurate calculations, successive trials may be used employing closer values for both $(h_c + h_r)$ and k.

5. Incremental cost of insulation, $0.50/(in.)(sq ft).

6. Total annual operating time, 8,320 hr or 95% of total annual time.

7. Incremental fuel cost, $0.25/MM Btu.

8. Temperature of pipe, 400°F.

The basic constants will be developed for an illustrative case and applied to the computation of one of the values in Table 8-5.

$$a = 8{,}320 \times 0.25 \times 10^{-6} \times (400 - 80) = 0.665$$

$$b = (\tfrac{1}{2} + 0.05)0.50 = 0.275$$

In order to compute R, a value of D_m will be required. If a 4-in. pipe with 1 in. of insulation is assumed, D_m = 5 in. and D_o = 6 in.

$$R = \tfrac{1}{2} \times \tfrac{5}{6} = 0.417$$

Then

$$x_{opt} = \left(\frac{0.665 \times 0.5}{0.275}\right)^{1/2} - 0.417 \times 0.5 = 0.89 \text{ in.}$$

To the nearest 1/2 in., 1 in. thickness would be specified.

Checking the assumptions made for k and $(h_c + h_r)$, it is noted that the total resistance (neglecting inside film and pipe resistance)

$$\frac{x}{k} + \left(\frac{1}{h_c + h_r}\right)\frac{D_m}{D_o} = \frac{0.89}{0.5} + \frac{1}{2} \times \frac{4.89}{5.78} = 2.20$$

The insulation contributes $1.78/2.20 \times 100 = 81\%$ of the total resistance. Its temperature is therefore at an average of 270°F, and, from McAdams [9], k = 0.45 Btu/(hr)(sq ft)(°F/in.). This is sufficiently close to the basic assumption made. Similarly, the surface temperature will be 141°F and $(h_c + h_r) = 2.3$ instead of 2.0 as assumed originally. Recalculation of x_{opt} gives

$$x_{opt} = \left(\frac{0.665 \times 0.45}{0.275}\right)^{1/2} - 0.362 \times 0.45 = 0.88 \text{ in.}$$

The effect of variations in pipe temperature, payout time, and value of heat are given in Table 8-5.

Actual commercial practice for insulation of pipes appears to be more conservative in the sense that usually somewhat thicker insulation is specified. Thus for 4-in. pipe at 400°F, 2 in. of 85% magnesia is usually employed. It can be seen that, insofar as heat-saving economics is concerned, such a thickness would correspond to a high maximum acceptable payout time and (or) a high value for incremental heat.

Equation (8.31) should be used only for approximation purposes because insulation does not vary continuously in price but in sharp steps as additional standard thicknesses are added. However, it should serve as indication of economical design.

With low-temperature operations, in particular, special problems other than heat saving will enter into economic design of insulation. Often the economical thickness from the standpoint of refrigeration saving will not be sufficient to prevent excessive condensation on the exterior surface, especially in locations where the humidity is high. The other factor that may influence the thickness needed is adequate control of the process to which insulation is applied. Knowledge of the process will dictate tolerable heat losses for each unit of equipment to which insulation must be applied. The greatest thickness of insulation required by any of these considerations will be the one specified for design.

Process Plant Components

8.04 THE OPTIMUM DESIGN OF DISTILLATION COLUMNS

The economic design of distillation equipment is of special interest because this is the unit operation that is most frequently employed for separating products from a reactor system, preparing raw materials for reactor feed streams, and for purifying finished products. Because of the existence of simplified methods for approximating the performance of distillation equipment, it is possible to treat this problem in reasonably concise form. The method presented here follows that originally derived by Colburn [13].

For a given separation, the greater the reflux, the fewer the plates or transfer units required, but the more steam and condenser water, the larger the diameter of the column, and the greater the size of the condenser and reboiler. A general solution for the establishment of optimum trays and reflux ratio is desirable to facilitate design calculations as well as to assist in developing convenient approximations.

It will be remembered that in Chapter 4 a distillation optimization problem was solved using two independent variables. However, the solution was complex and the calculations difficult and time-consuming. In the present case only one variable, the reflux ratio, will be considered.

The cost of a distillation operation is made up of three items:

1. The annual fixed charges against the installed cost of the distillation column and its internals (sieve trays, bubble cup trays, packing, or others). The incremental cost of a distillation column will be that of one tray plus the section of the column containing that one tray.

2. The annual fixed charges against the cost of the heat-exchange surface in the condenser and reboiler.

3. The annual cost of the heat needed to produce the vapor stream in the column plus the cost of the water (or air cooling) needed to condense the vapor stream at the top of the column.

As is well known, in distillation it is convenient to express vapor and liquid flow rates in terms of moles per unit time so as to take advantage of the fact that the molal latent heats of vaporization of many compounds are essentially equal. This usage will be adopted here.

The cost of a distilling column per pound mole of distillate can be expressed as

$$C_1 SN/EYD \tag{8.32}$$

where C_1 = annual incremental unit investment cost [$/(sq ft)(plate)(yr)]

S = tower cross sectional area (sq ft)

N = number of theoretical plates

E = fractional plate efficiency

Y = hours/yr operation

D = distillate rate (lb moles/hr)

But

$$S = V/G_a = D(1+R)/G_a \tag{8.33}$$

where V = vapor throughput rate (lb moles/hr)

G_a = allowable vapor velocity [lb moles/(hr)(sq ft)]

R = reflux ratio (moles of overhead liquid product/mole of distillate)

Then, the above cost of a column becomes

$$C_1 N(1+R)/EYG_a \tag{8.34}$$

In addition to the fixed charges for unit incremental column cost, there will be condenser and reboiler costs; this is item 2 mentioned above. The condenser and reboiler costs per pound mole of distillate can be written as

$$C_2(A_c + A_b)/YD$$

where C_2 = annual incremental unit investment cost in condenser and reboiler equipment [$/(sq ft)(yr)]

A_c = area of condenser (sq ft)

A_b = area of reboiler (sq ft)

From a consideration of heat transfer in both reboiler and condenser the combined area, $A_c + A_b$, can be expressed as

$$2[D(1+R)L_v]/(U\Delta T) \tag{8.35}$$

Process Plant Components

where L_v = latent heat of vaporization of substance being distilled (Btu/lb mole)

U = overall heat-transfer coefficient in reboiler and condenser; assumed the same in both [Btu/(hr)(sq ft)(°F)]

ΔT = overall temperature difference; assumed the same in both (°F)

The term $D(1 + R)L_v$ represents the latent heat transferred into the condenser, and in an adiabatic column (as most are) this is also the heat absorbed from the reboiler.

As a matter of experience, it is known that U and ΔT are essentially the same in both reboiler and condenser. Then, the combined area of reboiler and condenser is given by Eq. (8.35).

Let G_b = vapor-handling capacity of the condenser and reboiler equipment combined, in lb moles/(hr)(sq ft of total transfer area). Then,

$$A_c + A_b = 2D(1 + R)L_v/U\Delta T = D(1 + R)/G_b \qquad (8.36)$$

$$G_b = U\Delta T/(2L_v)$$

So condenser and reboiler costs become

$$C_2(1 + R)/YG_b \qquad (8.37)$$

The cost of steam and coolant per pound mole of distillate, in the case of feed at the boiling point, will be

$$C_3(1 + R) \qquad (8.38)$$

where C_3 = cost of steam and coolant to vaporize and condense, respectively, 1 lb mole of distillate.

The total cost of the distillation operation is the sum of these three items.

Here, it should be noted that C_1 and C_2 are defined as incremental costs, yet they are substituted directly in the total cost equation. As mentioned in Sections 2.15 and 8.03, this is a convenience when the total cost equation is to be differentiated analytically, the results set equal to zero, and the equation solved algebraically. Upon differentiation the total cost expression for the equipment reduces to a constant value, and if this constant is used in the original (before differentiation) equation it will remain unchanged throughout the subsequent manipulation. Therefore, it is permissible to simplify matters by using symbols for incremental costs rather than total costs in the original total cost equation.

$$\text{Cost} = \frac{C_1 N(1+R)}{EYG_a} + \frac{C_2(1+R)}{YG_b} + C_3(1+R) \tag{8.39}$$

The optimum value of R will be that where $d(\text{Cost})/dR = 0$,

$$\frac{d(\text{Cost})}{dR} = \frac{C_1 N}{EYG_a} + \frac{C_1(1+R)}{EYG_a}\frac{dN}{dR} + \frac{C_2}{YG_b} + C_3 = 0$$

Solving for the value of R_{opt} for which the cost is a minimum gives

$$1 + R_{opt} = \frac{N + [C_2/(YG_b) + C_3](YEG_a)/C_1}{-(dN/dR)} = \frac{N+F}{-(dN/dR)} \tag{8.40}$$

where F is designated as the "cost factor."

There is no convenient expression for N in terms of R to permit evaluation of dN/dR for the general case. Therefore the solution of Eq. (8.40) first involves preparing a plot of N vs. R for special cases. This may be accomplished by means of the curve of Gilliland [14]; Figs. 4.5 and 4.6 are examples of such plots. From the plot of N vs. R, values of slopes are obtained at a number of points, and a curve is drawn of dN/dR vs. R. Then Eq. (8.40) is solved by cut and try. That is, at chosen values of R, both N and dN/dR are found and the two sides of Eq. (8.40) are evaluated. This procedure is repeated until the two sides of the equation are equal.

A number of solutions, some made by Colburn, according to the procedure outlined above, have been plotted in Fig. 8.4. Table 8-6 has been prepared to illustrate the values for optimum reflux ratio and number of trays for a set of conditions such as might obtain in typical petroleum distillate fractionations near 1 atm pressure. For preparing this table, the following assumptions were made:

1. Payout time, 2 years.

2. Steam cost, $0.50/M lb.

3. Average incremental cost of condensers and reboilers, $16/sq ft.

4. Total annual operating time, 8,320 hours—95% of total annual time.

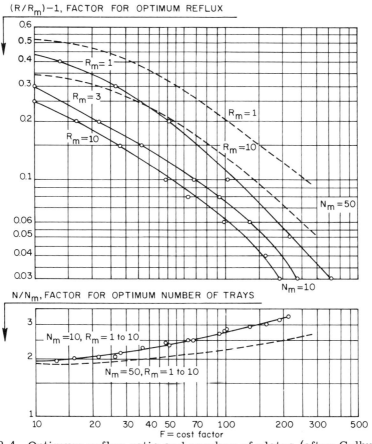

Fig. 8.4. Optimum reflux ratio and number of plates (after Colburn; Ref. [13]).

5. Heat-transfer coefficient for condensers and reboilers, 100 Btu/(hr)(sq ft)(°F).

6. Cooling water cost, $0.06/1,000 gal.

7. Cooling water enters at 80°F and leaves at 115°F.

8. Maintenance charges, 5%/yr.

9. G_a = 15.0 lb moles/(hr)(sq ft). Based on approximately 2 ft/sec superficial velocity for hydrocarbon vapor; the velocity for materials such as water will be higher.

Table 8-6

	Factor for optimum reflux $f = (R_{opt}/R_m) - 1$ $R_{opt} = (1+f)R_m$									Factor for optimum trays N_{opt}/N_m			F Cost factor
	$N_m = 10$			$N_m = 20$			$N_m = 50$			$N_m = 10$	$N_m = 20$	$N_m = 50$	
	R_m			R_m			R_m			R_m	R_m	R_m	
	1	3	10	1	3	10	1	3	10	1 to 10	1 to 10	1 to 10	
Base case	0.20	0.12	0.10	0.24	0.17	0.16	0.31	0.21		2.4	2.3	2.1	53
Payout time 1 yr	0.24	0.14	0.12	0.28	0.20	0.17	0.37	0.24		2.2	2.1	2.0	40
Payout time 5 yr	0.13	0.09	0.07	0.17	0.13	0.10	0.22	0.15		2.7	2.5	2.2	86
Steam cost $0.30/M lb	0.22	0.13	0.11	0.27	0.16	0.14	0.35	0.22		2.3	2.1	2.0	45
Steam cost $0.75/M lb	0.18	0.11	0.09	0.21	0.13	0.11	0.29	0.19		2.5	2.3	2.1	60
$G_a = 50$ lb mole/(hr)(sq ft)	0.06	0.04	0.03	0.08	0.06	0.05	0.13	0.08		3.1	2.8	2.4	177

Process Plant Components

10. For the completely installed column, the cost of one additional plate with the section of tower that it occupies is assumed to be $90/sq ft; 1970 costs--see Chapter 5. This cost applies to towers of carbon steel construction with sieve trays spaced 24 in. apart, the tower diameter is approximately 6 ft, and the tower height is not more than 120 ft. The costs of distillation towers are discussed more fully in Chapter 5.

11. The plate efficiency is taken as 0.95. This is reasonable for hydrocarbon fractionation, but not for other systems.

12. G_b = 0.10 lb mole/(hr)(sq ft) of condensing and reboiling surface totaled. This corresponds to a molecular weight of 100 and a latent heat of vaporization of 150 Btu/lb. Using a heat-transfer coefficient of 100 and an average temperature difference of 30°F in either condenser or reboiler, the value of G_b = (100)(30)/(2)(100)(150) = 0.10 lb mole/(hr)(sq ft).

To illustrate the application of these assumptions, the basic constants will be developed for Eq. (8.40), using the above information.

$$C_1 = r(90) = \left(\frac{1}{T_m} + m\right)90 = \left(\frac{1}{2} + 0.05\right)90 = 0.55(90)$$

$$C_1 = \$49.50/(\text{sq ft})(\text{plate})(\text{year})$$

$$C_2 = r(16) = (0.55)(16) = \$8.81/(\text{sq ft})(\text{yr})$$

Cost of steam = (100)(150)(0.50)/(900)(1,000) = $0.00832/lb mole vapor

Molecular weight of vapor = 100

Latent heat of vaporization of distillate = 150 Btu/lb

Latent heat of vaporization of steam = 900 Btu/lb

Cost of water = (100)(150)(0.06)/(35)(8.35)(1,000) = $0.003075/lb mole vapor

C_3 = 0.00832 + 0.003075 = $0.0114/lb mole vapor

C_2/YG_b = 8.81/(8,320)(0.10) = $0.01064/lb mole

C_1/YEG_a = 49.50/(8,320)(0.95)(15) = $0.000418/lb mole

From these the cost factor, F, becomes

$$F = (C_3 + C_2/YG_b)(YEG_a/C_1) = 53$$

For the case discussed above, it is instructive to make some calculations which will show the change in the optimum value of the reflux ratio as the minimum reflux ratio and the number of theoretical plates at total reflux change. Figure 8.4 may be used for this purpose.

	Cost factor = 53	
	$N_m = 10$	$N_m = 50$
	$R_m = 3$	$R_m = 10$
R_{opt}	$1.12 R_m$	$1.21 R_m$
N_{opt}	$2.4 N_m$	$2.1 N_m$

From these, and from more extensive calculations of the same type, it is fairly easy to draw the conclusion that reasonably large changes in the important costs and operating conditions affecting a distillation column will result in only modest changes in the optimum reflux ratio and the optimum number of plates. For practically all realizable conditions the optimum reflux ratio will vary between 1.1 and 1.4 times the minimum, and the optimum number of plates will be between 2 and 3 times the number at total reflux.

As a useful approximation design method, it may be said that the operating reflux ratio of a fairly ideal system may be 1.2 times the minimum value, and the number of theoretical plates will be twice the number at total reflux. This conclusion applies despite rather wide changes in the cost of steam, the cost of cooling water (or the use of air cooling), changes in payout time, and changes in the nature of the system being distilled.

Calculations for minimum reflux depend on vapor-liquid equilibrium data, and, if the accuracy of such data is questionable in a given case, a higher reflux ratio than that indicated by economic balance should be employed. Often the total cost is not increased greatly by moderate increases in the reflux ratio (up to 1.5 times the minimum), and an additional safety factor is thus provided at little additional cost.

Distillations of an unusual nature often occur in industry. Systems involving nonideal vapor-liquid equilibria, azeotropes, pinch regions, two liquid phases, and three or more components are often encountered. In such cases the Gilliland correlation cannot be used. However, a graphical solution, such as the McCabe-Thiele method, is always possible.

Pyle [15] has pointed out that when process considerations involve an easy separation, so that only a small reflux ratio is required, the flow of liquid across the plate in a column will be low. At low liquid rates, splashing of liquid over the weir and leakage around the vapor riser or around the edges of the weirs may all result in a reduced liquid seal, with lowered efficiency. The reflux rate under such conditions may not be set by the economics of heat saving, but by the minimum liquid rate at which the plate will operate satisfactorily.

No generalization between number of transfer units and reflux ratio has been developed similar to Gilliland's curve for plate columns. If, in the absence of an exact relationship, the curve for theoretical plates is employed for transfer units as a reasonable approximation, it is possible to utilize Eq. (8.40) and Fig. 8.4 by substitution, for the cost factor $F = (C_2/hG_b + C_3)(hEG_a/C_1)$, the following group $F = (C_2/hG_b + C_3)(hG_a/C_4)$ where C_4 = amortized unit investment cost, \$/(sq ft)(transfer unit)(yr).

Setting aside the problem of selecting the optimum reflux ratio, it is instructive to consider the total cost of conducting a distillation operation and the relative importance of the components of this cost.

Equation (8.39) gives the total cost of operating a distillation column. The answer is expressed as \$/lb mole of distillate.

$$\text{Cost} = \frac{C_1 N(1 + R)}{EYG_a} + \frac{C_2(1 + R)}{YG_b} + C_3(1 + R) \quad (8.39)$$

capital investment charges on column — capital investment charges on condenser and reboiler — cost of steam and cooling water

For the base case discussed previously, and allowing $R = 3$ and $N = 20$

Cost = 0.0334 + 0.0423 + 0.0456 = \$0.1213/lb mole distillate

If the molecular weight of the distillate is 100—say toluene (approximately)—this is 0.12 ¢/lb of distillate. Here, the separate quantities are essentially equal.

It is apparent that the reflux ratio is the most important variable, and that N, which is dependent on R, is also significant. In Example 4.4 the effect of the independent action of the two variables, N and R, is examined.

In general, for ordinary distillations, the following statements can be made:

1. The total cost of distillation is the sum of three essentially equal costs.

2. Despite the greater size and expense of the column and the trays, the low capacity of the reboiler and the condenser causes the operating cost in dollars per pound mole of distillate to be approximately the same for both units.

3. Distillation is a cheap method of separation. Even for relatively high cost steam and cooling water, the cost of a distillation separation is approximately 0.1¢/lb of distillate.

Optimum Bottoms Concentration

There are two problems in distillation column optimization:

1. Where valuable components of the system are retained in the process plant and not discarded to waste. In cases of this type the economic balance is concerned only with the costs of the column, the condenser and reboiler, and the steam and cooling water. This problem has been considered; see Eq. (8.39).

2. Where valuable components of the system are discarded to waste or may be destroyed by subsequent chemical processing of one or both of the top and bottom streams. In this case the economic balance must include a charge for the value of the lost material.

An example of Class 1 is the separation of ethylbenzene from styrene in the manufacture of styrene; see Bodman [1, pp. 131-149] and Example 4.4. Here the ethylbenzene (the top stream) is recirculated to the dehydrogenation reactor and the styrene (the bottom stream) is further processed to purify it from tars and polymers. Some styrene is lost in the tar and this second separation would be of Class 2.

An example of Class 2 is the separation of methanol from water following a solvent recovery operation. Some methanol will be lost when the water is rejected from the system. The economic balance is made between the cost of operating the column and the cost of the valuable material recovered.

An exact solution for the optimum bottoms concentration would involve solving simultaneously for the optimum reflux ratio as well as the optimum bottoms concentration.

An equivalent answer consists in assuming that the reflux ratio is constant. Then, the economic balance is between the cost of the column and the value of the material recovered. The answer obtained is the number of plates for that particular reflux ratio as a function of the bottoms concentration. By checking back to the previous solution, Eq. (8.39), it is established whether the number of plates is also optimum with respect to the reflux ratio selected. If it is not, a new value for the reflux ratio is assumed. It is apparent that a plot similar to Fig. 8.4 can be prepared.

The above method of application and the partial derivative method of Example 4.4 both involve a considerable amount of hand calculation. This is so undesirable that it is improbable that such calculations would be done. However, at the present time, such calculations have been programmed for the electronic computer [1, pp. 209-213], and it is possible to vary the bottoms concentration over a fairly wide range and to calculate a total cost for each one of these values. In this way the effect of the bottoms concentration on the distillation economics may be calculated swiftly.

In practically every case it results that a very high recovery (> 99%) of the desired product is economically justified. Only in cases where the relative volatility is less than 2 (resulting in a high column cost) and the sales price of the desired material is low (resulting in a small benefit from the recovery of the material) will a recovery of less than, say, 95% be considered. This combination of circumstances almost never occurs.

In connection with bottoms concentration there are two other factors that must be considered.

1. In some systems several distillation columns may be operated in series, with a pure product being taken from the top of each column. It is apparent that any low boiler remaining in the bottoms stream from one column will appear in the top product from the succeeding column. The purity requirements for the top product from the downstream columns will influence the concentration of low boiler which can appear in the bottoms stream from the preceding column. Economic balance may not be as important. The separation

of benzene, toluene, and xylene by fractional distillation is an example.

2. The disposal of waste streams has now become a matter of great importance, subject to stringent public regulation and requiring careful attention to the design and operation of equipment for the prevention of air and stream pollution. It is entirely possible that the demands of the public health authorities will be such that a certain value of bottoms concentration cannot be exceeded. In such cases, the economic-balance principle is still valid, but the principal variable will become the reflux ratio or the number of plates.

8.05 THE OPTIMUM DESIGN OF GAS ABSORPTION COLUMNS

In the chemical industry many valuable materials are recovered from gas streams by countercurrent absorption in various solvents. The equipment used for this purpose is of the same kind as used in distillation: tall towers (usually made of metal), with either plates or packing filling the interior. In nearly every case, the gas that enters the bottom of the absorption column is quite dilute, usually not more than 2 or 3% of solute gas in the carrier gas. Also, the liquid leaving the bottom of the tower, containing the solute gas in solution in the solvent, is quite dilute. As a consequence, rather large quantities of inert gas and inert solvent are processed in order to recover rather small quantities of valuable material.

In gas absorptions the valuable material is usually recovered from the solvent by fractional distillation. This distillation operation must be considered when analyzing the economics of gas absorption. There are, in general, two different procedures for recovering the solute material from the solvent by distillation.

1. If the solute is not soluble in water and a nonaqueous solvent is used, steam stripping may be used to recover the solute from the essentially nonvolatile solvent. Upon condensation of the overhead vapors, an essentially pure product may be obtained. Substances such as propane and butane may be recovered from petroleum gases by this method.

2. If the solute is soluble in water it is probable that water will be used as a solvent in the absorption tower. The pure product is then recovered by fractional distillation between the solute and

water. A column having both stripping and rectifying sections must be used.

Figure 5.4 shows a flow sheet for the recovery of a solvent material such as methanol from an inert gas such as air by absorption into water followed by a fractional distillation to recover the pure methanol overhead and methanol-free water from the bottom of the distillation column.

The total cost of the absorption and distillation operation is made up of several parts:

1. The cost of the solute gas which is not recovered. It is apparent that 100% of the gas cannot be recovered because that would require an infinitely tall absorption column. However, some quite large percentage of the inlet solute gas can be absorbed.

2. The cost of any solvent that is lost. Some absorption solvent will be lost from the system as a vapor in the exit gas, some will be lost by deliberate purging from the bottom of the distillation tower, and some will be lost from leaks and mechanical handling. These losses are small and, unless the solvent is valuable (not water or oil), they may be neglected.

3. The cost of the power required to force the gas through the absorption column. Rather large quantities of gas are treated to recover a small quantity of solute and the power requirement may be significant. Power is also needed to pump the solvent over the top of the absorption tower and into the distillation tower. This is usually negligible.

4. The annual charges on the capital investment in the fan and motor required to force the gases through the absorption tower.

5. The annual charges on the capital investment in the absorption tower and its internals. For simple solvent recovery—as for methanol—the column may be made of carbon steel and operated at atmospheric pressure, thus fairly cheap. However, petroleum industry columns are operated at much higher pressures, necessitating heavier steel, and some absorptions, such as those involved in the production of nitric acid, require both heavy steel and corrosion resistance. In addition, there are many instances in the chemical industry where atmospheric absorptions are conducted in acid-resistant towers that are quite expensive.

6. The cost of the distillation operation required to recover the solute from the absorber bottoms stream. This problem has been discussed in detail in the preceding section.

If heat exchangers are used on the main process streams the capital and operating costs related to these exchangers must be included in the list of system costs. In the subject case of the simplest possible system there are no heat exchangers, and no costs.

It is apparent that an economic balance must exist. The absorption column may be built very high and so absorb almost all of the solute gas, thereby reducing the cost of the lost solute to a low level. However, a tall column will have a high cost and a large pressure drop, thereby consuming large amounts of power for forcing the gases through it, and having a powerful fan and motor, which are also expensive. On the other hand, if the absorption column is short, the solute gas loss will be high (and expensive), but the cost of the column, fan, motor, and power will be low. In addition to these considerations, the cost of the distillation will depend on the quantity and the concentration of the absorber effluent stream. The distillation is more expensive for a dilute absorber stream.

In the following treatment only dilute solutions are considered. With dilute solutions the rates of flow of gases and liquids in the column are essentially constant throughout the length of the column, and the presence of the solute in either liquid or gas phase has little effect on the properties of that phase. For dilute solutions, concentrations may be expressed as mole fractions or mole ratios without significant effect on the numerical answer.

The size of the absorber column is dependent on several things:

1. The quantity of gas that is flowing through the column. This will determine the diameter of the column because the column must be sized so as to operate at about 70% of the flooding velocity. The cost of the column and of the packing are dependent on the column diameter. The rate of gas flow is

$$W = (0.786)(D^2)(G')$$

where W = rate of flow of gas in the column (lb m/hr)

D = diameter of the column (ft)

G' = allowable gas mass velocity in column [lb m/(hr)(sq ft)]. G' is determined from flooding velocity correlations, the kind of column internals, and the temperature and pressure of the operation.

2. The ratio of the slopes of the operating line, L/G, and the equilibrium line, m. The term L/mG is called the absorption factor,

and, for dilute solutions, it may be regarded as constant throughout the column. The value of m is set by the temperature, pressure, and chemical nature of the system. The value of G is set by the flooding characteristics of the column internals, but L may be varied. Actually L is the principal variable, since m and G are usually fixed within rather narrow limits.

3. The fraction of the solute gas that is absorbed. This number is given by

$$f = (y_1 - y_2)/y_1$$

where   f = fraction of inlet solute gas that is absorbed by the solvent

y_1 = mole fraction of solute gas in the absorber gas inlet stream

y_2 = mole fraction of solute gas in the absorber gas outlet stream

The fraction that is absorbed can be varied independently of the absorption factor simply by building the column as high as desired. That is, by using more or fewer transfer units or plates.

For a packed column the height of the column is given by the expression

$$H_T = H_{OG} N_{OG}$$

where H_T = the height of the active section of the column; towers usually have 4 to 6 ft of additional height above and below the absorbing section

H_{OG} = the height of an overall gas-phase transfer unit (ft)

N_{OG} = the number of overall gas-phase transfer units

There is a large amount of knowledge concerning H_{OG} and its dependence on liquid and gas flow rates, and other tower parameters. For practical tower packings, such as Raschig rings and Berl saddles, the work of Cornell et al. [16] has shown that H_{OG} is almost constant and equal to 2.0 to 2.5 ft for all practical purposes. This simplification is quite useful. It reduces the complexity of the problem by a large factor while, at the same time, not introducing any appreciable error. A detailed calculation of H_{OG} is given by Treybal [17, p. 261].

N_{OG} may be calculated from the formula

$$N_{OG} = \frac{\ln[(y_1/y_2)(1 - mG/L) + mG/L]}{1 - mG/L}$$

Charts, such as Fig. 8.20 of Treybal [17], permit the rapid calculation of this number. Thus, although the formula is complex, its use is simple.

By combining H_{OG}, N_{OG}, and the cost of an absorption column plus the packing, a formula for the annual cost of the absorption column and the packing may be obtained. This formula is

$$rH_{OG}\left[\frac{\ln[(y_1/y_2)(1 - mG/L) + mG/L]}{1 - mG/L}\right][280D + 0.786D^2 C_p] \text{ \$/yr}$$

where r = fraction of total investment in the installed absorption column per year to allow for maintenance, depreciation, and an acceptable minimum profit

280D = installed cost of 1 vertical ft of the absorption column shell ($)

C_p = cost of 1 cu ft of 1-in. Raschig rings ($)

It should be noticed that this formula contains the three principal variables mentioned above.

The cost of the lost solute vapor in the absorber gas outlet stream is

$$(8,000)(0.786)D^2 G' y_2 \left(\frac{\text{mol wt of solute vapor}}{\text{mol wt of inert gas}}\right) C_s \text{ \$/yr}$$

where C_s = the cost of the solute ($/lb)

This formula contains two of the three variables.

In order to evaluate the power consumption of the fan and motor which force the gas through the packing it is necessary to have an expression for the pressure drop through the packing as a function of the packing, the gas rate, and the liquid rate. Expressions of this type exist, but they are fairly complex. An examination of the correlations of pressure drop in the Chemical Engineers' Handbook [18, pp. 18-27 to 18-29] show that an empirical equation of the form

$$\Delta P = 4.3 \times 10^{-3} [G'/\phi]^{0.8}$$

where ΔP = pressure drop (inches of water/ft of packing)

G' = gas mass flow rate [lb m/(hr)(sq ft)]

Process Plant Components

$$\phi = (\rho_G/0.075)^{0.5}$$

ρ_G = density of inert gas (lb m/cu ft)

gives an adequate representation of the pressure drop-flow rate relationship for gas flowing through 1-in. Raschig rings wet with water. It should be noted that the equation does not contain any term for the liquid flow rate. For 1-in. Raschig rings, and within the range of liquid flow rates likely to be used [500 to 1,000 lb m/(hr)(sq ft)], the correlations show that the liquid flow rate has a negligible effect on the pressure drop. Therefore, the cost of electrical power consumed in forcing the gas through the packing is (including an allowance for a 50% efficiency of the fan and motor)

Cost of electrical power =

$$\left[\frac{1.06 \times 10^{-6} G'^{1.8} D^2 H_{OG}}{\phi^{0.8} \rho_G}\right] \left\{\frac{\ln[(y_1/y_2)(1 - mG/L) + mG/L)]}{1 - mG/L}\right\} \text{\$/yr}$$

where the cost of electrical power is taken as \$0.02/kw-hr.

The fan and the motor necessary to pump the gas through the packing will have a capital cost and this will have an annual charge. The capital cost of fan and motor is given in Chapter 5 as 18.6 (cu ft/min)$^{0.68}$. For low pressures this is essentially independent of the column pressure drop, but is rather strongly dependent on the amount of gas to be handled.

The annual charge for the fan-motor combination is

$$r\{18.6[(G'/\rho_G)(0.786D^2)(1/60)]^{0.68}\} \text{ \$/yr}$$

Fan and motor combinations, with their attendant gear reducers and complicated installations, are expensive, and their annual charges must be considered.

The cost of the distillation to recover pure product from the absorber effluent stream is—using the result of Section 8.04

$$\frac{0.786 D^2 L'(8,000) x_1}{\text{mol wt of solution}} \left\{\frac{C_1 N(1 + R)}{EYG_a} + \frac{C_2(1 + R)}{YG_b} + C_3(1 + R)\right\}$$

where L' = liquid mass velocity in absorption column [lb m/(hr)(sq ft)]

x_1 = mole fraction of solute in absorber effluent stream

D = diameter of absorption column (ft)

The other symbols have the same meaning as in the preceding section.

Adding all these annual costs, there results

Total cost of absorption ($/yr) [Eq. (8.40)] = $rH_{OG} \left\{ \dfrac{\ln[(y_1/y_2)(1 - mG/L) + mG/L]}{1 - mG/L} \right\}$

$\times (280D + 0.786D^2 C_p)$

annual capital investment related charges on absorption column plus packing

$+ \dfrac{6,300D^2 G'y_2(\text{mol wt of solute})C_s}{\text{mol wt of inert gas}}$

annual cost of lost solute from top of absorption column

$+ \left[\dfrac{1.06 \times 10^{-6} G'^{1.8} D^2 H_{OG}}{\phi^{0.8} \rho_G} \right]$

$\times \left\{ \dfrac{\ln[(y_1/y_2)(1 - mG/L) + mG/L]}{1 - mG/L} \right\}$

annual cost of power required to force gas through absorption column

$+ r\{18.6[(G'/\rho_G)(0.786D^2)(1/60)]^{0.68}\}$

annual capital related charges on fan and motor required to force gas through absorption column

$+ \left[\dfrac{0.786D^2 L'(8,000)x_1}{\text{mol wt of solution}} \right] \left[\dfrac{C_1 N(1+R)}{EYG_a} + \dfrac{C_2(1+R)}{YG_b} + C_3(1+R) \right]$

annual cost of distillation operation

$+ 0.786D^2 L'(8,000)(0.04)/(8.35)(1,000)$

annual cost of discarded solvent (water @ $0.04/M gal)

Process Plant Components   403

Here it must be noted that the cost of the distillation column, as used in the above equation, was originally derived for a distillation column feed which was at its boiling point. In the subject case, it is assumed that the feed to the distillation column will be at a temperature well below its boiling point. An analysis of the economics of the distilling system shows that the thermal condition of the feed has almost no effect on the economics of the distillation operation. Therefore, in this case, the treatment of Section 8.04 is adopted without change.

Equation (8.40) has three independent variables: y_2 (or fraction solute gas absorbed), L' (liquid mass flow rate in absorption column), and D (essentially, the vapor velocity in the absorption column). It should be possible to differentiate Eq. (8.40) with respect to these variables, obtain three simultaneous equations, and solve for the optimum values of y_2, L', and D. This procedure would be cumbersome and would not present a clear picture of the effect of these variables on the economics of gas absorption. A better method would be to assume values of D (derived by fixing W and G') and mG/L (which would fix L') and calculate the cost as a function of y_2 (or fraction absorbed) for the various values of D and mG/L. A complete set of such calculations would be tedious and time-consuming. However, an electronic computer would make short work of them [19].

Example 8.3

As an illustration of the use of Eq. (8.40) the problem of Treybal [17, pp. 257-263] may be considered. In this problem ethanol is to be absorbed into water from an inert gas consisting of carbon dioxide. The inlet gas contains about 1 mole% ethanol, and the dilute ethanol-water solution from the absorption tower is to be distilled to produce the ethanol-water azeotrope, leaving the water essentially ethanol-free.

The details of both the absorption and distillation calculations for the ethanol-water system are given by Treybal [17]. These computations need not be repeated here. The student can consult the reference given.

In summary:

Gas flow rate = 1,640 lb m/hr; L/G = 1.0 moles water/mole gas; m = 0.73; allowable gas flow rate = 1,246 lb mole/(hr)(sq ft); this is approximately 65% of the flooding velocity;

Column cross-sectional area = 1,640/1,246 = 1.31 sq ft

Column diameter = 1.29 ft; H_{OG} = 2.4 ft; L' = 515 lb m/(hr)(sq ft) ρ_G = 0.109 lb m/cu ft; ϕ = 1.2; r = 0.6; cost of ethanol = $0.076/lb; y_1 = 0.01; cost of packing = $7.00/cu ft for 1-in. Raschig rings; mol wt of gas = 44; mol wt of liquid = 18.

The present calculation is started by assuming a value for L/G, and further assuming that rather small changes in L will have little effect on the flooding velocity (see Fig. 6.26 of Treybal [17]) so that the flooding velocity for this system may be calculated. Once the flooding velocity is known, the actual velocity is calculated by assuming an actual velocity of from 50 to 70% of the flooding velocity. This value of G is then held constant, which sets the value of D. The cost is calculated at constant L for various values of y_2 (the fraction absorbed).

The treatment of the distillation calculation is worthy of some detailed consideration. In almost all absorptions of water-soluble solvents the feed to the distillation column will be quite dilute--between x_1 of 0.005 and 0.01. This is really quite small. A glance at the McCabe-Thiele diagram for any of the materials, such as ethanol, being distilled from water, will show that the reflux ratio must be fairly high and strongly dependent on the feed concentration, x_1. After a detailed study of the effect of the value of x_1 on the reflux ratio and the number of plates, the following conclusions are reached:

1. The term $x_1(1 + R)$ is essentially constant at a value of 0.132 for a range of x_1 between 0.01 and 0.005 and for R between 12 and 25.

2. The number of plates for the ethanol-water separation may be considered as constant at 29 actual trays. This number is deduced from a study of the McCabe-Thiele diagram for ethanol-water, and from Treybal's calculation of the plate efficiencies. By using a constant number of trays the value of x_B, the distillation column bottoms concentration, will vary as x_1, the distillation column feed concentration, varies. However, in every case the value of x_B will be so low as to be negligible.

3. As a consequence of 1 and 2 above the distillation cost can be calculated from Eq. (8.40)

$$\frac{0.786 D^2 L'(8,000) x_1 (1 + R)}{(18)(0.894)} \left(\frac{C_1 N}{YG_a} + \frac{C_2}{YG_b} + C_3 \right)$$

Here, D and L' pertain to the absorption tower. The term 0.894 is the mole fraction of ethanol in the distillate. $N = 29$; $G_a = 27$; $G_b = 0.125$; $Y = 8,000$; and $C_3 = 0.0114$. G_a and G_b are different from those used in Section 8.04 because water is distilled in this example whereas hydrocarbons of much higher molecular weight were used in Section 8.04. Then,

$$\left[\frac{(0.786)(8,000)(0.132)}{(18)(0.894)}\right][(0.00665 + 0.0088 + 0.0114)]D^2L'$$

Distillation cost = $1.39D^2L'$ $/yr

These useful simplifications reduce the rather complex distillation calculation to a simple matter. It is, of course, specific to the ethanol-water system and only applicable for column feed concentrations near 0.01 mole fraction ethanol. In the case of more complex distillations it would be necessary to treat each case separately.

It is now possible to express the total cost of the absorption-distillation operation [Eq. (8.40)] in a simple way as

$$\text{Cost} = 402DN_{OG} + 7.9D^2N_{OG} + 502D^2G'y_2 + 2.0 \times 10^{-5}D^2G'^{1.8}N_{OG}$$
$$+ 2.66G'^{0.68}D^{1.36} + 1.39D^2L' + 0.03D^2L' \text{ \$/yr}$$

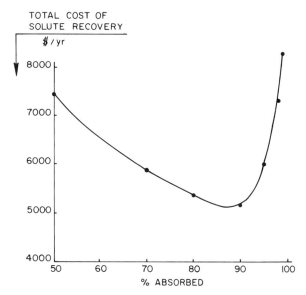

Fig. 8.5. Absorption of ethanol in water; % ethanol absorbed versus total cost. $L/D = 1.0$; $mG/L = 0.73$.

The cost as a function of the fraction absorbed, at a constant value of mG/L, may be easily calculated from the above equation. The solid line in Fig. 8.5 shows the variation of the total cost with the fraction absorbed for mG/L = 0.73 (L/G = 1). The curve goes through a minimum at a value of about 88% absorbed. The cost must increase rapidly as the absorption approaches 100%.

It is interesting that the answer obtained here is somewhat smaller than past experience might indicate. There are at least two reasons for this: (1) The scale of the subject operation is quite small--essentially pilot plant size--and the size of the equipment is small. Small size equipment is comparatively more expensive than larger equipment. (2) Present-day costs for equipment are substantially increased over such costs of a few years ago. As a consequence, the economic optimum has shifted in favor of smaller percent recovery.

The cost equation shown above can be differentiated with respect to y_2, and the optimum value of y_2 obtained for each L/G without having to calculate several points on the line. Thus,

$$\frac{d(\text{cost})}{dy_2} = rH_{OG}(280D + 0.786D^2 C_p) \left\{ \frac{-y_1}{y_2 \left[y_1(1 - \frac{mG}{L}) + \frac{mG}{L} y_2 \right]} \right\}$$

$$+ \frac{6,300D^2 G'(\text{mol wt of solute})(\text{cost of solute})}{(\text{mol wt of inert gas})}$$

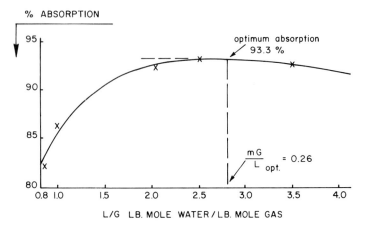

Fig. 8.6. Optimum value of mG/L: Absorption of ethanol by water.

Process Plant Components 407

It is apparent that when $d(cost)/dy_2$ is set equal to zero, a quadratic equation in y_2 results. Consequently, it is simple to solve for y_2 for various assumed values of L/G.

Figure 8.6 shows the result of the calculations for Treybal's problem. It is interesting to note that the optimum absorption is higher than that obtained in Fig. 8.5 (which was restricted to L/G of 1.0), and that the optimum value of L/G is unusually high: 2.8. However, as so often happens in process optimization work, the maximum is fairly flat, and essentially equal costs could be obtained by operating at L/G values ranging from 2.0 to 4.0.

The relatively low value for the percent absorption and the unusually high value for L/G reflect the present day high cost of equipment, the short payout time of 2 years, and the low cost of ethanol. In such cases the economic balance shifts away from a high solvent recovery towards a smaller capital investment. Equipment is reduced at the expense of material.

8.06 THE OPTIMUM DESIGN OF LIQUID-LIQUID AND SOLID-LIQUID EXTRACTION SYSTEMS

Liquid-liquid and solid-liquid systems may be optimized by the same general procedure employed for gas absorption. An analysis of these systems must include the capital investment related charges and the operating costs of the solute recovery and solvent purification portions of the operation. In liquid-liquid extraction, distillation is nearly always used for solvent recovery, but in solid-liquid extraction (leaching) other methods such as evaporation, precipitation and filtration, and steam stripping are used.

In liquid-liquid extraction the separation factor is frequently high and only a few stages are needed. Quite high recoveries (>99%) of solute are economically feasible.

In leaching, the separation factor is usually unity (simple dilution) and more stages (say 7 or 8) are used together with recoveries of 95 to 98% of the solute.

8.07 THE OPTIMUM DESIGN OF CONTINUOUS DIRECT CONTACT ROTARY DRYERS

In the chemical industry many different kinds of devices are used for the drying of solid materials. Drying is accomplished by the transfer of sensible heat from a gas to a solid, the evaporation of liquid from the solid, and the removal of the evaporated liquid as a vapor in the exit gas. Thus, a dryer is both a heat-transfer and a mass-transfer device. Because heat transfer is easier to study and better known than mass transfer, many drying devices are designed as heat-transfer equipment.

A widely used dryer that has been studied in this way is the direct contact continuous rotary dryer. Here, a rather finely divided particulate solid is caused to come into direct contact with a heating gas in a fairly long cylindrical device that rotates slowly while the solid and the gas pass continuously through the dryer. The gas and the solid can move through the dryer either cocurrently or countercurrently.

It is apparent that a great deal of gas can be used, causing the dryer to be fairly large in diameter and fairly short (and be cheaper), but using a large amount of heat to bring the gas to the desired inlet temperature. In addition, much of the heat in the gas will be wasted when the gas is vented to the atmosphere, or heat-recovery equipment will have to be installed. On the other hand, a small amount of gas may be used, resulting in a smaller diameter dryer which may be rather long (and more expensive), but having lower heating costs. It is important to know if there is an economic optimum point for minimum total cost of operation.

The cost of operation of a direct contact continuous rotary dryer is made up of several items:

1. The fixed charges on the dryer and its auxiliary equipment. Such dryers are quite expensive, and the fixed charges, particularly if the payout time is short, are high.

2. The cost of the heat supplied to the entering gas. All of the heat that is transferred from the gas to the solid must be put into the gas before the gas enters the dryer. The more gas that is used, and the higher the gas temperature, the greater the cost. If the temperature is to be quite high flue gases from an open flame may be used, but if the temperature is to be about 200°F (as is common for many chemical processes) a steam-heated exchanger is used.

Process Plant Components   409

3. The fixed charges on the gas heater. The higher the gas temperature and the more gas that is used, the higher the cost.

The hot gas leaving the dryer may be forced through heat-recovery equipment and some credit returned to the drying process. The economics of such heat-recovery units have been discussed in Section 8.02. For gases leaving the dryer at temperatures below 800° to 1,000°F it is usually not economically feasible to invest in heat-recovery equipment. In many cases in the chemical industry the gases leaving the dryer are at temperatures below 200°F and heat-recovery equipment is not economically justifiable. In such cases, the gases leaving the dryer will have no value.

As the temperature of the dryer increases, the recovery of heat from the dryer outlet gas stream will become more attractive. When heat-recovery equipment is installed, the size of the inlet heater will be reduced and so will the amount of steam required in the heater. These savings will be offset by the fixed charges on the heat-recovery equipment.

An approximate design method for rotary dryers is given in Appendix C. The assumptions used here will be the same as in the design method in Appendix C.

1. The gas will be assumed to be air, and it will enter the gas heater at 70°F and 50% relative humidity; this is an absolute humidity of 0.008 lb water/lb dry air.

2. The solid will be assumed to be always in the constant rate period of drying. Thus, the solid temperature will always be at the wet bulb temperature of the drying gas.

3. The allowable gas mass flow rate in the dryer will be 1,000 lb mole/(hr)(sq ft of dryer cross section). The gas mass flow rate is usually made as high as possible without causing excessive dust losses. The number given above is reasonable.

4. The gas and solid will flow countercurrently through the dryer. Under assumption (2) there is no difference between cocurrent and countercurrent flow with respect to the average temperature difference for heat transfer.

From an overall humidity or water balance the necessary amount of gas can be calculated,

$$W(X_i - X_o) = F(H_o - H_i)$$

where W = pounds of dry solids to be treated per hour

X_i = entering solid moisture content (lb water/lb dry solid)

X_o = leaving solid moisture content (lb water/lb dry solid)

H_o = humidity, or moisture content, of leaving gas (lb water/lb dry gas)

H_i = humidity, or moisture content, of entering gas (lb water/lb dry gas)

F = lb dry gas/hour

Thus,

$$F = W(X_i - X_o)/(H_o - H_i) \tag{8.41}$$

A typical design problem will specify: W, X_i, T_{Di}, X_o, and H_i. Thus, if values of H_o are assumed, values of F may be computed. For every value of H_o assumed, it is possible to find T_{Do}, the outlet gas dry bulb temperature, from the air-water (assuming that the gas is air—which is probably the case) humidity charts. This is done by following the adiabatic saturation line from the point T_{Di} and H_i to the intersection of the horizontal humidity line H_o and the adiabatic saturation line. Air-water humidity charts are printed in many easily accessible references. Thus, for every H_o assumed there will be a corresponding T_{Do}.

The assumption of H_o makes it possible to calculate F from Eq. (8.41) and ΔT_m, the logarithmic mean temperature difference for heat transfer, from

$$\Delta T_m = \frac{(T_{Di} - T_w) - (T_{Do} - T_w)}{\ln\left[\dfrac{T_{Di} - T_w}{T_{Do} - T_w}\right]}$$

In this equation T_w represents the wet-bulb temperature of the entering hot air.

Since G, the mass rate of flow of the gas in the dryer, has been set at 1,000 lb m/(hr)(sq ft),

$$D = [F/(0.786G)]^{0.5}$$

$$D = 0.0356(F)^{1/2} \quad \text{or} \quad 0.0356[W(X_i - X_o)/(H_o - H_i)]^{1/2}$$

Thus, the diameter of the dryer, D, is also set by the assumed value of H_o.

The rate of heat transfer in the dryer is given by the expression

$$Q = U_a V \Delta T_m$$

where Q = the rate of heat transfer in the dryer (Btu/hr)
U_a = the overall volumetric heat-transfer coefficient between gas and solid [Btu/(hr)(cu ft)(°F)]
V = the volume of the dryer (cu ft)
ΔT_m = overall logarithmic temperature difference between gas and solid (°F)

$$Q = W(X_i - X_o)(1{,}000) \quad \text{Btu/hr}$$

Here the sensible heat load of the solid is neglected. The latent heat of vaporization of the water is taken as 1,000 Btu/lb.

The work of Friedman and Marshall [20] gives an equation for U_a:

$$U_a = 20\, G^{0.16}/D$$

or

$$U_a = (20)(1{,}000)^{0.16}/0.0356\, F^{1/2}$$

$$U_a = 1{,}700/F^{1/2} \tag{8.42}$$

The volume of the dryer may now be calculated from

$$V = 1{,}000\, W(X_i - X_o)\, F^{1/2}/(1{,}700\, \Delta T_m)$$

or

$$V = \left\{ 1{,}000\, W(X_i - X_o) \left[\frac{W(X_i - X_o)}{H_o - H_i} \right]^{1/2} \right\} \bigg/ (1{,}700\, \Delta T_m)$$

$$V = \frac{0.589\, W^{1.5} (X_i - X_o)^{1.5}}{(H_o - H_i)^{0.5}\, \Delta T_m} \tag{8.43}$$

Thus, from an assumed value of H_o it is possible to calculate the volume of the dryer, its diameter, and its length.

The installed cost of a rotary dryer may be expressed as a function of its peripheral area, $A = \pi DL$, where L is the length of the dryer expressed in feet.

Since
$$V = 0.786\, D^2 L$$
$$L = V/(0.786\, D^2)$$
and
$$A = \pi D(V/0.786\, D^2) = 4V/D$$

Substituting the expressions for V and D in the above equation for A gives

$$A = 66.2\, W(X_i - X_o)/\Delta T_m \qquad (8.44)$$

From the correlations of the costs of rotary dryers (see Chapter 5) it can be found that the installed cost of a carbon steel dryer for rather low temperatures (200°F) is

$$\text{Cost} = 585\, A^{0.8}\ \$$$

where A is the peripheral area of the dryer. The installed cost of the dryer is then

$$\text{Cost} = 585 \left[\frac{66.2\, W(X_i - X_o)}{\dfrac{(T_{Di} - T_w) - (T_{Do} - T_w)}{\ln\left[\dfrac{T_{Di} - T_w}{T_{Do} - T_w}\right]}} \right]^{0.8}$$

The annual fixed charges on the dryer will be r times the above expression for the installed cost.

The heater required for the heating of the inlet air may be assumed to be a standard steam-heated exchanger of the usual shell and tube type, and made of carbon steel. Of course, extended surface heat exchange might be used and might prove to be cheaper.

The overall heat-transfer coefficient for this heater may be taken as 5 Btu/(hr)(sq ft)(°F). For the present example it will be assumed that the heating air will be raised from 70°F to 200°F before entering the dryer; condensing steam at 250°F will be used as the heating medium. Then,

$$\Delta T_H = \frac{(250 - 70) - (250 - 200)}{\ln\left(\dfrac{250 - 70}{250 - 200}\right)} = 102°F$$

Process Plant Components

The heating load on the exchanger will be

$$Q = \frac{W(X_i - X_o)}{H_o - H_i}[0.24(200 - 70) + 0.5H_i(200 - 70)]$$

$$Q = UA_H \Delta T_H = 5A_H(102) = 510 A_H$$

Then

$$A_H = \frac{W(X_i - X_o)}{(H_o - H_i)(510)}[(31.2 + 65 H_i)]$$

The installed cost of the heat exchanger may be determined from the material in Chapter 5; installed cost = $346 A^{0.62}$.

Thus, the fixed charges on the air heater will be

$$r\left\{346\left[\frac{W(X_i - X_o)}{510(H_o - H_i)}(31.2 + 65 H_i)\right]^{0.62}\right\} \text{\$/yr}$$

The cost of the steam used to heat the air before it enters the dryer will be

$$C_s\left[\frac{W(X_i - X_o)}{H_o - H_i}\right][0.24(200 - 70) + 0.5H_i(200 - 70)]8,000$$

where 8,000 = number of hours of operation/year
C_s = cost of steam \$/10^6 Btu

If $C_s = 0.80$ the annual cost of steam becomes

$$0.202[W(X_i - X_o)/(H_o - H_i)]$$

The gases leaving the dryer will be at a fairly low temperature and will be assumed to have no value.

With the cost terms evaluated as functions of H_o or T_{Do} it is possible to calculate the total cost of a rotary drying operation by assuming various values of H_o, finding T_{Do}, and then computing the total cost. There should be a minimum somewhere between 90% and 5% relative humidity of the exit gas.

As in other examples of optimum findings discussed before, it would be possible to write a complex analytical expression for the total cost, differentiate this with respect to T_{Do}, set this expression equal to zero, and so solve the problem analytically. This would be quite complex, and not as simple or as clear as a plot of costs against T_{Do}.

Example 8.4   A Rotary Dryer Problem

Consider the schematic flow sheet.

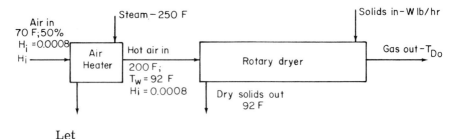

Let

X_i = inlet solid moisture content = 10%

= 0.10/0.90 = 0.111 lb water/lb dry solid

X_o = outlet solid moisture content = 1%

= 0.01/0.99 = 0.0101 lb water/lb dry solid

The gas inlet temperature is set at 200°F and the inlet relative humidity at 50%. From this it can be found that the gas wet-bulb temperature is 92°F; this is T_w.

W = lb dry solid entering (and leaving) the dryer per hour

= 1,000 lb m/hr

The quantity of air required is then

$$W(X_i - X_o)/(H_o - H_i) = 1{,}000(0.111 - 0.0101)/(H_o - 0.008)$$

$$= 100.9/(H_o - 0.008)$$

Assuming various values of the relative humidity of the exit gas, the results in Table 8-7 are obtained.

Process Plant Components

Table 8-7 Temperatures and Humidities in Direct Contact Drying

Relative humidity of exit gas (%)	H_o $\left(\dfrac{\text{lb water}}{\text{lb dry air}}\right)$	T_{Do} (°F)	F $\left(\dfrac{\text{lb dry air}}{\text{hr}}\right)$	ΔT_m (°F)	N_T (Number of heat transfer units)
90	0.0325	95	4,080	29.3	3.6
60	0.0300	105	4,540	44.8	2.1
40	0.0275	115	5,120	55.0	1.5
20	0.0230	132	6,660	69.0	1.0
10	0.0200	148	8,320	80.0	0.55
5	0.0135	174	18,150	93.6	0.28

Note that as the relative humidity of the exit gas decreases (that is, as the gas is allowed to pick up less and less moisture) several things happen:

1. The exit gas humidity decreases markedly.

2. The exit gas dry-bulb temperature increases and approaches the entering temperature.

3. The logarithmic mean temperature difference for heat transfer changes as H_o changes.

From these basic considerations it is possible to calculate the three costs of the rotary dryer operation. This analysis assumes that the value of the heat in the exit gas is negligible.

The total annual cost of the drying operation is

$$\text{Cost} = r\left\{585\left[\frac{66.2W(X_i - X_o)}{\Delta T_m}\right]^{0.8}\right\} + r\left\{346\left[\frac{W(X_i - X_o)}{510(H_o - H_i)}(31.2 + 65H_i)\right]^{0.62}\right\}$$

the fixed charges on the dryer · · · · · · · · · · · the fixed charges on the air heater

$$+ 0.202\,W(X_i - X_o)/(H_o - H_i)$$

the cost of steam to heat the inlet air

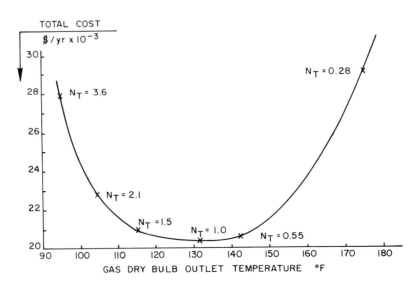

Fig. 8.7. Direct contact rotary dryer; optimum number of heat-transfer units.

Let r = 0.50; then

$$\text{Cost} = 308{,}000/(\Delta T_m)^{0.8} + 591/(H_o - 0.008)^{0.62}$$
$$+ 20.2/(H_o - 0.008) \quad \$/\text{yr}$$

The total cost may be plotted against the corresponding value of T_{Do} and the optimum value of T_{Do} found. The curve is shown in Fig. 8.7. It is apparent that a minimum does exist, that it is fairly flat, and that any T_{Do} between 115° and 150°F would be an acceptable operating point.

It is also of interest to calculate the number of gas-phase heat-transfer units, as defined by the equation

$$N_T = (T_{Di} - T_{Do})/\Delta T_m$$

This number is shown in the last column of Table 8-7 and is written beside each plotted point in Fig. 8.7. The minimum cost occurs between 0.5 and 1.0 transfer units. This finding is somewhat at variance with the statement [20, p. 20-20] that the economic number of transfer units is between 1.5 and 2.5. However, because of the rather flat minimum it is not too greatly different. The total cost at 1.5 transfer units is only 2.5% higher than the total cost at the minimum. This is an insignificant difference.

8.08 THE OPTIMUM DESIGN OF CHEMICAL REACTORS

A chemical reactor is a device which provides a space, sealed off from the surroundings, where the chemical reaction may take place under the desired conditions of temperature, pressure, and concentration.

Chemical reactions as carried out under industrial conditions are nearly always complex. In a reactor system there will be the main, or desired, reaction, plus one or more competing reactions. The competing reactions produce undesirable by-products and so reduce the yield of the reaction.

The reactor product stream must be treated so as to separate and purify the unreacted raw material and the desired product. The unreacted raw material may be recycled to the reactor, the desired product is further processed, and the undesired by-product recovered and disposed of in some way.

The design and operation of a chemical reactor system can be subjected to an economic analysis and an optimum found. There are several considerations:

1. If the reactor is made small, the reaction time will be short, the conversions will be low, and (usually) the yield of the desired product will be high; only a small quantity of by-product will be made. The cost of the reactor will be low, as will the associated capital investment-related charges. On the other hand, a low conversion leads to high costs for reactant separation and recycle and a greater investment in separation equipment.

As the reactor size tends toward zero the production costs will increase towards high values.

2. If the reactor is made large, the reaction time will be high and the reaction will proceed toward high conversions with accompanying lower yields and high production of undesirable by-products. In addition, the cost of the reactor will be high with high capital investment-related charges. On the other hand, a high conversion means lower separation and recycle costs.

As the reactor is made larger the cost of the operation will tend toward high values because of poor yields and large reactor capital investment cost.

Clearly, there is an optimum point.

A chemical reaction has several important variables:

1. The kinetics of the reactions. The speed with which the chemicals react is the most important characteristic of a reactor design. The rates of the various reactions are usually strongly dependent on temperature and composition. The nature of this dependence is fairly well known and is described in detail in many texts (see Ref. [18], pp. 4-1 to 4-25; and Ref. [21], Chapter 4).

The reaction rate is usually written in the form $r = c_A^a c_B^b$ and $k = Ae^{-E/RT}$. Many different forms of the general reaction are known (see Ref. [18], p. 4-4). An exact mathematical analysis quickly becomes quite complex and, usually, such analyses are restricted to the simpler forms. In many cases it makes little difference as to the exact form used, and a simple analysis is usually sufficient.

In order to determine the constants A and E in the equation for k it is necessary to have reaction rate data at two or more temperatures. These are seldom available, and when they are it is often difficult to know the exact reaction mechanism. As a consequence, it is not often that accurate knowledge of A and E exists. For many cases of chemical reactor design such knowledge is not necessary. However, reaction rate data at two or more temperatures are very good to have. Jordan [21, Chapter 4] gives an extended discussion of methods of obtaining reaction rate data and of using the experimental results.

2. The heat effect of the reaction. Almost all chemical reactions have a heat effect. Frequently, this heat effect is significant and a reactor must be designed so as to supply or accept the heat. The product of the reaction rate and the reaction heat effect is a reaction characteristic of great importance because this determines the size and the shape of the heat-transfer area associated with the reactor.

3. The corrosive nature of the reacting materials. Obviously, this determines the kind of material from which the reactor is made, and this, in turn, has an important effect on reactor design and cost.

4. The relative economic value of the reactants, products, and by-products.

Process Plant Components   419

Any economic evaluation of a chemical reactor must take all these factors, and several others, into consideration.

A simple example will help.

Example 8.5

Consider the chemical reactions

$A + a \rightarrow B$

$B + a \rightarrow C$

	Molecular weights			
	A	B	C	a
	90	130	170	40

where A, B, and C are liquids of moderate molecular weight (such as toluene) and a is a gas, such as chlorine, which reacts readily with both the raw material, A, and the desired product, B. The by-product C has no value, but can be disposed of by burning.

The desired production rate for B is 10,000,000 lb/yr, and the plant is to operate 8,000 hr/yr.

Laboratory studies have shown the following:

1. The two reactions are first order irreversible with respect to the concentrations of A and B expressed as mole fractions. Thus,

$-dx_A/dt = k_1 x_A$ $\qquad +dx_B/dt = k_1 x_A - k_2 x_B$

$k_1 = 2.0 \times 10^{12} e^{-19,120/RT}$ $\qquad k_2 = 3.0 \times 10^{20} e^{-32,700/RT}$

k_1 and k_2 are expressed in 1/hr. R = 1.99 cal/(gram-mole)(°K)

T = temperature, °K. The heat of reaction for both reactions may be taken as -30,000 cal/gram-mole A or B.

The reaction is to be conducted continuously in a single (one)-stage well-stirred vessel. The vessel may be made of low-carbon steel and must be equipped with a water-cooled jacket, an internal coil of 1-in. tubing carrying cooling water, and a turbine-type agitator.

Under these conditions the following equations apply [21, Chapter 4]:

$$M(dx_A/dt) = Fx_{Ai} - Fx_A - r_A M \qquad (8.45)$$

where M = lb moles of liquid in vessel

F = rate of flow of feed material (lb moles/hr)

x_{Ai} = mole fraction of A in the entering material (in this case x_{Ai} = 1.0)

x_A = mole fraction of A in the vessel and in the exit stream

r = rate of disappearance of A by the chemical reaction [lb moles/(hr)(mole of solution)]

At the steady state $dx_A/dt = 0$; then

$$Fx_{Ai} - Fx_A = r_A M$$

$$r_A = k_1 x_A$$

Then,

$$x_A = Fx_{Ai}/(F + k_1 M) = F/(F + k_1 M) \tag{8.46}$$

Also,

$$M(dx_B/dt) = Fx_{Bi} - Fx_B + r_B M \tag{8.47}$$

where x_{Bi} = concentration of B in the entering stream

x_B = concentration of B is the vessel and in the exit stream

r = rate of appearance of B [lb moles/(hr)(mole of solution)]

At the steady state $dx_B/dt = 0$; also, $x_{Bi} = 0$. Then,

$$x_B = k_1 MF/(F + k_1 M)(F + k_2 M) \tag{8.48}$$

and

$$x_C = 1 - x_A - x_B \tag{8.49}$$

From these equations the composition of the liquid in the reactor (and the product liquid) may be computed.

It is apparent that the product composition is a function of the feed rate F, the reactor volume M, and the temperature T. These three variables are not independent because the production rate must be satisfied; that is

$$Fx_B d = 10{,}000{,}000 \text{ lb/yr} = 9.6 \text{ lb moles/hr}$$

Process Plant Components

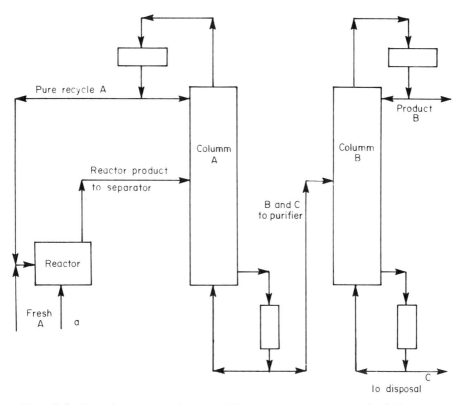

Fig. 8.8. Reactor-separator-purifier sequence. Example 8.5.

where d is the fraction of B produced in the reactor which is recovered in the purification system. Studies of the economics of distillation systems (see Section 8.05) show that d is almost always greater than 0.99. For present purposes d may be taken as 1.00.

The reaction system will consist of the reactor and the accompanying distillation columns. The flow plan will be as shown in Fig. 8.8. The reaction will occur in the reactor and the product stream will be separated in the distillation columns. Pure A will be taken from the top of column A and recycled to the reactor feed stream. The bottom stream from column A will be fed to column B. Pure product B will be taken from the top of column B, and the by-product C will be taken from the bottom of column C.

This system is one of three variables: temperature, reactor size, and flow rate. It would be possible to use the methods of

Chapter 4 and solve the problem for the optimum values of each variable. However, this would prove to be difficult and time-consuming. It is just as instructive to reduce the problem to one of a single variable by making some valid assumptions.

1. Assume that the temperature will be 60°C. Then, $k_1 = 0.60$ hr^{-1} and $k_2 = 0.12$ hr^{-1}.

2. Assume a value for F. Then, $x_B = 9.6/F$.

Using these values, M may be computed from Eq. (8.48), and x_A may be computed from Eq. (8.46). Then, $x_C = 1 - x_A - x_B$.

This calculation gives one set of operating conditions and a reactor size. Various values of F may be assumed and a group of flow rates, reactor sizes, and liquid compositions found. Thus, the results given in Table 8-8 are obtained.

The heat effect in the reactor must be evaluated. There are three aspects to the reaction heat effect.

1. The reactor system must be designed with sufficient heat-transfer surface to contain the reaction heat release.

2. The cooling water used in the heat-transfer system has a cost, and this cost is a part of the cost of the reaction system.

3. The control dynamics of the reactor system must be acceptable. The amount of heat-transfer area and the temperature of the cooling water are important items in this respect. In this example the control characteristics of the system will not be considered. For a detailed discussion see Bodman [1, p. 42].

The heat released in the reactor (Table 8-9) will be

$$q = \left[\frac{1 \times 10^7}{130}\right]\left[\frac{(1.8)(3 \times 10^4)}{8,000}\right] + (Fx_C)(1.8)(3 \times 10^4)$$

$$q = 519,000 + 54,000\, Fx_C \quad Btu/hr$$

The heat absorbed by the cold reactor feed stream will be

$$F(90)(0.45)(60 - 20)(1.8) = 3,240\, F \quad Btu/hr$$

The heat to be absorbed by the cooling water through the heat-transfer surface is then

$$q = 519,000 + 5.4 \times 10^4 Fx_C - 3,240F \quad Btu/hr$$

Process Plant Components

Table 8-8 Flow Rates and Reactor Sizes

F (lb moles/hr)	M (lb moles)	x_A	x_B	x_C	Reactor size (gal)
136	17.5	0.925	0.071	0.004	223
44	22.0	0.768	0.218	0.014	298
31	27.0	0.657	0.314	0.029	386
23	40.0	0.489	0.430	0.081	620
21	50.0	0.412	0.457	0.131	820
20.4	60.0	0.363	0.471	0.166	990

Each reactor will have a jacket and an internal cooling coil. The size of the coil can be designed so that it is sufficient to contain the anticipated heat release. Questions of the dynamics of temperature control will not be examined here.

Table 8-9 Reaction Heat Effect

F (lb moles/hr)	x_C	Heat release (Btu/hr)
136	0.004	108,200
44	0.014	401,500
31	0.029	467,500
23	0.081	545,500
21	0.131	599,300
20.4	0.157	625,800

Cost of the Operation, Example 8.5

The cost of the reaction system is made up of six separate items: (1) the raw material cost; (2) the capital investment-related charges for the reactor; (3) the cost of the cooling water for the reactor; (4) the cost of separating, purifying, and recycling unreacted A to the reactor; (5) the cost of separating and purifying B; (6) the cost of labor and supervision.

1. The raw material cost. For a production of 10,000,000 lb/yr of B:

lb A required to make B: $10,000,000(90/130) = 6,930,000$ lb/yr

lb a required to make B: $10,000,000(40/130) = 3,080,000$ lb/yr

lb A required to make C: $8,000(90)(Fx_C)/170 = 4,240\, Fx_C$ lb/yr

lb a required to make C: $8,000(2)(90)(Fx_C)/170 = 3,770\, Fx_C$ lb/yr

Cost of A = $0.06/lb; Cost of a = $0.03/lb

Annual raw material cost in dollars per year:

$(6.93 \times 10^6 + 4{,}240\, Fx_C)0.06 + (3.08 \times 10^6 + 3{,}770\, Fx_C)0.03$

2. The capital investment-related charges on the reactor. From Table 8-8 and Chapter 5, the data in Table 8-10 are obtained.

Table 8-10   Capital Investment-Related Charges

F (lb moles/hr)	M (gal)	Cost of reactor[a] and agitator ($)	Yearly charge on reactor[b] ($/yr)
136	223	7,250	4,000
44	298	8,700	4,800
31	386	10,450	5,750
23	620	14,400	7,940
21	820	17,920	9,880
20.4	990	20,260	11,130

[a] Cost includes cost of jacket and internal coil.
[b] $r = 1/T_m + m$; $T_m = 2$ yr; $m = 0.05$; $r = 0.55$.

Process Plant Components   425

3. Cost of cooling water.

Heat effect in reactor = 519,000 + 54,000 Fx_C Btu/hr

Cooling water cost = $0.05/1,000 gal

Cost of cooling water to reactor:

$$\left\{ \frac{5.19 \times 10^5 + 0.54 \times 10^5 Fx_C}{(1.8)(10)(8.35)(1,000)} \right\} (0.05)(8,000) \quad \$/yr$$

Table 8-11 Cooling Water

F (lb/moles/hr)	x_C	Fx_C (lb moles/hr)	54,000 Fx_C	Cooling water cost ($/yr)
136	0.004	0.54	29,200	1,460
44	0.014	0.62	33,500	1,478
31	0.029	0.90	48,600	1,510
23	0.081	1.87	101,000	1,650
21	0.131	2.75	148,400	1,780
20.4	0.157	3.20	173,000	1,842

4 and 5. The cost of separating and recycling unreacted A and purifying B. In this case it is assumed that the separation will be done by fractional distillation. From Eq. (8.39),

Cost of distillate ($/lb mole distillate) = $\dfrac{C_1 N(1 + R)}{EYG_a} + \dfrac{C_2(1 + R)}{YG_b} + C_3(1 + R)$

In this case the separation of A from B is fairly easy since the two compounds are similar in structure and are 40 points different in molecular weight. Accordingly, the relative volatility will be high, the reflux ratio will be low, and the number of plates small.

Approximate calculations show that N = 6, R = 1, E = 0.45, G_a = 20, G_b = 0.10. From Section 8.04, C_1 = 49.50, C_2 = 8.81, C_3 = 0.0114.

Cost of distillation of A from B =

$$\frac{(49.50)(6)(2)}{(8,000)(0.45)(20)} + \frac{(8.81)(2)}{(8,000)(0.10)} + (0.0114)(2)$$

= $0.053/lb mole of distillate

The cost of recovery of B from C should be approximately the same. Then, the cost of distillation may be calculated (Table 8-12).

Table 8-12 Cost of Distillations

F $\left(\frac{\text{lb moles}}{\text{hr}}\right)$	Fx_A $\left(\frac{\text{lb moles}}{\text{hr}}\right)$	Fx_B $\left(\frac{\text{lb moles}}{\text{hr}}\right)$	Cost of recovery A ($/yr)	Cost of recovery B ($/yr)	Total cost ($/yr)
136	126	9.6	53,500	4,100	57,600
44	34	9.6	14,400	4,100	18,500
31	20	9.6	8,480	4,100	12,580
23	11	9.6	4,670	4,100	8,770
21	8.7	9.6	3,690	4,100	7,790
20.4	7.4	9.6	3,140	4,100	7,240

This calculation is approximate and neglects some fine points. However, for present purposes it will suffice. For a detailed example of such a calculation using an electronic computer, see Ref. [22].

6. The cost of labor and supervision.

Direct labor: 0.5 operators/shift, 4 shifts, $4.50/hr

Supervision: 15% of direct labor

Labor and supervision cost =

(1.15)(0.5)(8,000)(4.50) = $20,750/yr

The cost of labor and supervision is a constant and therefore has nothing to do with the location of the optimum point. However, the labor charge is fairly heavy and does influence the total cost of the operation significantly.

The total cost of the operation is the sum of the six items (Table 8-13).

Table 8-13   Total Cost of Reaction System

F (lb moles/hr)	M (gal)	x_A	x_B	x_C	Fraction A converted
136	223	0.925	0.071	0.004	0.075
44	298	0.768	0.218	0.014	0.232
31	386	0.657	0.314	0.029	0.343
23	620	0.489	0.430	0.081	0.511
21	820	0.412	0.457	0.131	0.588
20.4	990	0.363	0.480	0.157	0.637

Raw material cost ($/yr)	Capital cost ($/yr)	Cooling water cost ($/yr)	Distillation cost A ($/yr)	Distillation cost B ($/yr)	Labor cost ($/yr)	Total cost ($/yr)
508,600	4,000	1,460	53,500	4,100	20,750	592,410
508,600	4,800	1,478	14,400	4,100	20,750	554,128
508,700	5,750	1,570	8,480	4,100	20,750	549,350
508,700	7,940	1,650	4,670	4,100	20,750	547,910
508,900	9,880	1,780	3,690	4,100	20,750	549,100
508,900	11,130	1,842	3,140	4,100	20,750	549,862

A plot of the total cost against x_B shows that there is a minimum at x_B = 0.40 and total cost of \$547,600/yr. This minimum is quite flat; between x_B = 0.22 and x_B = 0.48 the change in the total cost is approximately 1%. This means that the reaction vessel may be between 300 and 900 gal in size, and the fraction conversion of the feed material may be between 23% and 60%, and yet the total cost of the operation will remain essentially constant.

The results of Bodman [1, p. 38] are similar. The minima are flat; the range of results is of the order of 6% or less.

REFERENCES

1. S.W. Bodman, The Industrial Practice of Chemical Process Engineering, The M.I.T. Press, Cambridge, Massachusetts, 1968, pp. 100-179.
2. T.K. Sherwood, A Course in Process Design, The M.I.T. Press, Cambridge, Massachusetts, 1963, pp. 84-96.
3. K.M. Guthrie, Chemical Engineering, March 24, 1968, p. 114.
4. R.M. Braca and J. Happel, Chemical Engineering, 60, 180 (1953).
5. A.S. Foust, L.A. Wenzel, C.W. Clump, L. Maus, and L.B. Andersen, Principles of Unit Operations, John Wiley and Sons, Inc., New York, 1960.
6. Chemical and Engineering News, March 8, 1971, p. 11.
7. D.Q. Kern, Process Heat Transfer, McGraw-Hill Book Co., New York, 1950, p. 106.
8. H. Ten Broeck, Industrial and Engineering Chemistry, 36, 64 (1944).
9. W.H. McAdams, Heat Transmission, 3rd ed., McGraw-Hill Book Co., New York, 1954, p. 193.
10. C.L. Williams and R.D. Damron, Hydrocarbon Processing and Petroleum Refiner, 44, 139 (1965).
11. M. Brooke, Chemical Engineering, December 14, 1970, p. 135.
12. W.H. Dodge, Chemical Engineering, May 23, 1966, p. 182.
13. A.P. Colburn, Division of Chemical Engineering Lecture Notes, University of Delaware, Newark, Delaware, 1943.
14. E.R. Gilliland, in The Chemical Engineers' Handbook, 4th ed., McGraw-Hill Book Co., New York, 1963, pp. 13-43.
15. C. Pyle, in Distillation in Practice (C.H. Nielsen, ed.), Chapter 1, Reinhold Publishing Co., New York, 1956.
16. D. Cornell et al., Chemical Engineering Progress, 56, No. 7, 68 and No. 8, 48 (1960).

Process Plant Components 429

17. R.E. Treybal, Mass Transfer Operations, 2nd ed., McGraw-Hill Book Co., New York, 1968.

18. The Chemical Engineers' Handbook, 4th ed., McGraw-Hill Book Co., New York, 1963.

19. J.C. Kim and M.C. Molstad, Hydrocarbon Processing and Petroleum Refiner, $\underline{45}$, 107 (1966).

20. W.R. Marshall and S.J. Friedman, in The Chemical Engineers' Handbook, 4th ed., McGraw-Hill Book Co., New York, 1963, p. 20-19.

21. D.G. Jordan, Chemical Process Development, Interscience Publishers, John Wiley and Sons, Inc., New York, 1968.

22. T. Umeda, A. Shindo, and E. Tazaki, Industrial and Engineering Chemistry, Process Design and Development, $\underline{11}$, 1 (1972).

PROBLEMS

8.1. A simple atmospheric-pressure vertical cylindrical tank is to be employed for the storage of a liquid product. Assuming that the wall thickness is constant at the minimum value required for stiffness and does not vary with diameter, compute the optimum diameter-to-height ratio.

8.2. (a) A process line in a petroleum refinery conveys liquid with specific gravity 0.9 at a rate of 750,000 lb/hr. Pumps in the plant are driven by electricity, and a maximum acceptable payout time of 2 years is required on incremental investment. What is the optimum pipe diameter?

(b) It has become necessary to pump 25,000 lb/hr of toluene from one point in a chemical plant to another. The equivalent length of pipe between the two points is 1,200 ft and the static pressure drop is 30 psi. There exists in the plant a suitable quantity of 1-in. pipe and a suitable centrifugal electric motor-driven pump. The pump and motor are now permanently installed but not in operation. The pump and motor are completely depreciated, and the value of the 1-in. pipe may be taken as zero. Thus, the only capital charge for this system will be the installation charge for the pipe.

The question arises as to whether it is more economical to utilize the existing 1-in. pipe and the pump-motor installation or to purchase and install new pipe of the optimum size with a proper size pump and motor.

8.3. For a given pressure, the thickness of a heater tube wall determines the stress to which the metal may be subjected. The stress that an alloy will stand varies with the time and temperature. The higher the stress at a given temperature, the shorter the tube life will be.

As a specific example, it is desired to determine the optimum thickness for a typical case involving 4-6 Cr, 0.5 Mo steel at a pressure of 640 psig, outside tube diameter of 5.9 in., and temperature of 1,095°F. The following cost data were assembled:

Case	Thickness (in.)	Service (years)	Capital investment ($)
A	0.35	1.2	12,200
B	0.46	6.0	16,100
C	0.52	12.0	18,200
D	0.58	24.0	20,400

Assuming that the current earning rate on capital is 15% and that the minimum attractive rate of return after taxes is 25%, determine which of the above alternatives is most attractive. Assume that the life of the heater in which the tubes will be installed is 24 years.

8.4. A refinery stream of 100,000 lb/hr of hot oil at 600°F is to be used to heat a cold stream of 200,000 lb/hr at 100°F by means of a shell and tube exchanger, consisting of one shell and two tube passes. The equipment is to be installed at an East Coast location in the United States, and a maximum acceptable payout time of 1 year is desired on optional investment. Calculate the optimum approach temperature and duty of the exchanger to be specified for this service.

Note:

Average specific heat of hot oil = 0.57 Btu/(lb)(°F)

Average specific heat of cold oil = 0.50 Btu/(lb)(°F)

Process Plant Components

8.5. A chemical plant has a distillation column that is being rebuilt to increase its capacity by 30%. The new rate of overhead vapor flow will be 25,000 lb/hr. The original reflux condenser is now too small, it is not in good condition, and must be replaced. The suggestion has been made that the new condenser be of the extended surface air-cooled type.

Calculate the capital investment and operating costs of new air-cooled and new water-cooled condensers.

The following data apply: Boiling point of condensate, 220°F; latent heat of vaporization of condensate, 160 Btu/lb.

8.6. (a) In a new plant, a vertical cylindrical vessel of large diameter will operate at a temperature of 600°F. It is desired to determine the optimum thickness of insulation for this tower.

Assume that management expects money invested to show a maximum payout time of 3 years. Maintenance may be taken at 6% of investment. The incremental installed cost of insulation will be \$1.00/sq ft of area and per inch of thickness. The plant will operate at 90% time efficiency. Combined $h_c + h_r$ for the outside coefficient of heat transfer to air may be taken at 2.0 Btu/(sq ft)(hr)(°F). Thermal conductivity of insulation is 0.4 Btu/(hr)(ft^2)(°F/in.). The value of heat may be taken at 60¢/MM Btu. Outside temperature averages 70°F.

(b) In a chemical plant it is desired to transport 50,000 lb/hr of steam at 1,100°F and essentially 1 atm pressure from a steam superheater to a chemical reactor through an insulated steel pipe 24 in. in diameter. The minimum allowable temperature of the steam entering the reactor is 1,080°F. What is the maximum allowable distance between the superheater and the reactor?

The data of Problem 8.6a will apply except that the cost of the superheated steam will be taken as \$1.00/10^6 Btu, and the number of hours of operation each year will be 8,000.

8.7. (a) Construct a table indicating the optimum reflux ratio and the optimum number of plates for typical close hydrocarbon fractionations, assuming a separation of 99% purity in overhead and bottoms for binary mixtures with a range of relative volatilities as shown in the following tabulation:

Difference in boiling point of components (°C)	Relative volatility at 100°C
1.0	1.03
2.0	1.06
3.0	1.09
5.0	1.16
8.0	1.27
10.0	1.35
15.0	1.56
20.0	1.82

(b) In Problem 7.1 the separation of ethylbenzene from p-xylene was discussed. In this case the difference in boiling points is 2.3°C. A column using 228 theoretical plates and a reflux ratio of 149 was called for. It was decided, as a practical matter, to use three distillation columns connected in series. Each column would be 200 ft tall and 21 ft in diameter. A very large installation indeed. The table calculated as the answer to part a above indicates that 306 theoretical trays would be needed with a reflux ratio of 39. The difference in the answer is largely due to the feed concentrations used in the two problems. This difference is not important to this question. Obviously such distillations require many trays and high reflux ratios.

The further separation of p-xylene from m-xylene and of m-xylene from o-xylene might be contemplated. Referring to the work above, discuss the practicality of these separations by fractional distillation.

	Boiling point
p-xylene	138.5°C
m-xylene	139.3°C
o-xylene	144 °C

8.8. Consider a distillation column producing a distillate of 80 mole % ethanol-water from a feed of 30 mole % ethanol-water at its boiling point. The bottoms concentration will be 0.01 mole % ethanol.

Process Plant Components   433

Compute the optimum reflux ratio, the optimum number of plates, and the cost of the distillation operation.

8.9. A chemical plant has an absorption column which is 2.7 ft in diameter and is packed with 30 ft of 1-in. Raschig rings. The column is used to absorb benzene from an inert gas of molecular weight 29.3. The gas has a benzene content of 3.0 mole %. The liquid absorbent is a benzene-like material with a molecular weight of 150. This liquid is a by-product of the chemical reaction wherein benzene is one of the original raw materials. The absorbing liquid and the absorbed benzene can be separated by fractional distillation in a large column which is in existence in order to process the reactor product.

Find the optimum conditions for the operation of the absorption column.

The following data apply:

1. The absorption column is to process 80,000 cu ft/hr of inlet gas, measured at 90°F and 1 atm.

2. The equilibrium relationship between solute and solvent is ideal and can be expressed as $y^* = 0.13x$, where y^* is the mole fraction of benezene in the vapor phase and x is the mole fraction of benzene in the liquid phase.

3. The height of an overall gas-phase transfer unit is 2.5 ft.

4. The density of the inlet gas is 0.09 lb m/cu ft.

5. The distillation column has 10 actual trays.

6. The cost data of Example 8.3 will apply.

Appendix A

SUMMATION OF TIME SERIES

In the first edition of this book Happel discussed the application of the calculus of finite differences to the summation of the time series normally encountered in economy studies. The details of Happel's mathematical derivations will not be given here. The interested student may consult references A[1] and A[2] given at the end of Appendix A.

In this section a summary of these expressions, together with some examples, will be given. In this discussion, the interest rate will be considered constant at i fraction per year, and the interest payments will be compounded annually at the end of each of n years. Then,

1. A sum of money S will result from n years of compound interest payments at the interest rate i, when the original sum of money was P.

$$S = P(1 + i)^n \qquad (A.1)$$

The term $(1 + n)^n$ is designated as the single payment compound amount factor. Or, the future worth of a single sum P earning compound interest at the rate i will be $P(1 + i)^n$.

2. The reciprocal of the single payment compound amount factor is $1/(1 + i)^n$. This term is called the single payment present worth factor. It is used to find the payment at the beginning of a given period which would result in the development of a predetermined sum. Thus,

$$P = S/(1 + i)^n \qquad (A.2)$$

This term is widely employed throughout this book to find the present value of a sum which will exist n years into the future.

3. Various economic situations, instead of calling for the evaluation of single individual payments, involve payments periodically over a period of time. For convenience, such payments are usually considered as occurring annually with interest also compounded annually, though any other period might be assumed with obvious elaboration of the formulas involved. In general, all such systematic payments will be brought to a common date, usually the present. This may be expressed mathematically as

$$\sum_{1}^{n} \frac{1}{(1+i)^k} = \text{present worth} \qquad (A.3)$$

where a series of payments for each of k years is to be summed, each payment being appropriately discounted.

In the usual application of compound interest it is often desired to find the total amount at the beginning of n years resulting from a series of annual uniform payments. That is, the same amount of money is invested for each of n years, at the end of each year. More generally, the summation may start with the m-th year instead of the first. If it is assumed that the individual payment is unity, the appropriate $f(k) = 1$, and the desired formulation is

$$\sum_{m}^{n} \frac{1}{(1+i)^k} = \left[-\left(\frac{1+i}{i}\right)\left(\frac{1}{1+i}\right)^k \right]_{m}^{n+1}$$

$$= \frac{1}{i(1+i)^{m-1}} - \frac{1}{i(1+i)^n}$$

Multiplication by $i(1 + i)^n$ and division of the product by $i(1 + i)^n$ gives

$$\sum_{m}^{n} \frac{1}{(1+i)^k} = \frac{(1+i)^{n-m+1}}{i(1+i)^n} \qquad (A.4)$$

If this formula is specialized by letting m = 1, there results

$$\sum_{1}^{n} \frac{1}{(1+i)^k} = \frac{(1+i)^n - 1}{i(1+i)^n} \qquad (A.5)$$

Summation of Time Series

Expression (A.5) is called the **uniform annual-series present worth factor**. Its reciprocal, which gives the value of each of a series of uniform payments required to equal a prescribed present worth, $[i(1+i)^n]/[(1+i)^n - 1]$ is called the **capital recovery factor**.

4. When it is desired to obtain the value of a series of payments at the end of the time instead of the beginning, the general formulation will be

$$\sum_{1}^{n} f(k)(1+i)^{n-k} = \text{future worth}$$

In the case of a uniform series, this reduces to $f(k) = 1$, and, since n is not a variable of summation, there results

$$(1+i)^n \sum_{m}^{n} \frac{1}{(1+k)^k} = \frac{(1+i)^{n-m+1} - 1}{i} \qquad (A.6)$$

obtained by simply multiplying Eq. (A.4) by $(1+i)^n$. Similarly, when $m = 1$,

$$(1+i)^n \sum_{1}^{n} \frac{1}{(1+i)^k} = \frac{(1+i)^n - 1}{i} \qquad (A.7)$$

Expression (A.7) is called the **uniform annual-series compound factor**. Its reciprocal $i/[(1+i)^n - 1]$ is called the **sinking-fund deposit factor**, since a fund established by means of a series of annual payments over a period of time is commonly designated as a sinking fund.

5. Another type of time series which appears to have interesting possibilities in engineering economy studies involves the situation where a series of payments, instead of remaining constant, is subject to a linear variation. That is, each payment is larger or smaller than the one preceding it by a constant amount. The various possibilities are derived by utilizing previous results for uniform series in combination with one involving an annual gradient.

Thus,

$$\sum_{m}^{n} \frac{k}{(1+i)^k} = \left[-\left(\frac{1+i}{i}\right)\left(\frac{1}{1+i}\right)^k \left(k + \frac{1}{i}\right) \right]_{m}^{n+1}$$

$$= -\frac{1}{i(1+i)^n}\left(n + 1 + \frac{1}{i}\right) + \frac{1}{i(1+i)^{m-1}}\left(m + \frac{1}{i}\right)$$

After further algebraic manipulation,

$$\sum_{m}^{n} \frac{k}{(1+i)^k} = \frac{1+i}{i}\left[\frac{(1+im)(1+i)^{n-m} - 1}{i(1+i)^n}\right] - \frac{n}{i(1+i)^n} \quad (A.8)$$

For the special case where $m = 1$, the following is obtained:

$$\sum_{1}^{n} \frac{k}{(1+i)^k} = \frac{1+i}{i}\left[\frac{(1+i)^n - 1}{i(1+i)^n}\right] - \frac{n}{i(1+i)^n} \quad (A.9)$$

Note that the expression in brackets in Eq. (A.9) is the <u>uniform annual-series present worth factor</u>, given in Eq. (A.5).

Example A.1

The return from a venture starts at zero but rapidly climbs in 2 years to $10,000/yr at a uniform rate of increase. It then remains at $10,000/yr for the next 5 years. For the remaining 8 years of the duration of this venture, the return drops uniformly to zero at the end of 15 years. (Note: As discussed in Chapter 3, unless other considerations were involved, a venture would not be operated beyond the time when it would give a minimum acceptable return rather than a zero return.)

If money is worth 10%, it is desired to compute the series of uniform end-of-year payments that would be equivalent to the above schedule of payments.

Solution. For the first 2 years, the present worth of the return will be

$$5{,}000 \sum_{1}^{2} \frac{k}{(1+i)^k}$$

After this, at the ends of years 3 to 7, inclusive, the series of payments of $10,000 will result in the following present worth:

$$10{,}000 \sum_{3}^{7} \frac{1}{(1+i)^k}$$

Finally, during the period of decline, there will be, for years 8 to 15, inclusive, a series that starts at year 8 at $10,000 less $1,250,

Summation of Time Series

and finally becomes zero after 8 such deductions. Thus the series will be

$$1{,}250 \sum_{8}^{15} \frac{15-k}{(1+i)^k}$$

Summarizing,

$$\sum_{1}^{15} \frac{f(k)}{(1+i)^k} = 5{,}000 \sum_{1}^{2} \frac{k}{(1+i)^k} + 10{,}000 \sum_{3}^{7} \frac{1}{(1+i)^k} + 1250 \sum_{8}^{15} \frac{15-k}{(1+i)^k}$$

$$= 5{,}000 \left\{ 11 \left[\frac{(1.1)^2 - 1}{0.1(1.1)^2} \right] - \frac{2}{0.1(1.1)^2} \right\}$$

$$+ 10{,}000 \left[\frac{(1.1)^{7-3+1} - 1}{0.1(1.1)^7} \right] + 1{,}250 \times 15 \left[\frac{(1.1)^{15-8+1} - 1}{0.1(1.1)^{15}} \right]$$

$$- 1{,}250 \left\{ 11 \left[\frac{1.8(1.1)^{15-8} - 1}{0.1(1.1)^{15}} \right] - \frac{15}{0.1(1.1)^{15}} \right\}$$

$$\sum_{1}^{15} \frac{f(k)}{(1+i)^k} = \$57{,}840$$

This is the present worth of the schedule of payment as outlined above. The uniform annual-series present worth factor for $i = 0.10$ and $n = 15$ is given by Eq. (A.5) as

$$\frac{(1+i)^n - 1}{i(1+i)^n} = \frac{(1.1)^{15} - 1}{0.1(1.1)^{15}} = 7.60$$

Therefore $57,840/7.60 = $7,610 is the equivalent uniform annual end-of-year payment required.

An alternative method to the use of the special summation formulas is simply to evaluate the present worth of each term f(k) year by year (Table A-1).

Table A-1 Evaluation of Present Worth, Year by Year

Year, k	Payment ($/yr)	Present worth factor $(1.1)^k$	Present worth
1	5,000	1.10	$ 4,550
2	10,000	1.21	8,270
3	10,000	1.33	7,500
4	10,000	1.46	6,850
5	10,000	1.61	6,210
6	10,000	1.77	5,650
7	10,000	1.95	5,130
8	8,750	2.14	4,080
9	7,500	2.36	3,180
10	6,250	2.59	2,410
11	5,000	2.85	1,750
12	3,750	3.14	1,190
13	2,500	3.45	725
14	1,250	3.80	330
15	0		
			$57,825

6. It is possible to develop formulas for more complicated expressions than the above. Thus, for example, for a series of payments in which each one differs as the square of the preceding one, there results the following for the special case of m = 1:

$$\sum_{1}^{n} \frac{k^2}{(1+i)^k} = -\left(\frac{1+i}{i}\right)\left(\frac{1}{1+i}\right)^k \left[k^2 + \frac{1}{i}(2k+1) + \frac{1}{i^2}\right]_{1}^{n+1} \quad (A.10)$$

After substitution and appropriate algebraic manipulations,

$$\sum_{1}^{n} \frac{k^2}{(1+i)^k} = \frac{2+3i+i^2}{i^2}\left[\frac{(1+i)^n - 1}{i(1+i)^n}\right] - \frac{2n(1+i)}{i^2(1+i)^n} - \frac{n^2}{i(1+i)^n} \quad (A.11)$$

Summation of Time Series

7. A situation that is sometimes encountered occurs when the function f(k) noted in Eq. (A.3) is itself an exponential. Thus, if $f(k) = (1/b)^k$, then by analogy to Eq. (A.4)

$$\sum_{m}^{n} \frac{1}{(b+bi)^k} = \frac{(b+bi)^{n-m+1} - 1}{(b+bi-1)(b+bi)^n} \tag{A.12}$$

A distinction must be made between the cases where a function is summed continuously or stepwise. Although the compound-interest formulas developed here assume a stepwise summation at stated intervals, it is not necessary for the function f(k) itself to be a discontinuous function. If it is continuous, only the values it takes at the summation points are of interest. Example A.2 illustrates a case in which it is desired to sum a continuous function, and illustrates Eq. (A.12).

Example A.2

A royalty interest returns an average of 5 bbl/day of oil having a price of $2.50/bbl. It is assumed that production will decline at a rate of 10%/yr continuously. It is desired to know the present worth of future income over a 10-year period, discounted at 5%/yr, on the assumption that the present price and constant percentage decline will apply over the 10-year period. This example is after Woods [3].

Solution. Designate the variables involved as follows:

P_0 = initial production rate (bbl/yr)
P_1 = production rate at end of 1st year (bbl/yr)
S_1 = production during the 1st year (bbl)
S_2 = production during the 2nd year (bbl)
t = time from start of operation (years)
c = rate of decline (fraction/yr)

The decline is assumed to be first order

$$\frac{dP}{dt} = -cP$$

whence upon integration

$$P = P_0 e^{-ct}$$

Now, to calculate the production during the first year,

$$S_1 = \int_0^1 P\, dt = \int_0^1 P_0 e^{-ct}\, dt = \frac{P_0}{c}(1 - e^{-c})$$

In a similar manner production for succeeding years can be computed so that, for any year k, the production will be

$$S_k = \frac{P_0}{c}(e^c - 1)e^{-ck} \text{ bbl/yr}$$

and its value $2.5 S_k$ \$/yr.

In order to obtain the required present worth, the values of $2.5 S_k$ are discounted and summed,

$$\sum_1^n \frac{2.5 S_k}{(1+i)^k} = \frac{2.5 P_0}{c}(e^c - 1) \sum_1^n \left[\frac{1}{e^c(1+i)}\right]^k$$

and employing Eq. (A.12)

$$\sum_1^n \frac{2.5 S_k}{(1+i)^k} = \frac{2.5 P_0}{c}(e^c - 1) \frac{(e^c + e^c i)^n - 1}{(e^c + e^c i - 1)(e^c + e^c i)^n}$$

In the present case $i = 0.05$, $c = 0.10$. Therefore

$$\sum_1^n \frac{2.5 S_k}{(1+i)^k} = \frac{2.5 \times 365 \times 5}{0.1}(1.1051 - 1)\frac{(1.105 + 0.055)^{10} - 1}{(1.105 + 0.055 - 1)(1.160)^{10}}$$

$$= 45{,}600(0.1051)\frac{(1.16)^{10} - 1}{(0.16)(1.16)^{10}} = \$23{,}200$$

Alternative forms are possible by using, for e^c, the approximation $(1 + c)$. If instead of assuming a continuous decline in production, which is obviously the proper way to evaluate the production decrease, a stepwise falling off had been assumed, two alternative choices would be possible. It could be assumed that the initial rate represented production for the entire year previous to the start of the problem, in which case production for the first year would be 90% as great, etc.; or, alternatively, it could be assumed that the initial rate given represented production for the first year, and that the second year's production would be 90% as great, etc.

Summation of Time Series

The first assumption would give

$$S_k = P_0(1 - c)^k$$

and

$$\sum_1^n \frac{2.5 S_k}{(1 + i)^k} = 2.5 P_0 \sum_1^n \left(\frac{1 - c}{1 + i}\right)^k = 2.5 P_0 \sum_1^{10} \left(\frac{1}{1.165}\right)^k$$

$$= 2.5 \times 365 \times 5 \times 4.75 = \$21,700$$

The second assumption would give

$$S_k = P_0(1 - c)^{k-1}$$

and

$$\sum_1^n \frac{2.5 S_k}{(1+i)^k} = 2.5 P_0 \sum_1^n \left(\frac{1-c}{1+i}\right)^{k-1} = 2.5 \times P_0 \times \frac{1+i}{1-c} \sum_1^{10} \frac{1}{1.165}$$

$$= 2.5 \times 365 \times 5 \times 1.17 \times 4.75 = \$25,300$$

A final interesting alternative results from the assumption of both continuous decline in production and continuous compounding of interest. In this case

$$dS/dt = P$$

Designating T as the present worth of production

$$dT = 2.5 \, P e^{-it} \, dt$$

Noting that $P = P_0 e^{-ct}$,

$$T = \int_0^n dT = 2.5 P_0 \int_0^n e^{-t(i+c)} \, dt = \frac{2.5 P_0}{i + c}[1 - e^{-n(i+c)}]$$

and, with the numerical data assumed,

$$T = \frac{2.5 \times 365 \times 5}{0.15}[1 - e^{-1.5}] = \$23,500$$

REFERENCES

1. H.S. Mickley, T.K. Sherwood, and C.E. Reed, Applied Mathematics in Chemical Engineering, 2nd ed., McGraw-Hill Book Co., New York, 1957, pp. 320-340.

2. John Happel, Chemical Process Economics, John Wiley and Sons, Inc., New York, 1958, pp. 223-246.

3. R.W. Woods, "Present Worth of Profits for Constant Percentage Decline," Oil-Gas Journal, p. 107 (April 5, 1951); p. 113 (April 12, 1951).

Appendix B

INFLATION

The amount of goods and services that the basic unit of a nation's currency may be exchanged for is called the purchasing power of that currency. Fundamentally, people work for purchasing power, not money. They wish to exchange the products of their own labor and time, measured in terms of the nation's currency, for the products of other people's labor and time, measured in terms of the same currency.

It is a matter of historical record that the quantity of goods and services that a unit of the nation's currency will buy has been decreasing steadily all over the world for many years. This decline in purchasing power is called inflation. The name probably derives from the greater number of units of currency necessary to provide constant purchasing power, or because of the rise in the level of prices for goods and services. In the United States the purchasing power of $1 of currency has been declining rather irregularly, but none the less steadily, for over 100 years. Figure B.1 shows a plot (1) of the U.S. dollar's purchasing power against time for the 115-year period from 1850 to 1965. Since 1965 the rate of decline has become even greater. Figure B.2 is another plot showing the change in the purchasing power of the dollar. This graph shows the Wholesale Price Index, as compiled by the U.S. Department of Commerce, plotted against time. A higher value of the WPI indicates a lower purchasing power of the dollar. There are other indices which relate the cost of equipment, or construction, or raw materials to time; these are widely reproduced and studied. Some of these indices are shown in Fig. 5.1. All of these graphs show quite plainly a fact that nearly everyone in the United States

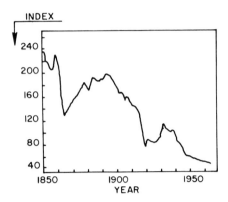

Fig. B.1. The decline in the purchasing power of the dollar, 1850-1965 [1].

knows from personal experience: moderate to severe inflation has almost always been present in U.S. economic life, the rate of inflation is now increasing, and there is every reason to believe that at least a moderate rate of inflation will continue into the foreseeable future.

An effect of inflation, which is of interest to chemical engineers, is described in Chapter 5. There it is pointed out that the costs of chemical plant equipment have been increasing with time

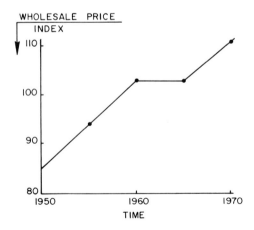

Fig. B.2. Wholesale Price Index versus time; a measure of inflation.

Inflation

for many years. This rate of increase has varied somewhat, but, despite the small variations, the trend has been toward higher prices for the same machine; inflationary. The well-known <u>Engineering News Record Index</u> for construction is now rising even more rapidly than in the past (Fig. 5.1), indicating a swift inflation in the cost of chemical plant construction.

Inflation also affects many other aspects of the chemical-process industries. The prices of raw materials, utilities, operating labor, and other goods and services have been increasing steadily. The interest rates which industry must pay for borrowing money have doubled in recent years, and competition from foreign producers has become more intense. In general, the existence of inflation has made it more difficult to obtain profitable operations in the chemical-process industries.

In times of inflation industrial corporations attempt to raise their selling prices as their costs of raw materials, labor, and machinery increase. If they are successful in so doing it is possible for them to offset the effects of inflation and to operate with a satisfactory profit margin and a good return on investment. A greater quantity of cash will flow but the general level of profits will be essentially the same; see Ref. [1, p. 191].

However, it is seldom that the effects of inflation can be fully offset in this simple manner. There are transient effects and imbalances which distort the economic situation, nearly always to the disadvantage of the industrial company. It often happens that companies cannot find relief from rising prices because of intense competition for sales, restrictive government regulations, and high interest rates on borrowed money. In recent years it has often occurred that chemical companies have been caught between the two forces of inflated costs and constant (or perhaps diminishing) prices. The result has been a sharp drop in the profitability of the chemical industry.

In this book the primary concern is the economic evaluation of new projects in the chemical industry. In order to make these evaluations a number of economic indicators or evaluation indices are calculated, and the various projects are compared on the basis of the numerical values of one or more of these indices. In this book the heaviest emphasis is on the venture worth method. This method involves the prediction of business activities some 5 to 15 years into the future. Sums of money presumed to be collected in the future are discounted to the present by means of the compound interest factor using the average rate of return realized by the company on its investments. This discounting of future sums to the present is an expression of the time value of money.

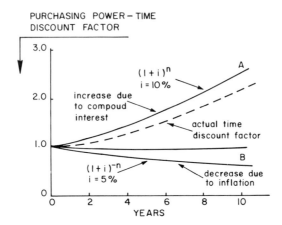

Fig. B.3. Purchasing power of one U.S. dollar versus time.

It would seem reasonable that, if the time value of money is to be invoked, then the fact of yearly monetary inflation should also be considered. The desirability of accounting for inflation in the economic studies and reports of industry has been stated by several workers [2,3]. Jordan [3] explains a simple method for accounting for inflation in the construction of the cash balance position diagram. Reference [2] discusses the changing of financial reports so as to account for inflation. Few corporations consider inflation when reporting the results of their operations, or specifically account for inflation in their economic evaluations.

In any existing business organization the effects of the time value of money and monetary inflation exist simultaneously and exert opposite effects on the purchasing power of future dollars. Figure B.3 shows the purchasing power of $1 as a function of time. Line A shows the increase in the purchasing power as a result of investment at compound interest in the absence of inflation. Line B shows the decline in the purchasing power as a result of monetary inflation in the absence of investment. The two forces are constantly at work, and depending upon the relative magnitudes of the two rates the resultant line will be somewhere in between lines A and B—as shown by the dotted line. The dotted line is the result of straightforward business operations in an economic climate that includes inflation. The dotted line is the line that is "seen" by corporations in the conduct of their business. The interest rate associated with the dotted line is that reported by corporations in their annual financial statements. It may be deduced from a study of such reports. This is the value of i which may be used in the

Inflation

venture worth equation. Using the venture worth equation as written will automatically consider the effect of inflation on future cash flows.

The venture worth equation also contains the term I, the new capital invested in the project. Part of the restriction placed on the project is that it must generate sufficient funds to recover I at the end of the projected life of the plant. The sinking fund deposit factor, $i/[(1 + i)^n - 1]$, is used to calculate the yearly charge against the project so that I will be recovered at the end of n years; see Eq. (3.8). However, if the project is to have a life of n years, the monetary value of I will not be I at the end of n years, but will be $I(1 + a)^n$ because of inflation at the average annual rate of a.

Therefore, if inflation is considered, the usual venture worth analysis, Eq. (3.9), will not provide sufficient funds to recover the purchasing power of the original investment. As a result, when inflation is present, one of the objectives of the investor is not realized; I will not be recovered at the end of n years. It is only if $I(1 + a)^n$ is recovered that the investor's objectives will be met.

The same situation exists for the investor in bonds, and, in order to counteract it, the investor insists on an interest rate that is high enough so that the effects of inflation are offset and a useful rate of return is achieved. Thus, in periods of serious inflation, as in the 1970s, the interest rate on bonds is high.

In connection with the venture worth analysis in an inflationary period there are several considerations:

1. No change is made in the formulation of W; I is considered a constant for all the project life.

2. A higher than usual value of i_m may be required so that only projects that are quite profitable (high R_k) will be approved.

3. The payout times are kept quite short, 2 to 4 years. In this case the inflationary forces do not have much chance to change I. Consequently, the question of the recovery of I dollars of purchasing power is not a serious one.

4. The venture worth is an index of the profitability of an investment and is used to compare the attractiveness of the investment in comparison to others which are possible. If all such investments are compared on the same basis their relative attractiveness will be apparent whether an allowance is made for inflation or not. All dollars will be inflated at the same rate.

REFERENCES

1. H. Sauvain, Investment Management, 3rd ed., Prentice-Hall, Inc., Englewood Cliffs, New Jersey, 1967, p. 193.
2. Reporting the Effects of Price Level Changes, Accounting Research Study No. 6, American Institute of Certified Public Accountants, New York, 1963.
3. D.G. Jordan, Chemical Process Development, Interscience Publishers, John Wiley and Sons, Inc., New York, 1968, p. 51.

Appendix C

RAPID APPROXIMATION METHODS FOR THE
DESIGN OF CHEMICAL-PROCESS EQUIPMENT

The subject matter of this book centers on the venture worth concept. It has been argued that this concept is the best method for providing a quantitative measure of the economic worth of a particular idea. An inspection of the venture worth equation shows that an accurate evaluation of the required new capital investment and the manufactured cost of the product are essential to the use of the venture worth concept. Furthermore, even if the venture worth is not used, no commercial enterprise can be seriously considered before the capital investment and product cost have been estimated.

In industrial work, it often happens that estimates of new capital investment and manufactured cost must be made in a very short time—say 8 hours. At first thought it seem ludicrous that serious decisions concerning a multimillion dollar chemical plant would be made on the basis of an economic and technical evaluation made in so short a time. However, this situation occurs frequently, and engineers must cope with it in the best possible manner. For a detailed discussion of economic and technical evaluation see Jordan [1, pp. 26-53].

It is apparent that in order to produce a plant design plus a technical and economic evaluation in a very short time, engineers must have a set of equipment design procedures which are not difficult or time-consuming. They must be able to design a distillation column or a chemical reactor in 15 minutes. Furthermore, the design must reflect modern chemical engineering knowledge and modern economic data. Despite the fact that the calculation is simple and quick, it must be correct.

The lack of time in which to complete the work is often accompanied by a lack of knowledge concerning the physics and chemistry of the process. This lack must be made up by a combination of experience, knowledge, and judgment.

An economic and technical evaluation, whether rapid or not, centers on the calculation of the process heat and material balances, and on the design and estimation of the costs of the separate pieces of equipment and their utilities requirements. If it is assumed that the process heat and material balances can be computed, it can be seen that the most important part of the cost estimate is the compilation of a list of the major items of equipment together with their size, materials of construction, utilities requirements, and costs; see Chapter 5.

In the following discussion rapid approximation methods of equipment design are presented. In Chapter 8 economic-balance methods were presented for a number of unit operations. The quick approximate rules given below were derived from considerations such as those discussed in greater detail in Chapter 8, and from a study [1,2] of characteristic operating parameters of chemical-process equipment.

PUMPS, COMPRESSORS, AND PIPING

Pumps

Pumps are selected on the basis of the kind and quantity of material to be pumped, the process pressures and temperatures, and the power required for pumping. Perry [2, pp. 5-14 to 5-58, 6-2 to 6-15, and 6-35 to 6-65] and Rase and Barrow [3] give detailed discussions.

The horsepower required by the pump may be computed from

$$Hp = (\Delta H)(G)(s)/3,960 \tag{C.1}$$

where Hp = pump horsepower required

G = fluid flow rate (gal/min)

s = specific gravity of fluid

ΔH = differential head (ft lb f/lb m)

Rapid Approximation Methods

The approximate brake horsepower for a given pump may be obtained from suitable efficiency data. For centrifugal pumps it will usually be about 60%. An estimate of the discharge pressure for a pump must include allowances for the pressure at the tower or vessel to which the fluid must be delivered, the line pressure drop, the static head due to the elevation of the discharge point above the pump, and the drop through a control valve ($\sim 30\%$ of the total variable pressure drop in the system). The differential pressure to use in computing the hydraulic horsepower is obtained by subtracting suction pressure from discharge pressure. For a recent discussion see Simpson [4].

Compressors

For information on types of compressors and their selection Perry [2, pp. 6-15 to 6-28] and Rase and Barrow [3] are recommended. For preliminary purposes the horsepower required of the compressor is calculated on the basis of a reversible adiabatic compression with intercooling between stages to the gas inlet temperature. In this case,

$$\text{Hp} = \frac{0.085 n T_1}{a} \left[\left(\frac{P_2}{P_1} \right)^a - 1 \right] \quad \text{(C.2)}$$

$$\frac{T_2}{T_1} = \left(\frac{P_2}{P_1} \right)^a \quad \text{(C.3)}$$

where Hp = theoretical horsepower required to compress an ideal gas

n = number of million std cu ft/day, 70°F, 1 atm

T_1 = initial temperature, °F abs (°F + 460)

T_2 = final temperature, °F abs (°F + 460)

P_1, P_2 = initial and final absolute pressures

$a = \dfrac{C_p/C_v - 1}{C_p/C_v}$, values for which are given in Table C-1 for various gases

Table C-1   Values for Specific Heat Ratio Function
$(C_p/C_v - 1)/(C_p/C_v) = a$

Average temperature (°F)	H_2	Air	Steam	Methane	Ethylene	Propane	Butane
0	0.29	0.29	0.25	0.25	0.21	0.15	0.10
100	0.29	0.29	0.25	0.23	0.19	0.13	0.09
200	0.29	0.28	0.25	0.21	0.17	0.11	0.08
300	0.29	0.28	0.24	0.19	0.15	0.10	0.07
400	0.29	0.28	0.24	0.18	0.13	0.09	0.06
500	0.29	0.27	0.24	0.17	0.12	0.08	0.05

Approximately (not accounting for temperature)
For monatomic gases (He, Ne)   a = 0.40
For diatomic gases (H_2, N_2, CO)   a = 0.29
For more complex gases (CO_2, CH_4)   a = 0.23
For other gases: a = $1.99/C_p$

Table C-2   Theoretical Horsepower for Compression of 1,000,000 std cu ft/day of Gas at 100°F

Compression ratio P_2/P_1	Value of a			
	0	0.10	0.20	0.30
1.5	19	20	20	21
2.0	33	34	36	37
3.0	52	55	58	62
4.0	66	71	76	82
5.0	76	83	90	98
6.0	85	93	102	113
7.0	93	102	113	126
8.0	99	110	123	138

Rapid Approximation Methods 455

Table C-2 tabulates the theoretical horsepower required to compress adiabatically 1,000,000 std cu ft of gas per day as a function of a and the compression ratio for an initial gas temperature of 100°F.

The column in Table C-2 for a = 0 refers to isothermal compression. For a gas at a different initial temperature than 100°F = T_1, the value from the table is multiplied by the absolute temperature ratio $(T_1 + 460)/560$. Except for permanent gases (air, hydrogen, methane) these values (in Table C-2) are only approximate because of gas-law deviations. For heavier hydrocarbon gases the theoretical horsepower will be less and the temperature rise higher than indicated.

The overall mechanical efficiencies of gas compressors may vary from 70 to 80%; large sizes and high C_p/C_v values give higher efficiency.

1. Cooling Water Requirement.
(a) Gas-engine drive.

	Btu/bhp/hr
Compression cylinder jackets	500
Compression intercooler	1,000
Compression aftercooler	1,000
Power cylinder jackets	4,000
Oil cooler	1,500

(b) Motor drive. Cooling water is required only in the compression end.

	Btu/bhp/hr
Jacket water cooling	500
Intercooler	1,000
Aftercooler	1,000

With both gas-engine and motor drive, allowable water temperature rise across cylinders is 25°F and in aftercoolers is 15°F.

Fuel-gas requirement of engine drive for gas engines is approximately 8,500 Btu/bhp/hr for large units, based on lower or net heating values of the gas employed.

Piping

The optimum sizing of piping is often accomplished by short-cut methods because detailed economic balances are not usually justified in selecting the numerous sizes that occur in a typical process plant; see Section 8.01.

Details regarding special types of piping and designs are available in handbooks [2, pp. 6-35 to 6-55] and codes [5] should be consulted for standard construction requirements.

1. Allowable Friction Loss. The economic balance between pipe diameter and total piping cost was presented in Chapter 8. The most useful form of the optimum solution is an optimum velocity, with the minimum allowable friction loss. Table 8-2 gives some results of a detailed calculation. For present purposes, Table C-3 shows some economic velocities that may be used.

Table C-3 Pipe Sizing Factors

Line type	Reasonable velocity[a] (ft/sec) in terms of pipe diameter, d (in.)	Pressure drop (lb/sq in.) Average per 100 ft	Pressure drop (lb/sq in.) Maximum, Total
Pump discharge lines	$\frac{d}{3} + 5$	2.0	–
Pump suction	$\frac{d}{6} + 1.3$	0.4	–
Steam or gas	20d	0.5	5.0

[a]Maximum liquid velocities should not exceed 20 ft/sec to avoid erosion.

Rapid Approximation Methods

Most trouble in practice is found on suction lines to pumps and compressors. The total pressure drop should be checked to be certain it is no more than one-half the total available head. For lines outside battery limits, the total pressure drop available must be used as the basis for calculating the line sizes. For more detailed economic balances see Table 8.2.

HEAT-TRANSFER EQUIPMENT

Shell and Tube Heat Exchangers

Shell and tube equipment is employed in heat exchangers, condensers, and coolers. This equipment is often purchased from manufacturers, and the process engineer must provide the necessary information for detailed design. Almost all engineering departments have elaborately detailed design check lists which, when completed, should furnish all the information that the manufacturer needs.

The manufacture of heat-transfer equipment is a difficult technique with many problems involving mechanical engineering, corrosion, metallurgy, and fluid pressure drop as well as the transfer of heat at certain rates and certain temperature levels. A practical economic design is the product of intensive cooperation between the process engineer and the equipment manufacturer. In many instances the ideal optimum heat exchanger cannot be used because of practical engineering, manufacturing, and construction difficulties. In such cases the final design will be the product of many compromises.

The detailed design of exchangers involves a number of factors that have been discussed in several books, among which the treatment of Kern [6] is recommended. The process engineer should be familiar with tube sheet layouts, average fouling factors, and corrections to the log mean temperature difference for multipass exchangers. These and other subjects are discussed in various references [2, pp. 11-1 to 11-49; 6; 7].

These matters are not really a part of rapid approximation designs. For the sake of speed and simplicity the following rules will be adopted:

1. Assume true countercurrent flow; no temperature difference correction. However, Ten Broeck's nomograph [8] is to be used.

2. All tubes are to be 3/4 in. in OD by 8 ft long.

3. No more than 3,500 sq ft of heat-transfer area in one shell.

1. Countercurrent Process Stream Exchangers. The economic basis for this form of heat exchanger has been described in detail in Section 8.02. The use of the Ten Broeck nomograph, Fig. 8.1, makes the calculation quite simple. Despite an apparently complicated relationship the time consumed in calculation is small and the answers obtained are much better than a guess. This nomograph requires several pieces of information. For first approximation purposes (unless the engineer has better information) use:

For liquid-liquid heat exchange, U = 50 Btu/(hr)(sq ft)(°F)

For gas-liqud or gas-gas heat exchange,
$$U = 5 \text{ Btu/(hr)(sq ft)(°F)}$$

Cost of heat, 80¢/million Btu

Cost of heat-exchanger area,

Area (sq ft)	Average purchased cost ($/sq ft)	
	Carbon steel	Stainless steel
1,000	7.9	23.7
10,000	3.1	9.0

2. Stream Coolers. An economic optimum calculation is not helpful because calculated water outlet temperatures are higher than desirable from the standpoint of scaling, tubing corrosion, and temperature of water feed to a cooling tower. Use

U = 150 Btu/(hr)(sq ft)(°F)

Water inlet temperature 90°F

Water temperature rise 30°F

3. Condensers. Use the same conditions as for stream coolers in 2 above.

4. Reboilers. A distinguishing characteristic of reboilers is that of film boiling and the critical temperature difference. In order to avoid vapor blanketing of the tubes the operating difference and the overall heat-transfer coefficient are deliberately restricted to

Rapid Approximation Methods

low values. This, in turn, sets a maximum on the heat flux, Btu/(hr)(sq ft), which is equal to $U \times \Delta T$.

For present purposes use

$U = 250$ Btu/(hr)(sq ft)(°F)

$\Delta T = 45$°F

Then, $q/A = U \times \Delta T = 11{,}250$ Btu/(hr)(sq ft).

5. Waste Heat Boilers. Assume that all the heat is transferred from a hot gas stream which decreases in temperature to boiling water at a constant temperature. Use

$U = 5$ Btu/(hr)(sq ft)(°F)

Let exit gas temperature approach to within 60°F of the steam temperature.

6. Air-cooled Exchangers and Condensers. In many cases cooling water is in limited supply and air-cooled condensers or exchangers, both employing finned tubes, are widely used in such cases. The detailed design of such exchangers is a matter for specialists, but approximately,

$U = 70$ Btu/(hr)(sq ft)(°F)

based on bare nonfinned tube surface.

Let the temperature of the exit air approach to within 40°F of the temperature of the entering high temperature fluid.

Assume the entering air temperature to be 90°F; this will be lower at night and in the winter, and provision for automatic adjustment must be made.

The quantity of air used is large, and fan capacities and power consumption may be high. For most chemical-process industry applications use 20 hp per 1,000 sq ft of bare tube area.

Direct-Fired Heaters

The combustion of fuels in a furnace supplies most of the high-temperature heat required by chemical processes. This heat may be used for the distillation of high-boiling materials, the preheating of the feed streams before a high-temperature reaction, or the supplying of the heat for an endothermic reaction.

For approximation design purposes it is only necessary to know the desired heat input, the necessary heat-transfer area (both radiant and convective sections), and the temperature of the exit heating gas. Details of furnace design need not be investigated; for such information the work of Kern [6, p. 674] or Perry [2, p. 10-32] should be studied.

Table C-4   Typical Heat-Transfer Rates in Direct-Fired Heaters
Btu/(hr)(sq ft)

	Radiant section	Convective section
Clean liquid, not vaporized	15,000	4,000
Dirty liquid, not vaporized	10,000	4,000
Oil or steam vapor superheater	8,000	4,000
Clean liquid, partly vaporized	12,000	4,000

These rates refer to the heat input to the flowing stream and not to the heat released in the firebox. Approximately 50% of the heat duty will be transferred in the radiant section and 50% in the convection section.

Exit flue gas temperatures usually range from 250° to 350°F higher than the feed inlet temperature. The temperature of the stack gases is usually between 650° and 950°F.

The fluid velocity in the tubes, for a nonvaporizing liquid, can be taken as 5 ft/sec.

DISTILLATION COLUMNS

Phase Equilibria

Many references to binary, ternary, and some more complex systems are given by Hala et al. [9]. If more conventional references

[2, pp. 13-2 to 13-9] fail to produce suitable data Hala should be consulted. This will not produce data, only a reference (perhaps not in English) to the original work. When examining vapor-liquid equilibria special attention should be paid to (1) the presence of azeotropes or relative volatilities lower than 1.2; (2) the presence of heat-sensitive compounds, particularly if they are less volatile and will appear in the reboiler of a column; (3) the corrosive nature of the materials.

The Thermal Properties of the System

The thermal properties of the streams must be known. There are many references to the thermodynamic properties of pure substances (for instance, Ref. 2 , pp. 3-147 to 3-195). The methods described by Reid and Sherwood [10] may be used to estimate these properties if none can be found in the literature.

At present (1970) it is common practice in the chemical engineering contracting companies that design most of the distillation columns to use the electronic computer, together with some of the complicated correlations that exist, to calculate the phase equilibria and the thermal properties of systems. This is done because the very high speed of the computer enables engineers to investigate many distillation situations in a very short time.

The Stream Concentrations

The concentrations of the feed, and tops and bottoms streams must be carefully considered. Very pure tops and bottoms streams will require tall columns, high reflux ratios, and more expensive distillations. As a first approximation use a top stream purity of 99.5% and 0.1% low boiler in the bottoms stream.

This generalization applies only to conventional systems where the relative volatility is essentially constant. For azeotropic and extractive distillations, or for systems with a nonconstant relative volatility, these approximations would not be applicable, and other knowledge must be used.

Column Process Design

The minimum reflux ratio may be established by well-known procedures [2, p. 13-24]. A practical industrial distillation will use an actual reflux ratio of 1.2 times the minimum reflux ratio. The number of theoretical plates may be obtained from Liddle's modification [11] of Gilliland's correlation [2, p. 13-43]. At a reflux ratio of 1.2 times the minimum the number of plates will be twice the minimum number.

The plate efficiency may be obtained from O'Connell's correlation [1, p. 439]. As a rough conservative rule the plate efficiency may be assumed to be 50%. Thus, the actual number of plates in the column will be four times the minimum number of theoretical plates.

Column Size

The physical dimensions of the column may be estimated from the following:

1. For atmospheric pressure distillations use a superficial vapor velocity of 3 ft/sec and a pressure drop of 3 in H_2O/actual tray.

2. For pressures below 100 mm Hg (vacuum distillations) the velocity can be from 6 to 8 ft/sec, and, for pressures above 1 atm, the vapor velocity should be reduced from 3 to near 1 ft/sec as the pressure increases. For a vacuum distillation the designer must be careful not to allow the pressure drop to become so high that the bottom temperature becomes too great.

3. Use a tray spacing of 2 ft. If the resulting column is taller than 150 ft the spacing might be reduced to 1.5 ft.

4. Use perforated (sieve) trays. At quite low pressures (~20 mm Hg) special perforated trays may be used. These have proven superior to packing [12].

5. For packed columns use 1-in. Pall rings of suitable metal. Also, consider the following tabulation:

Total pressure	Superficial vapor velocity (ft/sec)	Pressure drop in H$_2$O/ft	H$_{OG}$ (ft)
1 atm	3	0.5	22
100 mm Hg	6-8	1.0	3

GAS ABSORPTION COLUMNS

Gas absorption columns are of two types: (1) those wherein the dissolving gas reacts chemically with the solvent or with some active ingredient in the solvent (chemical absorption); (2) those wherein the solute gas simply dissolves because of its solubility in the solvent (physical absorption).

Both types are widely used in industry and there is a large amount of information and experience on both types [2, pp. 14-2 to 14-40; 13].

Physical Absorption

This type is usually confined to solvent recovery such as acetone or methyl ethyl ketone in water, or to process recoveries such as benzene in light oil. Packed or plate columns may be used. Packing is often selected because of lower pressure drop and better corrosion resistance, but for high-pressure absorptions (as in the petroleum industry) plate columns are often employed. Table C-5 summarizes the design recommendations for atmospheric pressure physical absorptions.

Chemical Absorption

These are of two types: (1) Where the chemical reaction is irreversible. Here the equilibrium pressure of the absorbed material is zero and only one theoretical stage of absorption is required to

Table C-5 Design Recommendations for Atmospheric Pressure Physical Absorptions

Packing	% Absorption of solute gas	Superficial vapor velocity (ft/sec)	Ratio of slopes of equilibrium line and operating line	Pressure drop (in. H_2O/ft)
1-in. Pall rings metal or plastic	99.9	3	0.7	0.5

Number of gas-phase transfer units - N_{OG} = 20

H_{OG} = 2 ft

Height of packing = 40 ft

Rapid Approximation Methods

absorb all the solute gas. For instance, CO_2 in NaOH. (2) Where the chemical reaction is reversible and the equilibrium pressure of the solute gas is a function of concentration and temperature. In this case the phase equilibrium relationship must be known. For instance, H_2S into diethanolamine.

The first case is not often encountered except for air pollution control or for the deliberate formation of a chemical compound. In the latter case a stirred vessel is often used, and there is no distinction between absorption and reaction kinetics. Despite the zero equilibrium partial pressure, the rather low mass-transfer rates accompanying these absorptions will require that a fairly tall column be used. The heat effect may also be severe, requiring some form of liquid cooling during absorption.

Let the following discussion concerning reversible chemical absorption apply to irreversible chemical absorption.

For reversible chemical absorption:

1. Absorb 99.9% of solute gas.

2. The heat effect depends on the concentration of the solute gas in the entering stream. If the concentration is higher than 10% it may be necessary to remove the liquid stream from the tower, pass it through a heat exchanger for cooling, and return it to the tower. See Sawistowski and Smith [14, p. 12] for a heat balance calculation.

3. Use a superficial vapor velocity of 3 ft/sec.

4. Use three theoretical absorption stages.

5. Assume a plate efficiency of 20%. The actual number of plates will then be 15. The tray spacing may be 2 ft so the column will be 30 ft tall.

6. For a packed column the H_{OG} will be about 12 ft when the gas concentration is in the range of 2% to 15%. For three absorption stages this will give a packed height of 36 ft.

7. Use an $L/G = 1.25(L/G)_{min}$.

8. Use a pressure drop of 3 in H_2O/actual tray for a plate column and 0.5 in H_2O/ft of packing for a packed column.

SOLVENT EXTRACTION SYSTEMS

Liquid extraction is used for several reasons, but two reasons are most important: (1) Separations not possible by distillation can be made; the separation factors for solvent extraction can be quite high. (2) The transfer of a valuable solute from one solvent to another lowers the cost of final purification. Distillation is nearly always used as the final separation and purification step, and this distillation accounts for much of the cost of the extraction process.

Usually the concentrations of the solute in the feed and extract streams are fairly low—not more than 20%. In this case the simple distribution law describes the phase equilibria quite well. For this situation there are published treatments that may be used for guidance in analyzing extraction problems [2, pp. 14-40 to 14-69]. Also, it frequently happens that the feed solvent and the extract solvent are totally immiscible so that the extract stream will contain little or none of the feed solvent. Thus, there is an almost total exclusion of the feed solvent from the extract stream. This property makes for a very high separation factor.

It is frequently true that the concentration of the valuable solute in the exhaust or raffinate stream may be very low. In this case the raffinate stream leaving the system will be almost pure and may need no further treatment.

Treybal [15, pp. 541-549] has studied the economics of solvent extraction in detail. From this work several conclusions may be drawn:

1. 99.9% of the valuable solute may be extracted into the extract solvent.

2. The optimum value of the slope of the equilibrium line to the slope of the operating line is 1.5.

The number of ideal contacting stages may be calculated from the plot given by Treybal [15, Figs. 6.39 and 8.3]. Since separation factors are usually fairly high, it is seldom that more than five stages will be needed.

Phase Equilibria

There are several compilations of liquid-liquid equilibria available. Perry [2, pp. 14-45 through 14-55] and Francis [16].

Rapid Approximation Methods

These contain references to data on a large number of systems. Treybal [15, pp. 56-119] and Jordan [1, pp. 571-578] give estimation methods. Scheibel [17], Jordan [1, pp. 584-594] and Treybal [15, pp. 360-370] have described Scheibel's method for the laboratory batch simulation of a continuous extraction. This method is quite practical and should be considered.

Mechanical Equipment

Solvent extractions are usually carried out in packed columns, sieve plate columns, or mechanically agitated equipment.

1. Packed Column. If a packed column is used the mass-transfer characteristics are poor. H_{OC}, the height of an overall mass-transfer unit based on the continuous phase, will be about 10 ft for 1-in. packing. The rates of flow will be V_D = 35 ft/hr (the superficial velocity of the disperse phase) and V_C = 25 ft/hr (the superficial velocity of the continuous phase). It is apparent that if more than four transfer units are required, the packed column will become too high.

2. Sieve Plate Column. If a sieve plate column is used the plate efficiency is given by

$$E = (8.95 \times 10^4 H_C^{0.5})(V_D/V_C)^{0.42}/\sigma \qquad (C.4)$$

where E = plate efficiency, fractional

H_C = tray spacing (ft)

V_D = superficial velocity of disperse phase (ft/hr)

V_C = superficial velocity of continuous phase (ft/hr)

σ = interfacial tension between two liquid phases
  [lb m/sq hr; (dynes/sq cm) × 28,650 = lb m/sq hr]

The interfacial tension is an important variable in liquid extraction. For totally immiscible systems such as benzene-water, σ is high and solvent extractions are difficult; E (above) will be about 10%. For partially miscible systems, such as ketones-water, σ will be lower and solvent extractions easier; E will be closer to 40%. The interfacial tension may vary from 40 dynes/cm to about 5 dynes/cm. A method for predicting σ is given by Treybal [15, Fig. 4.10].

For a sieve plate column let V_D and V_C be 50 ft/hr; use H_C = 2 ft.

3. Mechanical Agitation. For liquid extractions which require more than four or five transfer units it is now common practice to use a mechanically agitated device. The agitated vessel mixer-settler extractor is widely used. Here, the stage efficiency will be high—say 90%. The power input to the agitator can be set at 1,000 ft lb f/(min)(cu ft/hr) of total liquid flow, and the power consumption may be set at 4 hp/1,000 gal of mixer capacity. The vessel size and power consumption can be calculated from this and the stream flow rates.

Other mechanical devices are also used. Details concerning these can be obtained from Treybal [15] and Jordan [1, pp. 661-703]. The mechanically agitated tower of Karr [1, p. 675] is particularly interesting. It is mechanically simple, has good efficiency, and high capacity. An HETS of 5 in. at a frequency × amplitude of the agitator (which moves vertically) of 200 in./min may be used. V_D and V_C of 50 ft/hr are typical.

CONTINUOUS CRYSTALLIZERS

The design of a continuous crystallizer is not a well-known procedure, but some fairly recent developments may be connected in order to give a reasonable design method that produces plausible results.

Using the work of Saeman it is possible to derive an equation [1, pp. 758-783] for the necessary volume of a circulating magma crystallizer [1, p. 739] in terms of the specific growth rate of the crystal, the supersaturation, and the required production rate.

$$V = [P^{1.33}/k S \rho_S](C/6\rho_C N)^{0.33} \qquad (C.5)$$

where V = volume of slurry (crystals and saturated solution) (cu ft)

k = specific crystal growth rate constant [(ft^4)(hr)/(lb m)]; assume equal to 5×10^{-6}

S = average supersaturation in well-mixed slurry (lb m/cu ft); assume equal to 0.01

ρ_S = density of magma (solution plus solid crystals) (lb m/cu ft)

P = crystallizer production rate (lb dry crystals/hr)

Rapid Approximation Methods

C = a shape factor; assume equal to 1.3

ρ_C = density of the solid dry crystal (lb m/cu ft)

N = number of seed crystals formed per hour

= $2.9 \times 10^8 \, P/\rho_C$

The specific crystal growth rate is best estimated from the correlation of Jenkins (see Jordan, Ref. [1], p. 724). The value varies with the viscosity of the solution, but values near 5×10^{-6} have been measured for sodium chloride and ammonium nitrate.

The supersaturation to be used is always quite low, for two reasons: (1) The amount deliberately produced in the circulating magma is kept low because experience teaches that a low supersaturation is best. (2) The fast mixing in the crystallizer disperses the supersaturated (recycled) stream into the large quantity of crystallizing magma. The well-stirred vessel effect lowers the entering supersaturation to a much smaller amount. As a first approximation use S = 0.01 lb m/cu ft.

The density of the slurry (magma) of solids and saturated solution will be between 30 and 40 lb m/cu ft. As a conservative estimate use 40 lb m/cu ft.

When designing continuous crystallizers careful attention must be paid to the means of producing the supersaturation (by cooling or by evaporation of solvent) and of recycling the magma. Utilities such as cooling water, electrical power, and steam may be required in rather large quantities for production of supersaturation and recirculation of magma. The entire installation may be quite large and expensive.

CONTINUOUS ROTARY VACUUM FILTERS

There are many devices for filtration, but for fairly large-scale production of solids a continuous machine such as the rotary vacuum drum filter is most widely used. These machines are designed from small-scale vacuum leaf tests on the slurry in question; see Jordan [1, pp. 828-836].

For rapid approximation purposes two different rates may be used:

1. For finely ground ores and minerals, cement slurry, or fairly soft precipitates a filtration rate of 1,500 lb m of wet solid per square foot of total drum area per day may be used. A drum rotation rate of 0.33 rpm and a vacuum of 18 to 25 in. of mercury are conventional operating conditions.

2. For coarser solids and crystals, such as salt, a filtration rate of about 6,000 lb m of wet solid per square foot of total drum area per day may be used. The drum rotation rate may be about 0.33 rpm and a vacuum of 2 to 6 in. of mercury may be used.

Information such as the above is of only marginal assistance, but it will aid in calculating an approximate machine size; see also Perry [2, pp. 19-76 to 19-86]. For a more complete discussion of rotary vacuum filters see Jordan [1, pp. 828-836].

CONTINUOUS ROTARY DRYERS

Drying is a frequently encountered unit operation and there are many elaborate devices for removing moisture from solids. For present purposes only the continuous rotary kiln dryer will be considered. It is not suitable for all problems, but it can be used for the continuous drying of small particulate solids. The drying of such solids represents a substantial fraction of all drying problems. For a more complete discussion on drying see Jordan [1, pp. 855-939].

The available design data for these dryers are based on the consideration of the dryer as a heat-transfer device. Measurements have been made of an overall heat-transfer coefficient. From this coefficient, the heat load on the dryer, and the mean temperature difference between gas and solid, the volume of the dryer can be computed. After assuming some reasonable value for the gas mass flow rate through the dryer, the cross-sectional area can be calculated, and then the dryer length established.

For rapid design purposes the following assumptions may be made:

1. The solid is always in the constant rate period of drying. Thus, the solid temperature will always be at the wet bulb temperature of the gas. The falling rate period can take a very long time to remove only a small amount of moisture. Drying in the falling rate period would make the rotary dryer too long.

Rapid Approximation Methods 471

2. The gas and solid move continuously through the dryer in cocurrent flow. Actually, the direction of gas flow makes little difference to the heat-transfer problem since the solid is always at one temperature.

3. The allowable gas mass flow rate (based on the empty cross section of the dryer) is assumed to be 1,000 lb m/(hr)(sq ft).

4. The number of heat-transfer units in the dryer is assumed to be 1.5.

$$N_T = (T_{Di} - T_{Do})/\Delta T_m \tag{C.6}$$

where N_T = number of heat-transfer units based on gas dry bulb temperature

T_{Di} = gas inlet dry bulb temperature (°F)

T_{Do} = gas outlet dry bulb temperature (°F)

ΔT_m = logarithmic mean temperature difference between the gas and the solid (°F)

5. The ambient air will be assumed to be at 70°F and 50% relative humidity. From this, and the inlet gas temperature T_{Di}, the inlet gas wet bulb temperature will be set.

6. The overall volumetric heat transfer coefficient is given by

$$Ua = 20 G^{0.16}/D \tag{C.7}$$

If G = 1,000 then Ua = 60/D, where D = dryer diameter (ft).

$$Q = (Ua)(V)\Delta T_m \tag{C.8}$$

where Q = dryer heat load (Btu/hr)

V = dryer volume (cu ft)

The actual design calculation proceeds in the following way:

1. Assume an inlet gas dry bulb temperature. This should be as high as possible, but not so high as to attack the solid or the dryer metal.

2. From the assumption of constant rate drying and 1.5 heat-transfer units, the gas outlet dry bulb temperature may be calculated.

3. Calculate the dryer heat load from the solids feed rate and the moisture to be evaporated.

4. Calculate the gas flow rate (lb m/hr) from a moisture balance.

5. There must now be agreement between the values of the gas rate calculated from the moisture balance and from the heat balance. If there is not, a new value of the gas inlet dry bulb temperature must be assumed, or some other value of N_T between 1.5 and 2 may be used. All heat and material balances must agree before the problem is solved.

6. With the gas rate known from the calculations above and a mass flow rate of 1,000 lb m/(hr)(sq ft) assumed, the cross-sectional area of the dryer can be calculated. Thus the diameter of the dryer is known.

7. Knowing D it is possible to calculate Ua, the volumetric heat-transfer coefficient.

8. From the known dryer heat load, the heat-transfer coefficient, and the logarithmic mean temperature difference, the volume of the dryer can be calculated.

9. From the volume and the cross-sectional area, the length can be calculated.

Some engineering judgment must be used concerning the ratio of length to diameter of the dryer. A rotation rate of about 4 rpm may be used.

CHEMICAL REACTORS

From the rather large amount of knowledge and experience concerning applied reaction kinetics now available, it can be shown that chemical reactions and chemical reactors are most conveniently classified according to the number and kind of phases present. This method of classification enables the designer to match the requirements of the reaction (residence time, temperature, pressure, agitation, resistance to corrosion) with the characteristics of the reactor that will enable it to supply these requirements. The various kinds of chemical reactors (short packed tubes, long empty pipes, stirred vessels, and fluidized beds) have quite different characteristics with respect to agitation, pressure drop, distribution of residence times, and materials of construction. By noting the special requirements of the reaction and the characteristics of the reactors it is possible for the designer to match the two and so choose the proper reactor for the particular reaction.

Table C-6 states that there are seven kinds of reactions and five kinds of reactors.

Table C-6   Chemical Reactors

Reaction type	Reactor suitable for reaction
Homogeneous: gas phase	Empty tube, continuous
Homogeneous: liquid phase	Empty tube or stirred vessel, continuous Stirred vessel, batch
Heterogeneous: liquid-liquid	Stirred vessel, batch or continuous
Heterogeneous: liquid-gas	Stirred vessel, semibatch or continuous Absorption tower, continuous
Heterogeneous: liquid-solid	Stirred vessel, batch or continuous Packed column, continuous
Heterogeneous: liquid-solid-gas	Small packed tubes (stationary solids, flowing gas), large packed (adiabatic) fixed bed (stationary solids, flowing gas), moving bed (large-sized solids moving down, gas up), fluidized solids (gas moving up, solids flowing but well mixed)

Before an approximate design of a chemical reactor can be started it is necessary that experimentation provide the designer with some facts:

1. The reaction conditions: Laboratory work must determine the temperature, pressure, flow rates, catalysts, concentrations, reaction times, conversions, and yields. This may come from only

one experiment, but experimental data must be available. It is not necessary to determine the kinetics or the mechanism of the reaction, but an experimental determination of reaction conditions, reactor geometry, conversion, and yield must be made.

2. The heat effect of the reaction: It is necessary that the heat released or absorbed by the reaction be known, or estimated fairly accurately. Usually the heat effect must be estimated (this is not difficult) because it is quite hard to measure. The heat of the reaction multiplied by the reaction rate will give the heat release (or absorption) per unit volume per unit time in the reactor. From this information the designer will obtain a fairly clear idea of the required heat-transfer capability of the reactor.

3. The agitation and mixing requirements of the reaction: The mixing of the reactor contents may have to be quite intense (in order to promote heat and mass transfer), or may not be needed at all (as for homogeneous mixtures of gases and liquids). Intense agitation may require rather elaborate mechanical equipment that may present many problems of cost and selection of materials.

4. The material of construction of the reactor and auxiliary equipment: If the corrosive nature of the reacting mass is mild the reactor may be built of conventional materials (steel) and in almost any desired shape. However, if the corrosion problem is severe, and ceramic materials must be used, the size and shape--together with the heat- and mass-transfer characteristics of the reactor--may be severely restricted.

The designer of a chemical reactor must combine the experimental facts listed above with his knowledge of agitation, heat-, and mass transfer, flow patterns in continuous equipment, and the properties of materials to design a mechanical device which will meet the requirements of the reaction. Even an approximate design must specify the following:

1. The volume of the reactor

2. The reactor geometry: length/diameter ratio, the diameter; the length, diameter, and number of tubes

3. The necessary amount of heat-transfer area and its arrangement and geometry

4. The type and the intensity of the agitation

5. The materials of construction

For a description of industrial reactors see Walas [18, pp. 267-294].

Rapid Approximation Methods

Using the classification of reactions given in Table C-6 the following numbers and design methods may be used for the rapid design of chemical reactors.

Homogeneous: Gas Phase

The reactor consists essentially of a multiplicity of small empty tubes connected in parallel and operated continuously. The reactions are usually fast, 1 sec or less, the flow velocities are high, and the heat effect is strong and usually endothermic (although exothermic reactions do occur). The heat transfer is poor, the flow turbulent (piston flow), and the temperature is nearly always high. Heat-resistant metals must be used, and the designer must be careful of small quantities of reactant gases (oxygen or chlorine).

Use:

Pressure drop = 0.2 psi/ft

2-in. diameter stainless steel tubes 20 ft long

Gas mass flow rate = 3,000 lb m/(hr)(sq ft)

Endothermic reactions may be heated by radiant heaters [use 18,000 Btu/(hr)(sq ft)] or by convection from hot gases [use 4,000 Btu/(hr)(sq ft)]. For both heating and cooling an overall heat-transfer coefficient of 5 Btu/(hr)(sq ft)(°F) may be used. The maximum safe skin temperature of the metal tubes must be known. This will depend on the internal pressure. For a first approximation use 1,200°F for stainless steel.

References: Jordan [1, pp. 176-178], Hougen and Watson [19, pp. 845-884], Walas [18, pp. 101-108]. For experimental work see Happel and Kramer [20].

Examples: Pyrolysis of light hydrocarbons to ethylene, propylene, and acetylene; nitration of paraffins (exothermic); thermal demethylation of toluene.

Homogeneous: Liquid Phase

Use well-stirred vessels for both batch and continuous reactions. Use a tubular or pipeline reactor only for continuous reactions.

Stirred vessels may be used singly or in series combinations of two, three, or four vessels. Nearly always four are used.

1. Stirred Vessels. Use baffled turbine-agitated vessels; power consumption 2 hp/1,000 gal; assume vessel contents completely mixed; heat-transfer coefficient to jacket or internal coil: 150 Btu/(hr)(sq ft)(°F). If the heat-transfer requirements are excessive, consider the use of an outside heat exchanger. The size of the vessel (the nominal holding time) must come from laboratory experiments.

2. Pipeline Reactors. The tubular or pipeline reactor may be used for slow reactions which have little or no heat effect. Here the flow may be laminar, the pipeline long, and temperature and pressures easily controlled. The distribution of residence times should be taken as that of piston flow despite the laminar flow model.

If the reaction is faster or the heat effect is larger (but not large) turbulent flow will be needed. A velocity of 5 ft/sec, a heat-transfer coefficient of 25 Btu/(hr)(sq ft)(F), and a pressure drop (psi/ft) = $0.7/D$ (where D is the pipe diameter in inches) may be used. The distribution of residence time will be that of piston flow.

Frequently the stirred vessel and the pipeline reactors may be connected in series. The earlier faster stages of the reaction may be conducted in the stirred vessel where the heat transfer is better, and then the less intense but longer part of the reaction may be done in a long pipe.

References. Jordan [1, pp. 193-225]; Walas [18, pp. 79-125].

Example. Continuous polymerization of styrene.

Heterogeneous: Liquid-Liquid

Here good phase dispersion and good heat transfer are needed. Use turbine-agitated baffled vessels with power consumption of 5 hp/1,000 gal; U = 150 Btu/(hr)(sq ft)(°F); provide outside chamber for phase separation, or, if fast separation is needed, use a centrifuge.

References. Jordan [1, pp. 193-214]; Levenspiel [21, pp. 384-481].

Example. Nitration of toluene with mixed acids.

Rapid Approximation Methods 477

Heterogeneous: Liquid-Gas

Use stirred vessels with baffles and turbine agitation; use 10 hp/1,000 gal power input to ungassified liquid; use gas superficial velocity of

0.2 ft/sec for gas which is mostly absorbed

0.1 ft/sec for gas which is 50% absorbed

0.05 ft/sec for gas which is mostly not absorbed

Let U = 100 Btu/(hr)(sq ft)(°F); assume that the fraction of the reacting volume that is gas is 0.20.

References. Jordan [1, pp. 151-165]; Levenspiel [21, pp. 384-481].

Examples. Air, or oxygen oxidation of p-xylene to terephthalic acid.

Heterogeneous: Liquid-Solid

Use turbine agitated baffled vessel; use agitator power input of 10 hp/1,000 gal; vessel length/diameter ratio = 2; U = 100 Btu/(hr)(sq ft)(°F); liquid phase mass transfer coefficient, k_L = 3.5 ft/hr; if small sized catalyst is used may use 5 lb solid/cu ft of slurry.

References. Jordan [1, pp. 135-138].

Examples. Dissolution of solid salts; extraction of ores with acids.

Heterogeneous: Liquid-Solid-Gas

1. Stirred Vessel. Turbine-agitated, fully baffled, 10 hp/1,000 gal; U = 100 Btu/(hr)(sq ft)(°F); L/D = 2; superficial vapor velocity of 0.05 to 0.2 ft/sec; solid content 5 lb solid/cu ft of slurry; fraction gas in slurry 0.2.

References. Jordan [1, pp. 135-165]; Satterfield and Sherwood [2, pp. 43-55].

Examples. Cottonseed oil hydrogenation; production of light hydrocarbons from CO and H_2.

2. Trickle Bed Reactor. Stationary solids, liquid descending, gas ascending. Design in same way as gas absorption tower. Use G = 1,000 lb m/(hr)(sq ft) or 3 ft/sec superficial velocity; L = 1,500 lb m/(hr)(sq ft). Heat effect should be low.

References. Lapidus [23].

Example. Hydrodesulfurization of petroleum fractions.

Heterogeneous: Solid-Gas

1. Small Packed Tubes. Solids stationary, with gas moving up or down. Use tube size of 1.5-in. dia; solid particle size 0.20 in.; gas pressure drop not more than 15% of upstream pressure; for atmospheric pressure operation use G = 1,000 lb m/(hr)(sq ft); $\Delta P/L$ = 15 in H_2O/ft of packed tube; number of tubes approximately 0.1 × feed flow rate (lb/hr); U = 15 Btu/(hr)(sq ft)($^\circ$F), for exothermic reactions a strong temperature rise may be expected.

References. Jordan [1, pp. 226-267]; Petersen [24].

Examples. Oxidation of o-xylene to phthalic anhydride; synthesis of vinyl chloride from hydrogen cyanide and acetylene.

2. Large Bed, Bulk Catalyst, Adiabatic Operation. No heat transfer; heat effect is absorbed or supplied by sensible heat change in gas. Frequently large quantities of inert diluent gas mixed with reactants. Pressure drop and gas flow rates same as for packed tubes. Bed diameter may be as large as desired. Gas distribution may be a problem.

Example. Vapor-phase hydrolysis of chlorobenzene to phenol.

3. Fluidized Bed.

1. Gas velocity, 0.5 ft/sec.

2. Gas flow pattern, perfect mixing. Treat fluidized bed reactor as a perfectly mixed vessel.

3. Overall heat-transfer coefficient to buried surface, 50 Btu/(hr)(sq ft)($^\circ$F).

4. Pressure drop per foot of bed height, 0.3 psi. Since large quantities of gas may be used, calculate necessary size and power of gas blower.

5. Provide 100% extra length of reactor for bed expansion and containment of solids elutriation.

Rapid Approximation Methods

6. Solids will be elutriated from bed. Provide cyclone separators and gas filters. Assume elutriation rate of 0.01 lb m/(sq ft)(sec).

PRESSURE VESSELS

This classification generally includes all cylindrical vessels in a process plant: reactors, reflux and separation drums, and fractionating towers. Such vessels are not usually purchased in standard sizes, but are specially designed for each duty required. Engineering contracting firms maintain special design groups whose duty it is to design and specify such equipment. Normally, such a group will have a great deal of experience and expertise, and they can answer the process engineer's questions quickly and accurately enough for estimation purposes. Also, the vessel fabricators will often give much information and advice without charge on short notice by telephone.

However, it often happens that the process development engineer has no such aid available or may not have the time to seek it. Almost always the student has no access to such information. This short section seeks to provide some assistance to those having to design and specify the size, shape, and the strength of process vessels.

In the U.S. the design of pressure vessels is governed by the procedures of the ASME code for unfired pressure vessels [26]. This code gives complete details of the design process. The application of these codes is discussed from the process viewpoint by: the Chemical Engineers' Handbook [2, Sections 23 and 24, and especially Table 24-10, Fig. 24-3 and p. 24-17. Section 23 discusses the corrosion resistance of many materials], and in books by Rase and Barrow [3] and Brownell and Young [27].

Pressure, Temperature, and Material of Construction

The mechanical design of a pressure vessel must start with the knowledge of the temperature, pressure, and chemical nature of the process stream. Obviously, a stream that is highly acidic, quite hot, and under some pressure cannot be contained in a carbon steel vessel, while on the other hand, carbon steel will be perfectly

Table C-7 Maximum Allowable Stress Values (psi) in Tension for Several Typical Vessel Metals

	Temperature (°F)			
	-20 to 750	750	850	1000
Carbon steel	11,250	10,250	7750	2500
Low alloy steel	18,750	15,650	9550	2500

	-20 to 100	200	400	700	1000
Stainless steel, type 302: 18 Cr, 8 Ni	18,750	17,000	15,450	14,800	12,500

suitable for a hydrocarbon stream at temperatures near 200°F and under pressure near atmosphere. There are many different materials for vessel fabrication (not all of them metal) and one will be most suitable for the task at hand. For the present discussion attention will be centered on metals and on process streams that are not excessively corrosive or abrasive.

Under these conditions the most important variables of the process stream to the design are the temperature and the pressure.

1. Temperature. It is terribly important to remember that the tensile strength of metals will decrease as the temperature increases to high values. Table C-7 gives a few values for the tensile stress of three metals as the temperature increases. When the temperature is quite low special alloys are required. [For a discussion of metals for low temperatures see p. 23-64 of 2. For aluminum and nickel alloys see p. 24-13 of 2.]

It is important to note from Table C-7 that different metals have different strengths and that these strengths decline as the temperature increases. Above about 750°F the strength declines sharply.

Between the temperatures of -20 and 750°F the design temperature is usually taken as 50°F greater than the operating temperature. Below -20°F and above 750°F it is important to set closer limits, based on a careful analysis of deviations from the design value owing to either control variations or changes in the process duty of the equipment.

2. Pressure. The vessel design pressure is generally specified as 10% or 10 to 25 psi over the maximum operating pressure, whichever is greater. The maximum operating pressure is determined by

Rapid Approximation Methods

taking into account normal variations owing to changes in vapor pressure, pump discharge head (considering shut-off cnditions), static head, and system pressure drops. Consideration should also be given to the desirability of basing design pressure on the highest pressure that might result from various operating failures or upsets. If exact data are lacking, the upper limit should be employed.

Assume that the maximum operating pressure equals normal operating pressure plus 25 psi.

For vessels operating under vacuum, design pressure of 15 psi gauge and full vacuum are specified.

Usually the thickness of the metal skin of a vessel is determined by the internal pressure. A modification of the hoop stress formula, which appears in the ASME code, is commonly employed.

$$t = PR/(SE - 0.6P) + C \qquad (C.9)$$

where t = thickness of shell, in.

P = design pressure, psi

E = joint efficiency, fraction (varies between 0.8 and 0.95, depending on the type of construction)

R = inside radius of shell, in.

C = corrosion allowance, in.

S = allowable working stress, psi

This formula and many variations of it are discussed on pp. 24-16 and 24-17 of [2].

The term "C," the corrosion allowance, is added to the thickness in order to give an additional amount of metal in order to compensate for wall weakness caused by corrosion. For situations that are known to be corrosive 0.25 in. are usually added; for noncorrosive streams an amount nearer 0.15 in. is used; and for known duties such as steam drums and air receivers allowances as low as 0.06 in. may be specified.

Structural rigidity is also a matter of some importance. A "good enough" schedule is as follows:

Inside diameter, in.	Minimum wall thickness, in.
42 and under	0.25
42 to 60	0.32
Over 60	0.38

3. **Conections, Openings, and Flanges.** All process vessels must have opnings and these must be connected to pipes, valves, and other vesels. The connections are usually short sections of pipe or tubin known as nozzles which are usually welded to the vessel and cannected to other devices by flanges. This is a very specialized ad detailed field of engineering which is usually not the province of the process engineer. [For a quick study the reader may consult p. 24-18 through 24-21 of 2.]

4. **Internally Insulated Vessels.** In the mechanical design of vessels operating above about 900°F, internal linings are often specified for heat insulation or erosion protection. In such cases the process design calculation will specify the maximum internal surface temperature of the lining, rather than the vessel temperature, which will be less.

The design temperature for the shell of a reactor vessel is determined by making a heat balance. The total heat loss to the atmosphere is assumed to be equal to the convection plus radiation loss from the external surface at 200°F. The maximum internal surface temperature is then determined by calculating the temperature drop between the maximum process fluid temperature and the inner surface, using the loss computed above, together with the convection and radiation coefficients between the fluid and the inner surface. All vessel internals will, of course, be designed to withstand the maximum process fluid temperature.

Drums

Vessels other than reactors and fractionating or absorbing towers are designated as drums. They are usually cylindrical in shape and either effect separation of entrained phases or provide surge capacity. Younger [28] lists factors that are useful in sizing such equipment.

1. **Dimensions and Position.** Any convenient length-to-diameter ratio in the economical range may be chosen. This usually lies between 2.5 and 5 to 1. As pressure is increased, the ratio becomes larger, but for most services a ratio of 3 to 1 is optimum. Settlers are specified at higher ratios, usually 4 and sometimes 5 to 1. As regards position, liquid drums are usually placed horizontally and gas-liquid separators vertically.

2. **Gravity Separators.** Frequently, the primary function of a vessel is to serve as a container in which solids and liquids may

Rapid Approximation Methods

settle, under the influence of gravity, away from gases and other liquids. The vessel size and shape is required to be such that there will be sufficient time for small drops or solid particles to move through the continuous phase, collect in a compartment of the vessel, and so be separated from the continuous phase. For a description of liquid-liquid separators, see p. 21-18 of [2]; for gas-liquid separators, see p. 18-82 of [2].

3. Liquid-Liquid or Liquid-Solid Separators. An attempt at designing a liquid-liquid or solid-liquid separator can start with the general relationship for a spherical particle moving through a continuous fluid under the force of gravity. This expression is given by equation 5-206 of [2].

$$V = 6.55(D_p/C)^{0.5}(\rho_s - \rho/\rho)^{0.5} \tag{C.10}$$

where V = particle velocity relative to the surrounding medium, ft/sec

D_p = particle diameter, ft

C = 24/particle Reynolds number

ρ_s = density of settling particle, lb m/cu. ft

ρ = density of continuous medium, lb m/cu. ft

For a water-organic liquid separation the actual size and shape of the particle will depend on the intensity of mixing and on the properties of the system involved, but for present purposes, let

D_p = 0.004 in.; C = 17; ρ_s = 62; ρ = 50.

Under these circumstances a settling velocity of 10 in./min is calculated. This value should never be exceeded.

To relate drum dimensions and stream flow rate it is assumed that the particle must fall (rise) through one half of the drum diameter in one drum nominal residence time. To be conservative one half of the drum verticle cross sectional area is used.

Then,

$$D = 0.22[(D_p/C)(\rho_s - \rho/\rho)]^{-0.25}Q^{0.5} \tag{C.11}$$

where D = drum diameter, ft

Q = process stream (both phases) flow rate, cu. ft/sec

length/diameter ratio = 4.

Equation C.11 gives results which appear reasonable when applied to flow rates that are intermediate in magnitude. Thus,

Table C-8 Approximate Sizes of Horizontal Drum Gravity Separators

Q (gal/min)	D (ft)	L (ft)
10	0.7	2.8
100	2.2	8.8
1000	7	28

For larger flows, which are commonly encountered in large chemical plants and petroleum refineries, the results appear too large.

4. Vapor-Liquid Separators. In vapor-liquid separators Eq. C.10 is also applicable. The equation is usually employed in a semiempirical fashion for entrainment calculations; thus,

$$V = k(\rho_s - \rho/\rho)^{0.5}$$

where V = allowable vapor velocity, ft/sec

$$k = 6.55(D_p/C)^{0.5}$$

The factor k is usually taken as constant instead of specifying a particle diameter and type of flow separately. Younger [28] suggests a value of k = 0.2 for vertical drums, if the height of the vapor space is over 3 ft. This corresponds to approximately D_p = 0.005 in. and C = 0.44.

For cases where low entrainment is desirable, as in compressor suction knockout drums, a value of k = 0.1 may be used. Severe entrainment will usually occur as values of k = 1.0 are approached. Disengaging area may be reduced by appropriate baffles and entrainment separators.

In the case of horizontal drums, a conservative basis for specifying area is to use the vertical cross-sectional area available above the liquid level in calculating allowable velocity (see Eq. C.11).

Rapid Approximation Methods

5. Surge Capacity. The volume required to provide sufficient holdup will depend upon several factors. The vessel should be sized on the basis of the one that results in the largest size. In cases where simultaneous separation of vapor occurs, the rules noted under "liquid separators" must also be considered.

6. Accumulators. Reflux drums and vessels containing product for intermediate storage are called accumulators. These vessels keep fluctuations from affecting smooth operation of the equipment which they feed. For normal conditions the holding time half-full should be 5 min for a reflux drum and 5 to 10 min for a product feeding to a subsequent tower.

Holdup times of 2 to 5 min half-full will be allowable for products moving to storage, but for drums upstream from furnaces or other services where failure of the flow would be undesirable, 30 min holding time half-full may be specified.

7. Knockout Drums. These vessels are used to separate liquid from vapor or gas streams, especially ahead of compressors. They should be designed so that the total surge capacity ahead of a compressor is sufficient to handle the flow of liquid that may become accidentally associated with the gas for a period of 10 to 20 min. The volume of such a drum should in no case be less than 10 times the gas volume per minute being passed through it.

REFERENCES

1. D.G. Jordan, Chemical Process Development, Interscience Publishers, John Wiley and Sons, Inc., New York, 1968.
2. Chemical Engineers' Handbook (R.H. Perry, C.H. Chilton, and S.D. Kirkpatrick, eds.), 4th ed., McGraw-Hill Book Co., Inc., New York, 1963.
3. H.F. Rase and M.H. Barrow, Project Engineering of Process Plants, John Wiley and Sons, Inc., New York, 1957.
4. L.L. Simpson, Chemical Engineering (Deskbook Issue), April 14, 1969, p. 167.
5. A.S.A. Code for Pressure Piping, American Standards Association, New York, 1951-1953.
6. D.Q. Kern, Process Heat Transfer, McGraw-Hill Book Co., Inc., New York, 1950.

7. A.S. Foust, L.A. Wenzel, C.W. Clump, L. Maus, and L.B. Anderson, Principles of Unit Operations, John Wiley and Sons, Inc., New York, 1960, pp. 223-246.

8. H. Ten Broeck, Industrial and Engineering Chemistry, 36, 64 (1944).

9. E. Hala, J. Pick, V. Fried, and O. Vilim, Vapor-Liquid Equilibrium, 2nd ed., Pergamon Press, Inc., New York, 1968.

10. R.C. Reid and T.K. Sherwood, The Properties of Gases and Liquids, 2nd ed., McGraw-Hill Book Co., New York, 1967.

11. C.J. Liddle, Chemical Engineering, p. 137, October 21, 1968.

12. J.C. Frank, G.R. Geyer, and H. Kehde, Chemical Engineering Progress, 65, 79 (1969).

13. W.S. Norman, Absorption, Distillation and Cooling Towers, John Wiley and Sons, Inc., New York, 1961.

14. H. Sawistowski and W. Smith, Mass Transfer Process Calculations, Interscience Publishers, John Wiley and Sons, Inc., New York, 1963.

15. R.E. Treybal, Liquid Extraction, 2nd ed., McGraw-Hill Book Co., New York, 1963.

16. A.W. Francis, Liquid-Liquid Equilibriums, Interscience Publishers, John Wiley and Sons, Inc., New York, 1963.

17. E.G. Scheibel, Industrial and Engineering Chemistry, 46, 43 (1954).

18. S.M. Walas, Reaction Kinetics for Chemical Engineers, McGraw-Hill Book Co., New York, 1959.

19. O.A. Hougen and K.M. Watson, Chemical Process Principles, 3rd ed., John Wiley and Sons, Inc., New York, 1947.

20. J. Happel and L. Kramer, Symposium on Applied Reaction Kinetics, June 13, 1966, American Chemical Society, Washington, D.C., 1967, p. 108.

21. O. Levenspiel, Chemical Reaction Engineering, John Wiley and Sons, Inc., New York, 1962.

22. C.N. Satterfield and T.K. Sherwood, The Role of Diffusion in Catalysis, Addison-Wesley Publishing Co., Inc., Reading, Mass., 1963.

23. L. Lapidus, Industrial and Engineering Chemistry, 49, 1000 (1957).

24. E.E. Petersen, Chemical Reaction Analysis, Prentice-Hall, Inc., Englewood Cliffs, New Jersey, 1965.
25. D. Kuni and O. Levenspiel, Fluidization Engineering, John Wiley and Sons, Inc., New York, 1969.
26. ASME Code for Unfired Pressure Vessels, American Society of Mechanical Engineers, New York, 1962.
27. L.E. Brownell and E.H. Young, Process Equipment Design: Vessel Design, John Wiley and Sons, Inc., New York, 1959.
28. A.H. Younger, Chemical Engineering, 62, 201 (1955).

INDEX

A

Abbott, L., 52, 89
Absorption
  of gases, 396-407, 463
  columns for, 396
    annual charges on, 397, 400
    design of
      approximate, 398-401, 463
      optimum, 396-407
    distillation systems in, 397, 401, 404
    economics of, 396-407
  factor, 398, 403
  heat exchangers in, 398
  total annual cost of, 402, 405
Academic work in chemistry, 22
Accounting
  depreciation charge for, 39-43, 92
  procedures, 36
Accounts receivable, 36, 271, 275
Acetylene, 14
Acid
  hydrochloric, 14
  nitric, 14
  sulfuric, 10, 13, 290-302
Acquisition of other companies, 25

Acrylic
  fibers, 19
  resins, 19
  rubbers, 19
Acrylonitrile, 19, 20
Acyclic systems in dynamic programming, 203-210
Advantage
  economic, 5
  technological, 5
Agitation (agitators) of vessels, 229
Air
  compression of, 455
  coolers, design of, 376-380, 459
  pollution, 14, 32
  as raw material, 10
Alcohol
  ethyl, 327-338, 403-407
  isopropyl, 291-303
  methyl, 236-240
Algorithm, the Simplex method, 198, 202
Allen, R. D. G., 72
Alternatives, choices between, 56, 61, 64, 79, 140, 143, 447
Alum, 10
Alumina, 10

489

Aluminum metal, 10
Ammonia, 13
Analogy, design by, 290
Analysis, economic,
  of financial investments
  of financial statements and securities, 272
  preliminary, 35-86, 87-148, 213, 289
  of projects (overall), 289-348
  theoretical and mathematical methods for, 149-211
  venture profit, 54
  venture worth, 91
Andersen, L. B., 352, 457
Approach, fractional, of stream temperatures, 363-376
Approximation, design, procedure for, 349, Appendix C
  of crystallizers (continuous), 468
  of distillation columns, 460
  of dryers, rotary continuous, 470
  of filters, continuous, 469
  of gas absorption columns, 463
  of heat transfer equipment, 456
  of pressure vessels, 479
  of pumps, compressors, and piping, 452
  of reactors, chemical, 472
  of solvent extraction systems, 466
Area
  cross sectional
    in columns, 173, 386, 398, 403
    in dryers, 410, 471
    in piping, 187, 351, 456
  for heat transfer, 54, 175, 358-377, 457
  extended surface, 376
Aries, R. S., 55

Ascents, steepest, method of, 180-184
A.S.M.E. Code for unfired pressure vessels, 479
Assets of corporations, 270-276
Atomic energy, 14
Attractiveness, economic, of a venture, 24, 56, 61, 91, 94, 101, 131, 136, 142, 289-348
  effect of inflation on, 95
Azeotropes, 403, 461

B

Balances
  economic, 53, 68, 289, 317, 324, 335
    capital investment, 46, 49, 111
    chemical reactor, 417
    distillation, 171, 394
    drying, 409
    fluid flow, 351-358
    gas absorption, 398
    heat transfer, 69-72, 357-381
    operating costs, 73
  energy, 294, 349
  heat, 175
  mass, 215, 294, 324, 330, 349, 409, 419
  water, in dryers, 409
Balance sheet, financial, 265, 275
Ball mills, cost of, 229
Banks, 26, 263
Barrow, M. H., 452, 479
Batch production of chemicals, 16
Battery limits of chemical plants, 241
Bauman, H. C., 90, 220, 222, 243
Bauxite as raw material, 10
Bazovsky, I., 322
Bell, J. E., 323
Bellman, R., 204
Benefit, economic, 263
Berl saddles, 399

INDEX 491

Bertetti, J. W., 318
Beveridge, C. S. G., 163, 181, 198, 202
Bleed (purge), 329
Board of directors, 26
Boas, A. H., 152, 163, 179, 181
Bodman, S. W., 179, 350
Boilers
  approximate design of, 459
  cost of, 228, 304
  waste heat, 68, 459
Bonds
  of corporations, 26, 264, 269-276
  of governments, 39, 269-276, 449
Borrowing of money, 7, 26, 88, 263, 269, 273, 278, 281
Bottoms from distillation columns, 394, 421
Braca, R. M., 352
Bresler, S. A., 336
Brooke, M., 378
Brownell, L. E., 479
Butadiene, 13, 75-79, 198-202
Butane, 10, 454
Butylene, 10, 198 202
By-products, 252, 417

C

Calcium carbide, 10
Calcium carbonate as raw material, 10
Calcium oxide, 10
Calculus, differential, 151-157
Canada, chemical companies in, 5
Capacity of production, 17, 73
Capital
  amount of new, 6, 31, 45, 93, 98, 258, 262, 449
  appreciation of, 88
  availability of, 79, 143, 263

[capital]
  charges related to, 64, 148, 176, 187, 322, 362, 367, 378, 385, 397, 413, 424
  incremental, 52
  interest on, 89, 263
  investment of, 6, 14, 31, 36, 45, 50, 93, 101, 212, 259, 261, 265, 269, 272, 279
  limitations on, 46, 79
  new, sources of, 6, 31
  productivity of, 49
  rate of earnings on, 92, 269-276
  recovery of, 39, 42, 90, 116, 266, 269, 273, 280, 449
  requirements, 4, 46
  return on, 45, 52, 55, 91, 269-276, 360
  spending, 7
  structure, of chemical companies, 27
  undepreciated, 132, 135
  use of, 31
  working, 36, 50, 55, 63, 79, 93, 100, 129, 135, 143, 253, 264
Carbon, 10
Carbon dioxide, absorption from, 403-417
Carbon monoxide, 14
Cash, 36, 47, 271
Cash flow
  of depreciation, 30, 42
  discounted, 96
  future, 94
    effect of inflation on, 94, 447
Cash position, diagram of, 47
Catalyst, 198, 260, 319
Cement, 10
Centrifuges, cost of, 228
Challenger, adverse minimum, for replacement, 309-317
Charges (see costs)
Chart
  air-water humidity, 410

[chart]
  breakeven, 74-78
  exclusion, 13
Chemical(s)
  basic, 13
  business, 5
  companies, 2, 3, 27
    changes in nature of, 3, 15, 32
    financial organization of, 26, 30
    foreign, in U.S., 5, 21, 447
    profitability of, 15, 22, 28, 46, 447
    size of, 27
  engineer, 3, 5
  Engineering magazine, 217
  equipment
    cost of, 6, 17, 21, 174, 188, 216-232, 242
    depreciation of, 38-43, 92, 96, 105, 113, 115-127
    design of, 170-179, 349-449, 451-485
    installation of, 218, 223-232
    maintenance of, 63, 251
    optimization of, 53, 148-210, 349-449
    replacement of, 35, 38, 62, 309-317
  heavy, 13
  industry
    characteristics of, 3-9
    definition of, 1, 2, 3
    employment in, 15
    future of, 15, 31
    growth rate of, 13, 31
    organic, 21, 32
    problems of, 4, 14, 21, 31, 46, 276, 281
    research in, 21-26, 33
    spending of money in, 31
    strengths of, 33
  manufacture, 1, 2, 3, 10, 12, 15
    cost of, 12, 14, 20, 46, 73-78, 102, 110, 244-254, 261, 300, 304, 324, 417

[chemical(s)]
  [manufacture]
    energy for, 14, 32, 250
    raw materials for, 10, 15, 21, 32
    pharmaceutical, 47
    sales, 12
    specifications for, 13
Chemistry, 1-3, 22, 33
Chemists, 1
Chilton, C. H., 216, 244, 252
Chlorine, 13
Choice between plans, 139
Clump, C. W., 352, 457
Coal
  as fuel, 15
  gasification of, 15
  as raw material, 10, 15
Cocurrent, 408
Coefficient
  algebraic, 196
  heat transfer, 69, 175, 358-384, 387, 391, 411, 458, 471
  of thermal conductivity, 381
Colburn, A. P., 385
Columns
  gas absorption
    capital cost of, 224, 400
    design of, 396, 463
    operating cost of, 400
    optimization of, 396-407
  distillation
    capital cost of, 169-179, 224
    design of, 169-179, 385-396, 462
    operating cost of, 173-179, 385-396
    optimization of, 169-179, 385-396
  extraction (liquid-liquid)
    capital cost of, 224
    design of, 467
    optimization of, 407
  packed (packing for), cost of
    Pall rings, 226
    Raschig rings, 226
Commerce, articles of, 1

INDEX 493

Commercialization, 4, 9, 22, 25
Competition
  between companies, 4, 6, 14, 21, 25, 34
  between products, 4, 6, 13, 23, 47
  from new processes, 135
  in equipment sales, 4
  international, 4, 6, 14, 21, 34, 447
Component parts of a chemical plant, 349-433
Composition, of process streams, 296, 330
Compounding of money
  annually, 89, 435
  continuously, 89
Compression of gases
  adiabatic, 152, 158, 181, 185-195, 206
  cost of, 188, 190, 454
  efficiency of, 188
  fuel for, 188
  isothermal, 455
  optimization of
    three stage, 154, 158, 181, 206
    two stage, 65, 152
  power for, 154, 156, 158, 188, 190, 454
  pressures in gas pipe lines, 185-195
  specific heat ratios for, 454
  work of, 152, 158, 181, 206
    contours of constant, 182
Compressors for gases
  approximate design of, 453
  centrifugal, 189
  cost of, 65, 68, 188, 190
  cooling water requirements for, 455
  mechanical efficiency of, 455
  reciprocating, 189
Computation
  continuous, 65

[computation]
  of physical properties, 461
  point-by-point, 65
Computers, use of, in economic evaluation, 149, 197, 204, 336
Condensers
  air cooled, 376
  approximate design of, 458
  cost of, 175, 376, 386
  water cooled, 71, 175, 378
Conditions for a maximum (minimum), 153, 184
Considerations, overall, in project analysis, 289-348
Constraints, mathematical, 152, 157, 162, 185, 195
Construction
  cost of, 6, 21, 25
  graphical, 167
  materials of, 215, 479
  spending for, 35
  time for, 103
Continuous
  compounding of money, 89
  operation of chemical plants, 16
Control
  automatic, of chemical plants, 16
  of chemical reactions, 7, 15, 422
  of industry by government, 26
Conversion in chemical reactions, 325, 417
Convection
  forced, in heat transfer, 357, 458
  natural, in heat transfer, 382
Coolers, 365-380, 458
Coordinates
  logarithmic, 216, 361
  rectangular, 105
Cornell, D., 399
Corporations, financial aspects of, 5, 26, 43, 259-282
Correlations
  Gilliland's, 171, 388
  Guthrie's, 174
  of costs, 222-232

Corrosion
  in absorption columns, 302
  in chemical reactors, 418, 474
  in heat exchangers, 359
  in pressure vessels, 479
Cost(s)
  advantage, 5
  of air cooled exchangers, 377-380
  capital related (see, capital, charges, related to)
  of chemicals, 5, 13, 16, 36, 41, 47, 96, 98, 104, 110, 244, 300, 324
  of chemical reaction systems, 417-428
  of compressing gases, 187, 453
  of construction, 17, 25, 447
  of dismantling, 93
  of distillation, 71, 385-396, 401, 417, 425
    factor for, 388
  of drying, 408-417
  effect of inflation on, 447
  of energy, 15, 63, 250
  of engineering, 5, 103, 237
  equation for, 64, 139, 178, 191, 218, 349-428
  of equipment, 17, 52, 216-231, 299, 349-428
  estimation of, 213-257, 293
    capital investment, 46, 80, 103, 213-244
    manufacturing, 41, 46, 96, 103, 244-254
    precision of, 80, 222, 243
    total plant, 103, 216, 232
  fixed, 79
  incremental, 52, 54, 64-71, 169, 361-376
  indirect, 247-252
  inflation of, 32, 95, 250, 446
  of insulation, 381
  of insurance, 251
  of interest, 263

[cost(s)]
  of labor (and labor related), 63, 232, 247-252, 426, 447
  of maintenance, 63, 251
  of manufacturing (production), 16, 32, 36, 41, 45, 63, 73-79, 97, 103, 110, 129, 139, 212, 244, 261, 300, 451
    superproduction, 73-79
  minimum acceptable return as, 44, 53
  minimum average annual, 64, 158
  operating, 32, 36, 63, 70, 149, 349
  preliminary estimates of, 214, 232, 243
  of process variables, 168-172, 196
  of pumping
    gases, 187-195
    liquids, 351-358, 378
  of raw materials, 17, 32, 36, 63, 260, 267, 424, 447
  of steam, 71, 250, 305, 387, 394, 413
  total yearly, before taxes, 64, 138, 150, 154, 158, 354, 360, 378, 381, 385, 402, 415, 427
  of transportation, 5, 11, 13
  unit, 168, 172, 175, 196
  unrecovered, 117
  of utilities, 250
  of water cooling, 378, 394, 422, 425
Cottle, S., 272
Courant, R., 154
Courter, M. L., 327
Credit
  financial, 264
  tax, 96-99, 102, 116-129, 133-139, 264, 280

INDEX 495

Criterion
  for investment attractiveness, 4, 44-79, 91-115, 129-133, 136, 138
  for shutdown (termination of operations), 133
Crowe, C. M., 336
Crystallizers
  cost of, 227
  design of, 468
Cumene, 12
Currency, 445
Customers
  sales to, 43
  work with, 24
Cycle of operations, 203, 317
Cyclohexane, 13

D

Damron, R. D., 376
Data
  for cost estimation, 215-232, 242, 250
  for process design, 454-485
DCF method of economic analysis, 112
Debentures, sale of, 265
Debt, financial, 27, 31, 45, 263, 271, 281
Defenders, adverse minimum, in machinery replacement, 310
Dehydrogenation, 170, 198
Delivery of chemical shipments, 13
Density, 352, 401, 404
Deposit of money, 89, 269, 278, 280
Depreciation
  calculation of
    declining balance, 38, 97, 99, 105, 107, 110, 116-127
    straight line, 38, 50, 55, 97, 102, 110, 116-127, 251, 311, 313

[depreciation]
  [calculation of]
    sum-of-years-digits, 38, 97, 110, 116-127
  cash flow, 43
  change in method, 117, 120-127
  charge for accounting, 39, 45, 55, 92
  in DCF method, 114
  effect on net profit, 47
  effect on payout time, 49, 65
  effect on taxation, 38, 45, 55, 79, 93, 99, 116, 120-128, 138, 176
  factors, 97, 99, 116-127
  of equipment, 36, 40, 45, 56, 63, 71
  in manufacturing cost, 36, 41, 46, 63, 104
  optimum method of, 116, 120-127
  special rates (accelerated), 38, 97, 110, 116-127
  summation and discounting of, 96, 99, 102, 108, 113
  time, 38, 94, 132-135, 266
  in venture profit, 55, 63, 93
  in venture worth, 94, 99
Derivative
  cross, 157
  first, 66, 152
  negative, 155
  non-vanishing, 152
  partial, 154, 181, 196
  second, 152, 156
Design
  optimum, criteria for, 62
  preliminary, 174, 215, 291, 451
  procedures, rapid approximate, 451-485
Deterioration of machinery, 40, 309
Development
  of chemical processes, 4, 18, 21, 24

[development]
  projects, 18, 24
  work, analysis by venture worth, 277
Diagram
  cash balance position, 48
  decision tree, 208
  McCabe-Thiele, 393, 404
  venture worth, components of, 98
  venture worth, sensitivity of, 110
Diameter
  of columns
    absorption, 398, 404, 465
    distillation, 391, 462
  of dryers, 410-417
  of piping, 185-195, 350-357, 381, 456
Differentiation
  analytical, 66, 72, 151, 168, 184
  graphical, 169
  numerical, 169
  partial, 72, 154, 168, 184, 351
Disbursements, 58
Discount
  of prices, 14
  rate, 92
Discounting of money
  annually, 90, 96, 436
  continuous, 90, 441
Discoveries, chemical, 22
Distance, optimum, in gas pipe lines, 185-195
Distillation, fractional
  columns
    approximate design of, 171-179, 385-396, 459-463
    cost of, 71, 174, 224, 385-396, 459-463
  of benzene, toluene, and xylene, 396
  with chemical reactor systems, 421

[distillation, fractional]
  cost of, 393
  in gas absorption systems, 396
  of isopropanol, 291
  of methanol, 394
  optimum bottoms concentration, 394
  of styrene, optimization of, 161, 169-179, 394
Dividends
  on corporate stock, 26, 40, 264, 279, 282
  of the Dow Chemical Company, 11, 28, 271
Dodd, D. L., 272
Dodge, W. H., 381
Dryers
  continuous rotary, 408-417
  cost of, 227, 408, 412
  design of, 408-417, 470
  dust losses in, 409
  gas and solids flow in, 410-417, 471
  heat transfer in, 410, 417, 470
  optimization of, 408-417
Dryden, C. E., 338
Drying, cost of, 408, 415
Duffin, R. J., 157
Dynamic programming, 203-210

E

Earnings
  corporate, 38, 40, 92, 264, 275
  after taxes, 40, 281
  average rate of, 91, 93, 276
  of the Dow Chemical Company, 29, 272
  net, 40, 272
  operating, 40, 360-367
Economic
  balance, 53, 64, 69, 72, 349-428
  criterion, 4, 44, 48-79, 91-116
  design, 289-338, 349-428

INDEX 497

[economic]
  evaluation, 4, 213, 277
  studies, 4, 267
Economy
  inflationary, 95, 264
  profit-oriented, 26
  state of, 264
Efficiency
  of calculation, 164
  of distillation trays, 386, 462
  of extraction plates, 467
  of extraction systems, 468
  of furnaces, 371
  mechanical, of gas compressors, 188, 455
  of metal joints, 481
  operating of plants, 35
  of time, 317
Emergencies, 17, 317
Employees in the chemical industry, 15
Energy
  atomic, 15
  balance, 215, 294, 349
  in compression, 66, 152, 181
  cost of, 14, 32
  electrical, 7, 14, 16, 250
  management of, 14
  shortage of, 14, 32
  use of, 14
Engineering
  chemical, 4
  design, 5
  research, 9
Engineers, 3
Equations
  algebraic, 72, 75, 154, 186, 192, 196
  differential, 153, 155, 168, 184
  economic evaluation
    cash position, 47
    DCF, 116
    net profit, 47
    payout time, 49
    return on investment, 48

[equations]
  [economic evaluation]
    total yearly cost, 64, 140
    venture profit, 55, 92
    venture worth, 94, 96, 135
  empirical, 75, 400
  Fenske's, 171
  with Lagrangian multiplier, 184
  linear, 196
  modified, for graphical differentiation, 168
  solution of, by trial and error, 169
Equilibrium
  chemical, 52
  phase
    in distillation, 52, 392, 460
    in gas absorption, 52, 399, 403
    in liquid extraction, 52, 466
  thermal, 52
Equipment, chemical, used, 91
Equity of stockholders, 271, 281
Estimates (estimation)
  of capital investment, 101, 213, 216-244
  of equipment cost, 216-232
  of manufacturing cost, 244-254
  of plant life, 39, 45, 260, 266, 280
  precision of, 243
  preliminary, 3, 213
  of sales price, 104
  of termination time, 133
  of venture worth, 94, 103
  of working capital, 253
Ethane from natural gas, 10
Ethyl benzene distillation, from styrene, 170-180
Ethylene, 10, 13, 17, 142, 327, 454
Evaluation methods
  economic, 35-82, 89-143, 149-210, 213-255, 259-282, 289-338, 349-428, 445
  technical, 35, 213-255, 289-338, 349-428, 451-485

Evaporator, tubes for, 315
Experience of process operations, 35
Expenses (see costs)
Extraction (liquid-liquid)
  columns for, 52, 225, 466
    approximate design of, 466
    economic evaluation of, 51, 407
  extract from, 407
  batch simulation of, 467
  mechanically aided, 468
  optimization of, 407
  phase equilibria for, 407, 466
  raffinate from, 466
  separation factor for, 407

F

Factor
  absorption, 398-403
  capital recovery, 96, 112, 140, 270, 311, 437
  charge, capital investment related, 64, 142, 352, 360, 372
  cost, in distillation, 388-393
  discounted tax credit, 121-129, 142
  friction, in pipes, 186, 352
  of Lang, in cost estimation, 234
  separation, 407
  single payment compound amount, 90, 435
  single payment present worth, 435
  sinking fund deposit, 92, 269, 279, 311, 437, 449
  uniform annual series compound, 310, 437
  uniform annual series present worth, 96, 100, 438
Failure, financial, 46, 133, 268

Fans
  in air cooling, 376-380
  cost of, 227
  power consumption of, 376, 397
Fees
  contingency, 237
  engineering, 3, 237
Fibers, synthetic, 7
Fibonacci numbers, 165
Films, 7
Filters
  cost of, 227
  design of, 469
Filtration, 407, 469
Financial
  analysis, 272
  investment, 33, 80, 213
    return on, 35
    maximization of, 35
  statements, annual, 272, 275
  structure of chemical companies, 26
Finite difference, calculus of, 435
Firms (business), engineering and construction, 3
Flanges for pressure vessels, 482
Flooding velocity in gas absorbers, 398
Flow
  of cash (money), 30, 41, 92, 96, 269, 281
  cocurrent, 408
  countercurrent, 362, 408, 457
  of fluids, 350-357
Flow sheet of processes, 215, 237, 291, 327, 421
Fluid mechanics, equations of, 186
Formula, hoop stress, 481
Fouling of heat transfer surface, 359
Foust, A. S., 352, 457

INDEX 499

Fraction
  of chemical reacted, 325, 417
  of gas absorbed, 399, 404
  of product recovered, 170, 421
  of solute recovered, 399, 407
Fractionation (see distillation)
Francis, A. W., 466
Frank, J. C., 170, 179, 462
Fried, V., 460
Friedman, S. J., 411, 416
Fuel
  availability of, 14
  cost of, 15, 250, 371, 383
    coal, 15
    gas, 15
    oil, 15, 250
Functions
  cost, 170, 179, 350
  mathematical, manipulation of, 152, 155, 180, 185, 196
  maximum (minimum) of, 153-158, 185, 195
  objective, 163, 196, 204
    gradient of, 163, 180
  profit, 73
  unimodal, 104-168
Funds
  corporate, 22, 30, 40, 263, 269, 279
  sinking, 92, 264, 269, 281
Furlow, R. H., 338
Furnaces, industrial, 53
  cost of, 228
  design of, 459
Future
  of the chemical industry, 31
  discounting of cash flow, 94
  predictions of, 97, 278
  sums, discounted to the present, 94, 120, 130, 135, 139

G

Gas
  absorption of material from, 396-407, 463
  natural, 14, 21
    as fuel, 14
    pipeline transportation of, 185-195
      total yearly cost of, 191
    as raw material, 10, 15, 21
  solute, 396
Gases
  stack, recovery of heat from, 53, 68
  specific heat ratios of, 454
Gass, S. I., 198
Geometric programming, 157-162
Geyer, G. R., 170, 179, 462
Gilliland, E. R., 171, 388, 392, 462
Glauz, R. L., 251
Golden mean, 164
Government
  regulation by, 26, 265
  taxation by, 37, 116, 265, 280
Gradient of functions, 180
Gradient methods for optimization, 179-183
Graham, B., 272
Grant, E. L., 56, 59, 90, 309, 318
Growth
  of a company, 26
  of the chemical industry, 4, 6, 25, 31
  of money, 89
Guthrie, K. M., 174, 219-228, 236, 240, 351

H

Hadley, G., 198

Hala, E., 460
Hammer mills, cost of, 229
Happel, J., 55, 91, 97, 232, 236, 239, 294, 352, 435, 475
Head, differential, in pumping, 352-357
Heat
  capacity, 363
  cost of, 15, 176, 250, 350, 357-384, 458
  for distillation, 170-179
  for drying, 408-417, 470
  exchangers
    air cooled, 226, 376-380
    cooler combination, 365-375
    cost of, 52, 68, 175, 223, 350, 361-380, 413, 458
    design of, 357-360, 457
    economic evaluation of, 51, 68, 360-374
    fouling in, 359, 457
    manufacture of, 359, 457
    operating costs of, 70, 360, 366
  flow of, resistance to, 376, 382
  recovery of, 53, 68, 80, 357-376, 469
  of vaporization, 175, 387, 391
Heneghan, W. F., 90
Henley, E. J., 336
Horsepower
  for compressors, 188, 400
  for pumps, 352-357
Hougen, O. A., 475
Humidity of air in drying, 409-415
Hvizdos, L. J., 154
Hydrocarbons, 10
Hydrogen
  as raw material, 10
  specific heat ratio of, 454
Hydrogen cyanide, 14, 19

I

Ilmenite as raw material, 10
Imports, 32

Improvements of products and processes, 22
Income
  deductions from, 38
  gross, 36, 39, 41, 47, 53, 63, 73, 93, 102, 130, 138, 263, 272
  effect of production rate on, 73
  tax, 37, 263, 271
  tax credit, 116, 133, 138
Indices
  economic, 47, 91, 263
  for equipment costs
    Chemical Engineering Plant Cost, 218
    Engineering News Record Cost, 216
    Marshall and Swift, 217
    wholesale price index, 218, 445
Inequalities in linear programming, 196
Inferiority, operating, of machinery, 309-315
Inflation, monetary, 91, 95, 445
  effect on costs, 32, 39, 95
  effect on interest rates, 39
  effect on planning, 32, 95
  historical record of, 39, 95, 445
Insecticides, research on, 24
Instrumentation, cost of, 235, 239
Insulation, optimum thickness of, 357, 380-384
Insurance, cost of, 251
Integration in chemical companies, 9, 11
Interest
  average rate of, 40, 92, 95, 139, 269, 435
  basic rate of, 263
  on bonds, 26, 39, 279, 449
  calculation of, 264
  compound, 40, 88, 268, 278, 435

INDEX
501

[interest]
 [compound]
  derivation of formulas for, 435-443
  effect of inflation on, 95, 447
  effective, 264
  incremental payments, 100
  on loans, 25, 89, 264, 270
  payments as business expense, 25, 89, 263, 271, 274
  prime rate, 265
  on salvage value, 136
  on undepreciated capital investment, 137
  on working capital, 135-139
Intermediates, chemical, 12
Internal Revenue Service, U.S. office of, 38, 115, 266
International
 companies, 4
 competition, 4
Inventories, 317
Investment
 average value of, 50
 capital, 16, 36, 46, 60, 79
 evaluation of, 44, 50, 214, 234
 fixed charge on, 64, 68, 73, 140, 176, 187
 incremental, 52, 60, 142, 169, 175, 223-230, 349, 387
  return on, 52, 56, 139, 176
 inflation of, 95, 265, 449
 initial, 46, 94, 99, 113, 279
 maximum, 80
 optional, 80
 recovery of, 40, 49, 92, 117, 129, 259-282
  effect of inflation on, 95, 265, 449
 restrictions on, 80
 return on, 4, 44, 48, 50, 54, 99, 141, 360, 447
 types of, 43
 undepreciated, 118, 135
Investor, 39, 52, 279

Ireson, W. G., 309, 318

J

Jaedicke, R. K., 272
James, W., 96
Jenkins, J. D., 469
Jelen, F. C., 316
Jordan, D. G., 35, 92, 103, 222, 230, 244, 278, 288, 319, 324, 418, 448, 451, 468, 475

K

Kapfer, W. H., 114
Karr, A. E., 468
Kehde, H., 170, 179, 462
Kermode, R. I., 157
Kern, D. Q., 360, 378, 457, 460
Key accounts in cost estimation, 232
Kim, J. C., 403
Kinetics of chemical reactions, 417-428, 472-479
Knowledge, scientific, 3, 23
Kramer, L., 475
Kuni, D., 479
Kuo, M. T., 336

L

Labor in chemical plants, 7, 15, 21, 71
 costs related to, 247, 294
Laboratory control, cost of, 249
Lagrangian multiplier, 184-195, 351
Land for chemical plants, 241
 value of, 93
Lang, H. J., 234, 236, 240
Lapidus, L., 478

Lavine, I., 243
Leaching, 407
Length
  of piping (equivalent), 352
  of processing runs, 317
Levenspiel, O., 476, 479
Liabilities, monetary, 26, 270, 274
Liddle, C. J., 462
Life
  of equipment, 39, 117, 138, 309, 360
  human, 1, 33
  of plants, 40, 92, 99, 117, 129, 141, 261, 309, 447
  of processes, 136
  of projects, 40, 45, 91, 266, 279
Line
  adiabatic saturation, 410
  equilibrium, 398
  operating, 398
List, check
  for cost estimation, 215
  of equipment, 235, 299, 333
Loans
  of money, 26, 263
  interest on, 89, 263
  maturity of, 263
Location of chemical plants, 14
Loss
  of money, 132
  tax, 133

M

Machinery and Allied Products Institute (MAPI), 313
Maintenance
  cost of, 36, 63, 134, 139, 142, 176, 187, 251, 314, 353, 361
  personnel, 15
Management, financial, 44, 53, 265

Manpower, 15, 17
Marchitto, M., 13
Margin of profit, 18, 41, 47, 53, 98, 300
Marginal operation of plants, 129
Market
  development of, 18, 23
  research, 4, 18
Marketing of chemicals, 18, 45, 246
Markets
  captive, 246
  open, 246
  securities, 27
Marshall, W. R., 411, 416
Materials of construction, 215, 298, 302
Materials
  handling of, 15, 17
  raw
    availability of, 10, 32
    cost of, 5, 10, 20, 32, 246, 261, 267, 300, 324, 334
    not recovered, 71, 170, 394
Mathematical
  advanced techniques, multivariables, 149-210
  area elimination, 163-167
  dynamic programming, 203-210
  Fibonacci search, 164-167
  geometric programming, 157-162
  golden mean search, 164-167
  justification for, 150
  Lagrangian multiplier, 184-195
  linear programming, 195-202
  partial differentiation, 154, 167-179
  use in industry, 150
  derivation of compound interest formulas, 435-444
  models, 149, 196
  simple techniques, one variable, 151

INDEX 503

Matrices, algebra of, 203
Maus, L., 352, 457
Maximum(s), 52, 60, 151, 155, 180, 184
McAdams, W. H., 370, 376, 382, 384
McCabe, W. L., 404
McCurdy, R. C., 21, 217
Melamine, 23
Metals, tensile strength of, 189, 481
Methane, specific heat ratio of, 454
Methanol
  distillation of, 237, 394
  recovery of, 237
Mickley, H. S., 66, 72, 435
Minimum(s), 72, 152, 155, 184, 194, 207, 408, 413, 427
  adverse, 310
Moisture in solids (drying), 408, 470
Molstad, M. C., 403
Money
  acquisition of, 26
  borrowed, 17, 26, 89, 264
  flow diagram of, 41
  income, 36, 41, 47, 63, 260
  purchasing power of, 445
  reward of, 26
  spending and recovery of, 49, 89
  time value of, 89, 447
Motor, electrical, 188, 352

N

Nelson, C. R., 327
Nitric acid, 14
Nitrogen, as raw material, 10
Nomograph for heat transfer, 364, 457
Nonlinear mathematical function, 197

Norden, R. B., 218
Norman, W. L., 463
Nuclear reactor, 15
Nylon, 23

O

Obsolescence
  economic, 40, 267
  technical, 38, 40, 267
O'Connell, H. E., 462
Officers, financial, of a corporation, 46, 134
Offsite, for plants, cost of, 242
Oil
  as fuel
    availability of, 14
    cost of, 15, 250
  as raw material, 5
  gasification of, 15
Optimality, principle of, 204
Optimum
  designs
    for absorption columns, 396-407
    for chemical plants, 289-308, 317-338
    for chemical reactors, 417-428
    for distillation columns, 169-179, 385-396
    for extraction columns, 407
    for heat exchangers, 68, 357-380
    for insulation, 68, 380-384
    for pipe lines, 68, 185-195, 350-357
    for rotary dryers, 408-417
  finding methods, 151-209
  general formulation of, 151
  points, 52, 65, 71, 152-157
  rate of production, 75
Overhead in plants, cost of, 249
Oxygen as raw material, 10

P

Packaging, cost of, 252
Painting, cost of, 314
Parts, component, of a chemical plant, 349-428, 450-485
Patents, 5
Peters, M. S., 38, 90, 202, 220, 243
Petersen, E. E., 478
Peterson, E. L., 157
Petroleum companies in the chemical industry, 5
Petroleum fractions as raw materials
  fuel, 14
  gas oil, 10
  naphtha, 10
  price of, 15, 21, 250
Phases in chemical reactors, 472
Phenol, 4, 20, 24
Phosphate rock, 10
Pick, J., 460
Pierce, D. E., 252
Pipe(ing)
  friction loss in, 186, 350, 456
  installed cost of, 189, 349, 351
  optimum design of, 189, 350, 456
  optimum diameter of, 354
  optimum velocity in, 356
  wall thickness of, 189
  yearly cost of, 190, 354
Plants, chemical
  business operation of, 51, 62, 72, 133, 259-283, 447
  cost of, 232-244
  design of, 62, 213-255, 290-303, 324-338, 451-485
  depreciation of (see depreciation)
  increasing capacity of, 304
  life of (see life)
  location of, 11
  maintenance of, 251

[plants, chemical]
  manning of, 248
  obsolescence of, 38, 267
  optimization of, 72, 289-308, 317-338
  petrochemical, 11
  production rate of, 17, 36, 72
  protection of, 249
  return on investment from, 43, 48, 111, 269
  sales from, 2, 4, 6, 13, 18, 36, 45, 72, 99, 129, 261
  shut down of, 92, 133
  venture worth of, 91-111, 303
  very large, 17
Plastics, 7
Plates
  for absorption, 225, 396, 465
  for distillation, 174, 225, 385, 462
  for extraction, 225, 467
  cost of, 174, 225
  efficiency of, 386, 462, 465, 467
Point
  breakeven, 74
  of maximum yearly profit, 74
  of minimum cost, 74
  optimum (see optimum)
  saddle, 155
Poisoning of catalysts, 319
Policies
  of corporations, 18, 40, 46, 259-283
  of governments, 32, 265
Pollution of air and water, 7, 14, 25, 32, 396
Polyethylene, 23
Polynomial, generalized, 159
Popper, H., 244
Position, cash, 47, 271
Power, electrical, cost of, 14, 250
  in air cooling, 377-380
  in fluid pumping, 352

[power, electrical, cost of]
  in gas absorption, 71, 397, 401
  in gas compression, 66, 188, 454
  in gas transport, 188
Prediction of future business, 97, 261, 266
Preliminary cost estimates, 35, 213, 289
Pressure
  in chemical processes, 328
  in gas compression, 66, 153, 181, 186, 206
    optimum interstage, 66, 155, 160, 181, 206
  in gas pumping systems, 185-195
Pressure drop
  in air cooling, 377-380
  in fluid pumping, 188-195
  in gas absorption, 400
  in heat exchangers, 358
  optimum, in gas pipe lines, 185-195
  through column packing, 463
  through control valves, 453
  through distillation plates, 462
Prices
  of raw materials, 10
  of chemicals
    current, 7, 34
    decline of, 7, 97, 105, 129
    list, 14
    purchase, 14
    sales, 4, 13-18, 36, 41, 47, 96, 104, 129, 261, 266
Principal
  of bonds, 26, 39
  of money, 264
Principles, chemical, 1
Process(es), chemical, 1, 4, 22, 135, 149, 267, 277
  design of, 149, 277, 289
  economics of, 22, 135, 266, 276, 317

[process(es), chemical]
  improvement of, 4, 22, 250
  poorly defined, 179
  units, components of, 138
Product, gross national, 7
Products, chemical
  competition between, 4, 262
  development of, 4, 9, 22, 32, 277
  introduction of, 4, 9, 22, 32, 268
  older (modification of), 24, 32, 260, 268
  profit margin on, 32, 74
  sales price of, 17, 73, 261
  sales rate of, 17, 73
Profit
  decline of (with size increase), 52, 282; (with time), 97
  function, 201
  gross, 63, 279
    discounting of, 98
  incremental, 51, 54, 60
  large, 17, 51
  maximum, 52, 323
  minimum acceptable, 59, 62, 289
  net, 40, 42, 47, 50, 55, 59, 73, 280
  production rate relationship, 17, 20, 74
  small, 13, 17, 20, 51
  of some chemical companies, 27
  spending for, 35
  taxation of, 40, 132, 280
  time record of, 8, 18, 97
  total, 54
  venture
    definition of, 54
    equations for, 55, 93
    maximization of, 64, 72, 79, 81
  present value of, 94
Profitability, 60, 91, 136, 154, 168, 262, 277, 282, 302, 317, 447

Programming
  dynamic, 203
  geometric, 157
  linear, 195
Programs for computers, 336, 349
Projects, analysis of
  economic, 55, 62, 80, 90-100, 199, 266, 289
  technical, 289, 290-303, 324-338
  use of computers in, 336
Propane, 101, 396, 454
Propylene (poly), 291
Pumps
  cost of, 223, 352
  selection of, 452
Pumping systems, cost of, 185-195, 349-357, 452
Purchasing power, 445
Purge streams in process plants, 328
Pyle, C., 393

Q

Quality, control of, 17, 249
Quantity
  of material produced, 16, 36, 41, 73, 103, 111, 245, 261
  of material sold, 36, 41, 49, 73, 77, 97, 103, 111, 261

R

Raschig rings for column packing, 226, 399, 404
Rase, H. F., 452, 479
Ratio
  of gas and liquid flow rates, 398, 404

[ratio]
  of heat capacities, 363-376, 454
  of recompression, 186-195, 454
  reflux, 170, 173-179, 385-396, 404, 462
    optimum value of, 388, 395
  of venture worth to original investment, 111
Reactions (reactors), chemical, 291, 319, 327, 417-428, 472-479
Reboilers for distillation, 173, 385-396, 458
Receipts of money
  gross, 58
  net, 58, 61
Recycle of process streams, 417
Redundancy of equipment, 322
Reed, C. E., 66, 72, 435
Reed, L. A., 154, 162, 183
Reid, R. C., 461
Refrigeration, mechanical, cost of, 230, 250
Region, elimination of, 163-167
Regulations, governmental, 26, 37, 116, 265
Rejection of new investment, 54
Reliability of systems, 322
Replacement of equipment, 35, 38, 62, 309
Reports, financial, 27, 275, 282
Research, chemical
  in corporations, 3, 9, 18, 22, 26, 33, 35
  decreasing effectiveness of, 33
  market, 18, 259
  kinds of, 23, 26
  on processes, 3, 9, 18, 23
  on products, 3, 9, 18, 23
  value of, in USA, 9, 21, 23

Restrictions
  on process life, 265
  on process variables, 184
Return, rates of, on investment
  annual gross, 48, 51, 56, 104, 111, 129, 141, 259-280, 295, 361, 447
    decreasing, law of, 52
      exponentially, 129-133
      linearly, 129-133
    in discounted cash flow method, 112-116
    incremental, 50-54, 60, 79, 92, 361
    marginal, 129
    maximization of, 35, 73
    minimum acceptable
      on incremental investment, 71, 79
      on total investment, 43, 45, 111, 260, 277
    present value of, 44, 51
    rejection of, 54
    venture worth, 111
Result, trial, 163
Reward of money, 26, 259
Risk
  degrees of, 267, 276
  financial, 18, 57, 80, 94, 113, 259-282
  on new products, 18, 268
  technical, 18, 260
Rosen, E. M., 336
Royalty, payments of, 253
Rubber, synthetic, 7
Rudd, D. F., 90, 204, 220, 322
Run, length of, 318, 322

S

Saeman, W. C., 468
Safety, 318
Sales, of the chemical industry
  dollar volume of, 18, 36, 260

[sales, of the chemical industry]
  marginal, 129
  prices, 4, 13, 36, 43, 97, 261
  profitable, 40, 45, 99, 129
  return, 54, 63, 70, 139
  trend with time, 6, 23, 32, 97, 129
Salvage value of equipment and plants, 39, 45, 55, 93, 100, 129, 134, 138, 259
  tax on, 93, 132
Satterfield, C. N., 476
Sauvain, H., 445
Savings of money
  incremental, 54
  investment, 89, 264
  operating, 349
Sawistowski, H., 465
Scale of production, economics of, 17, 20
Schechter, R. S., 163, 181, 198, 202
Schedules, production and withdrawal, 317
Scheibel, E. G., 467
Schcid, F., 198, 202
Schweyer, H. E., 317
Science, chemical, 1, 3, 33
Scientists, 3
Sections of systems, 290
Securities of corporations
  bonds, 26, 269, 278
  stocks, 26, 269, 278
    common, 26, 278, 280
    preferred, 26, 278
Shannon, P. T., 336
Shannon, R., 322
Sherwood, T. K., 66, 72, 77, 157, 187, 350, 435, 461, 476
Shindo, A., 426
Shutdown of plants
  planned, 136, 318
  premature, 92, 133, 358
  criterion for, 133
Silica as raw material, 10

Silicones, 10
Simplex algorithm in linear programming, 198
Simplification of designs, 62
Simpson, L. L., 453
Situation (business), special, 5, 221
Size
  of chemical companies, 27
  of chemical equipment, 218
  of chemical plants, 20
  optimum, 54
Smith, W., 465
Sodium, 10
Sodium carbonate, 10, 13
Sodium chloride as raw material, 10
Sodium hydroxide, 2, 13
Sodium silicate, 10
Solute
  in gas absorption, 396, 400, 463
  in liquid extraction, 407, 466
Solutions, dilute, in gas absorption, 398, 463
Solvent
  in gas absorption, 396, 400, 463
  in liquid extraction, 407, 466
  recovery of, 237
Source
  of company funds, 26, 30, 279
  of cost data, 216, 223-230, 242, 246
South America, chemical companies in, 5
Spacing of trays in a column, 171, 174, 385, 391, 462
Spares of equipment, 322
Sprouse, R. T., 272
Stages
  in an acyclic system, 203-210
  in a chemical reaction system, 419
  in a compressor, 155, 206

[stages]
  in a separation system, 171, 385-396, 462
Startup of process plants, 99
Statistics, financial, 30, 272, 275
Steam, cost of, 71, 176, 250, 268, 387, 394, 413
Stevens, W. F., 154, 162, 183
Stobaugh, R. B., 20
Stock
  preferred, 26, 264, 276
  common, 26, 270, 280
    dividends on, 27, 31, 264, 280
    of Dow Chemical Company, 29, 265
    earnings per share, 27
    sale of, 31, 279
    treasury, 271
  corporations, 26, 269
Stockholders, 26, 264, 271, 280
Stores and supplies, cost of, 249
Strauss, R., 105
Stress
  in pipes, 189
  in pressure vessels, 481
Stripping by steam, 396, 407
Study
  economic, 132
  preliminary, 288
Styrene
  distillation of, 169-179, 394
  production of, 13, 17, 23
Suboptimization in dynamic programming, 203-210
Success, business, 99
Sulfur
  liquid, pumping of, 350
  as raw material, 10
Summation
  continuous, 90, 440
  stepwise
    algebraic expressions for, 435-443
    of discounted gross income, 102

INDEX 509

[summation]
  [stepwise]
    of discounted tax credit, 102, 116
    of money, 94
    of time series, 47, 112, 435-443
Supersaturation in crystallization, 468
Supervision, cost of, 247
Surface, response, 180, 203
Surplus, earned, 7, 264, 271, 279
Survey, market, 4

T

Tanks, storage, cost of, 229, 349
Taxation, 4, 37, 63, 93, 116, 264
Taxes
  base for, 37, 56
  computation of, 4, 36, 47, 55, 93, 133, 139, 142, 263, 279, 361
  credit for, 4, 96, 116
  deduction of, 316
  effect of depreciation on, 116
  period, 141
  property, 251
  rate, in venture profit and venture worth, 55, 63, 96
Taylor, G. A., 90, 313
Tazaki, E., 426
Techniques, special mathematical, 148-210
Technology, chemical, 20, 94, 260, 268
Temperature
  adiabatic saturation, 410
  difference, 54
    arithmetic mean, 175
    logarithmic mean, 362-380, 410, 471
      corrections to, 377, 457
    optimum, 68, 72, 358-376

[temperature]
  dry bulb, 409-417, 471
  level of, 72, 357-385, 409-417, 421
  wet bulb, 409-417, 471
Ten Broeck, H., 363, 366, 370, 457
Tension, interfacial, 467
Terbough, G., 309
Terephthalic acid, 13
Tetraethyl lead, 23
Thermal properties of process streams, 461
Thiele, E. W., 404
Thuesen, H. G., 318
Time
  for construction, 45, 101
  operating, 101, 360, 370, 381
  payout, 4, 45, 49, 55, 91, 140, 255, 280, 295, 353, 407, 449
  series, summation of, 435-443
  start-up, 45
  value of money, 89, 140
Timmerhaus, K. D., 38, 91, 202, 220, 243
Titanium dioxide, 10
Towers, cooling, cost of, 230
Trays (see plates)
Treasury, U.S. Department of, 41
Transfer units
  in absorption, 399-407, 464
    height of, 399, 404, 464
  in distillation, 393
  in drying, 415, 471
  in liquid extraction, 467
    height of, 467
Transportation
  of finished products, 5, 13
  of gas, in pipelines, 185
  of raw materials, 5, 11
Trends
  of chemical prices, 19
  of chemical technology, 19
  of monetary inflation, 446

Treybal, R. E., 399, 403, 407, 466, 468
Turbines
 gas, for pipe lines, 188
 steam, for pumping, 352
Turnover of investment as a function of sales, 48

U

Umeda, T., 426
Utilities, cost of, 250

V

Valves, cost of, 352
Value
 limiting, 152, 196
 present, 90, 96, 100, 436
 of reactants and products, 418
 of shipments, 6, 7
 time, of money, 89, 269, 435
Variables
 in chemical reactions systems, 421
 design, 204
 multi, 154, 168
 process, 149, 168
  cost of, 168, 176
  dependent, 168, 177
  independent, 168, 171
 single, 152
 slack, 196, 202
 state, 204
Velocity
 economic, of fluid flow, 351-357, 456
 flooding, 398, 403
 mass, 386, 398, 403, 409, 471, 475
 superficial, 174, 462, 465, 467, 477

Venture, 37, 43
 attractiveness of, 54, 91, 96, 101, 113, 132
Vessels, process
 cost of, 229, 349
 design of, 479
Vilim, O., 460
Vinyl chloride, 20
Volatility, relative, 170

W

Wages, 7, 16
Waggoner, J. V., 18
Walas, S. M., 474
Water
 as coolant, 32, 71, 175, 365-380, 422
 pollution of, 32
 price of, 5, 32, 71, 176, 250, 365-380
 quality of, 32
 as raw material, 5, 7, 10, 32
 supplies of, 7, 32
Watson, C. C., 90, 204, 220, 322
Watson, K. M., 475
Weaver, J. B., 90
Wenzel, L. A., 352, 457
Wessel, H. E., 248
Williams, C. L., 376
Woods, R. W., 441
Work, of compression, 66, 153, 181, 206
Worth
 economic, 136
 future, 437
 present, 91, 96, 102, 134
 venture, 4, 91, 95-117, 136, 139, 154, 447
  calculation of, 96, 103-111
  components of, 96-100, 106
  effect of inflation on, 95, 449
  maximum in, 136

INDEX 511

[worth]
  [venture]
    sensitivity diagram for, 105, 110
    uses of
      in development work, 277
      in economic evaluations, 95, 103, 135, 139, 149, 168
      in termination of operations, 134

Y

Yard (of a chemical plant), cost of, 249
Yield from a chemical reaction, 319, 325, 424
Young, E. H., 479
Younger, A. H., 482, 484

Z

Zabel, H. W., 13
Zener, C., 157
Zimmerman, O. T., 243